CONSTRUCTION FOR INTERIOR DESIGNERS

CONSTRUCTION FOR INTERIOR DESIGNERS

Second Edition

ROLAND ASHCROFT MA, FCSD
Course Director, Interior Design, Duncan of Jordanstone College of Art, Dundee
Illustrated by the author

Longman
Scientific &
Technical

Longman Scientific & Technical
Longman Group UK Limited,
Longman House, Burnt Mill, Harlow,
Essex CM20 2JE, England
and Associated Companies throughout the world.

© Longman Group Limited 1985
This edition © Longman Group UK Limited 1992

First published 1985
Reprinted by Longman Scientific & Technical 1987, 1988, 1989
Second edition 1992

British Library Cataloguing in Publication Data
A CIP record for this book is available from the British Library

ISBN 0 582 08125 4

Set 4 in Compugraphic 10/11 pt Plantin
Printed in Hong Kong
NPC/01

Contents

CHAPTER 9   ENVIRONMENTAL ISSUES

Preface to Second Edition

The major revisions in this edition concern *The Building Regulations, 1991: 1992 Edition*. These regulations apply to England and Wales only. Readers in Scotland should refer to *The Building Standards (Scotland) Regulations, 1990: Technical Standards* and in Northern Ireland, *The Building Regulations (Northern Ireland) 1977*.

Other revisions concern materials and techniques that are becoming outdated. In particular, products including asbestos fibres are being superseded by non-asbestos alternatives due to fears of asbestos-related diseases. Generally the book reflects this trend although vinyl-asbestos flooring is included since it still offers a low-cost, widely used finish requiring no special precautions in its installation or in disposing of surplus or waste.

A new chapter is added which introduces two of the prominent environmental issues relevant to buildings and interiors.

The author is grateful to K.A. Rosier (Nuralite (UK) Ltd), K. Penny (Absestos Information Centre) and Eternite (UK) Ltd for technical advice in preparing this edition, and to Caroline Peters (School of Design, Duncan of Jordanstone College of Art) for wordprocessing the text and tables.

Preface to First Edition

Unlike 'architecture', 'interior design' is rather loosely defined both in terms of:
- what it actually is;
- who practises it.

For the purposes of this book, interior design might be considered as design work concerned with the interior of buildings. Consequently, the activity of interior design involves the solution of the aesthetic and functional problems posed by the designing of the interior of a building. This includes both the interiors of new buildings, and the remodelling of existing buildings.

The activity of interior design might be undertaken by people of various professional backgrounds, and at various 'levels'. At the more formal level, interior design might be practised by such people as:
- interior designers *per se*;
- architects;
- designers of related disciplines such as textile and furniture designers.

At a less formal level, craftsmen or decorators might undertake interior design work.

In order to solve an interior design problem satisfactorily, any of these practitioners should exhibit:
- a receptiveness to the spatial characteristics of both the interior, and the building as a whole;
- a capacity to evaluate a client's requirements and translate this into an aesthetically and structurally appropriate interior scheme.

This book is aimed at assisting those who practise, or are training to practise, the activity of interior design. The information presented in the book should enable the reader to:
- comprehend the principles involved in construction generally;
- recognise how these principles operate in relation to the external enclosure of a building;
- achieve a reasonably comprehensive understanding of the

structural elements and materials used in interior construction.

In order to achieve the second aim, this book generally uses small-scale and domestic-type buildings to illustrate construction principles in relation to external enclosure. The reason for this is twofold:

- it is not the purpose of this book to examine the whole of the immense range of building construction techniques;
- small-scale and domestic buildings are familiar to everyone, and consequently the construction principles should be more easily grasped.

This textbook is only capable of providing a framework for knowledge of interior construction. This framework should be augmented by such publications as:

- manufacturers' and suppliers' trade literature;
- specialist textbooks (see bibliography);
- documents such as the Building Regulations and relevant British Standards and Building Research Establishment Digests.

The author would like to thank the following for their valuable assistance in reviewing certain chapters and for their constructive criticism: David Upton – Chapter 1; Bob Baldwin – Chapter 2; Derek Crafts – Chapters 3 and 5; Bryan Dyer – Chapter 4; Tony Collard – Chapter 6; John Woolen – Chapter 7; Steve Constanti and Malcolm Rapier – Chapter 8. My thanks also to Avril Cornock for her skilful typing of the entire manuscript, and to Elinor Allenby for checking and correcting the text.

Acknowledgements

We are grateful to the following for permission to reproduce copyright material: Nuralite (UK) Ltd for references to 'Nutec Nuralite'; the Controller of Her Majesty's Stationery Office for various extracts from *The Building Regulations 1991: 1992 Edition*.

CHAPTER 1

Structural principles

Introduction

A structure might be described as a series of components or 'structural elements' which, when fitted together, are capable of withstanding the loads and forces to which they are subjected.

A building is such a structure, being composed of a wide variety of components of different sizes and materials which, when fitted together, are capable of withstanding such loads and forces as:

- wind;
- the furnishings, people and machinery in the building (referred to as 'superimposed' loads);
- the considerable loads imposed by the weight of the building components themselves (referred to as 'dead' loads).

Clearly then, in the designing of a building and its interior, it is necessary to understand the way in which structures behave in relation to the materials and the construction techniques used.

Much of building design utilises the knowledge of established materials and components (such as bricks, timber joists, doors, etc.), which are fitted together by established techniques (for example the typical techniques used for forming a window opening, or for installing a staircase). Designers need to be sufficiently versatile to adapt and elaborate on these established principles, in order to fulfil the design requirements peculiar to a particular building/interior scheme. Sometimes, however, the designer will use less familiar techniques and materials in the design of a building/interior structure. In such circumstances, the designer might use the services of a structural engineer, who is trained to:

- calculate the loads and forces to which the structure will be subjected;

● calculate and design structural components of suitable size and of suitable material to withstand these loads and forces.

This chapter explains some of the terminology used in the study of the behaviour of structures, and considers some of the structural behaviour that might be expected in the components of a building/interior.

Definitions of basic terms

For a building structure to function properly, it should be capable of remaining stable when it is subjected to the various forces imposed on it. This does not mean that the building structure should be totally rigid, but that the degree of movement should be kept within tolerable limits. In fact, the material of any structure will move or deflect to some extent when subjected to a force.

Force

A 'force' acting on a structure, is the effect of weight on the structure. For example, a person weighing 70 kilograms (70 kg) standing on a floor, will push down on the floor with a force of 70 kilograms force (kgf). A strong wind blowing against the side of a wall will also exert a force of so many kilograms.

Forces due to the 'mass' (or weight) of objects are different from forces due to factors such as the wind, since they are subject to the effect of gravity. Gravity has the effect of pulling objects downwards at a rate of acceleration of 9.81 metres per second, per second (m/s^2). Forces are measured in units of 'newtons' (N) or:

One newton is the force required to accelerate a weight of 1 kg by 1 m/s^2.

Consequently, an object of 1 kg weight which is subjected to gravitational force will exert a force of 9.81 N (or 1 kg × 9.81 m/s^2 = 9.81 N). In the case of the person standing on a floor, the force exerted will be 70 kg × 9.81 m/s^2 = 686.7 N

Structural stability relies on a structure, or part of a structure, resisting 'active forces' (imposed forces) by 'passive forces' (a balancing force produced within the structure).

For example, when a chair is placed on a floor, the chair legs exert a downwards active force on the floor. In order to resist the active force, the floor produces an upwards passive force of similar magnitude. If a person sits on the chair, the active force is increased, and consequently the floor will provide an increased, upwards passive force. Passive force is produced in a structure by the material of the structure

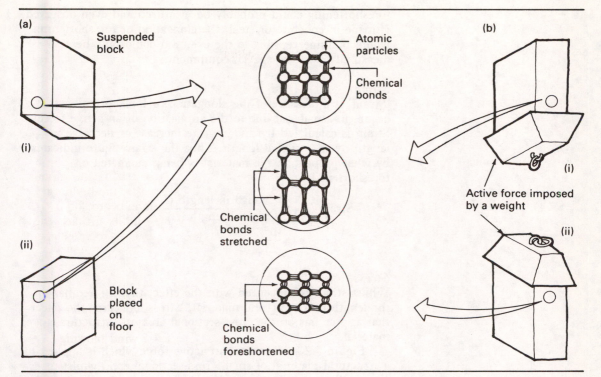

(a)

Suspended block

(i)

(ii)

Block placed on floor

Atomic particles

Chemical bonds

Chemical bonds stretched

Chemical bonds foreshortened

(b)

(i)

Active force imposed by a weight

(ii)

Fig. 1.1(a)
(i) *Suspended block at rest*
(ii) *Block at rest on floor*

Fig. 1.1(b)
(i) *Active force applied to suspended block*
(ii) *Active force applied to block on floor*

actually changing shape due to the exertion of an active force.

Figure 1.1(a) illustrates:

(i) a block of material which is securely suspended from a structure;

(ii) a similar block, which is placed on a secure floor structure.

In both cases, the blocks are free from applied active forces, and the material of the blocks is relatively 'at rest'.

The inset diagram of Fig. 1.1(a) illustrates diagrammatically a typical material at rest, where the atomic particles are held together by chemical bonds. In Fig. 1.1(b), an active force is applied to each of the blocks. In the case of (i) above, the active force pulls the material of the block, and a passive force is produced by the chemical bonds between the atomic particles being stretched. In the case of (ii) above, the active force pushes down on the block, and a passive force is produced by the chemical bonds between the atomic particles being foreshortened. The degree of elongation or foreshortening which occurs in the block will depend not only on the magnitude of the applied force, but also on the nature of the material of the block and its cross-sectional area. If made of rubber, for instance, the elongation and foreshortening could probably be measured and even observed. If made of concrete or steel, the elongation or foreshortening would be imperceptible to the eye, and could only be measured by use of special equipment.

Strain

'Strain' is a measure of the elongation or foreshortening that occurs in a material due to the application of an active force. Strain is calculated by dividing the increase or decrease in length by the original length. Thus the strain figure indicates by what proportion the material is being elongated or foreshortened.

$$\text{Strain } (e) = \frac{\text{Change in length } (l)}{\text{Original length } (L)}$$
$$e = \frac{l}{L}$$

Stress

Whilst strain is concerned with the effect that a force has on the length of a structural material, 'stress' concerns the effect that a force has on the cross-sectional area of a structural material.

Figure 1.2(a) illustrates an active force which is applied to a structural material of small cross-sectional area, whilst in Fig. 1.2(b), a similar force is applied to a structural material of much greater cross-sectional area. Clearly, the effect of the force will be much greater in Fig. 1.2(a) than it will in Fig. 1.2(b). Stress is calculated by dividing the active force applied by the cross-sectional area of structural material on which it is applied:

$$\text{Stress } (f) = \frac{\text{Force (or load) } (W)}{\text{Cross-sectional area } (A)}$$
$$f = \frac{W}{A}$$

In the case of Fig. 1.2(a), the stress in the material will be 100 kN divided by 5 mm^2 resulting in a stress of 100/5 or 20 kN/mm^2, whereas in Fig. 1.2(b), the stress will be 100/50, or 2 kN/mm^2.

A stress which is capable of causing a material to fail (for example by crushing, or by excessive deformation) is known as the 'failing stress', or 'ultimate stress'.

The stresses in the structure of a building must be kept well below the ultimate stress levels, otherwise the structure could collapse. In calculating suitable sizes for the materials used in a building structure, an engineer will use 'permissible'

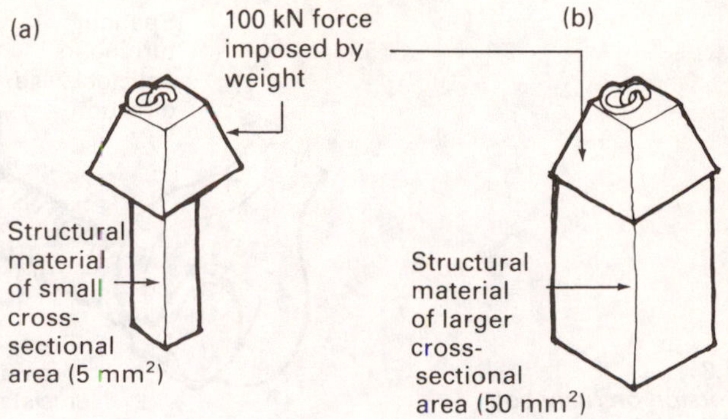

Fig. 1.2
(a) *Active force applied to small cross-section*
(b) *Active force applied to large cross-section*

or 'safe working' stresses, where the ultimate stress figures are reduced substantially, in order to ensure that the structural material will not be subjected to forces which could cause failure. If, for example, an active force applied to a structural material resulted in an ultimate stress of 20 N/mm^2, the engineer might work on a safe working stress of 5 N/mm^2. In this case the engineer would be adopting a 'safety factor' of 4, since the ultimate stress is four times the permissible stress. In order to achieve the permissible stress figures, it may be necessary to either increase the cross-sectional area of the material, or to decrease the active force.

The active forces illustrated in Figs. 1.1 and 1.2 have caused the structural material to either elongate or foreshorten due to the force being applied vertically to vertical members. Such stresses, which are caused by axial forces, are known as 'direct' stresses. Sometimes, an active force is applied at right angles to a structural member, such as a weight applied to a beam, a foot applied to the rung of a ladder, or heavy objects placed on a long shelf. In such circumstances the structural member concerned will bend, and the stress occurring in the structural member is referred to as a 'bending' stress.

Compression and tension

A force which causes a structural member to foreshorten (such as that illustrated in Fig. 1.1(b)(ii)) is known as a 'compressive force' and will result in a 'compressive stress'. Excessive compressive stress tends to cause crushing of a material.

A force which causes a structural member to elongate (such as that illustrated in Fig. 1.1(b)(i)) is known as a 'tensile force' and will cause tensile stress. Excessive tensile stress will tend to cause a material to stretch.

5

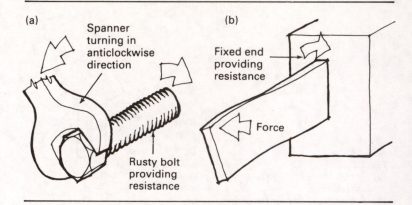

Fig. 1.3
(a) *Torsion on rusty bolt*
(b) *Effect of a torsional force*

Torsion

'Torsion' occurs as a result of twisting (Figs. 1.3(a) and (b)).
If, for example, a rusty bolt is turned by a spanner, the
spanner will exert a turning force on the bolt, but the rusty
bolt fails to turn. The force applied by the spanner is a
'torsional force' and the bolt will experience 'torsional stress'.

Excessive torsional stress will tend to cause a material to
twist, or eventually break.

Shear

Where forces act on a structural member in opposing
directions, a sliding tendency will occur (Fig. 1.4). Such forces
are referred to as 'shear force' and result in 'shear stress' in
the material. Excessive shear stress in a structural member will
cause it to either deform, or eventually to break, as shown in
the illustration.

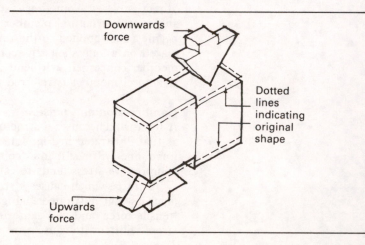

Fig. 1.4 *Effect of shear forces*

Elasticity

Most materials used in building and interior construction are 'elastic'. Elastic materials change their shape when subjected to a force (elongate if subjected to tensile force, foreshorten if subjected to compressive force), but when the force is removed the material will recover its original shape.

Some materials, such as soft putty, or plasticine, are 'plastic'. Plastic materials change shape when subjected to a force, but do not recover their original shape.

With elastic materials, strain is proportional to stress. If, for example, a particular sample of material were subjected to a tensile stress of 50 N/mm², and this resulted in a strain of 0.0001, then if the stress were increased to 100 N/mm², the strain would increase to 0.0002. A stress of 200 N/mm² would result in a strain of 0.0004, and so on. In each case, if the stress was divided by the strain, the same figure would result:

$$\frac{50}{0.0001} = 500\ 000\ \text{N/mm}^2$$

$$\frac{100}{0.0002} = \frac{200}{0.0004}\ \text{etc.}$$

In other words, the stress : strain relationship is a 'constant'.

For any material, the stress/strain constant is known as the 'modulus of elasticity', which is basically an indication of the stiffness of a material.

$$\text{Modulus of elasticity } (E) = \frac{\text{Stress}}{\text{Strain}}$$

Table 1.1 indicates the modulus of elasticity figures for a number of common building materials. Note that the higher the values of E, the stiffer the material.

The E values are extremely useful to engineers in enabling them to calculate the changes in dimension that might occur in a structural member due to particular compressive or tensile forces.

Table 1.1   Modulus of elasticity for building materials

Material	Average E (MN/m²)
Plastics (such as polythene)	1 400
Timber	10 000
Brick	10 000
Concrete	21 000
Aluminium	70 000
Steel	200 000

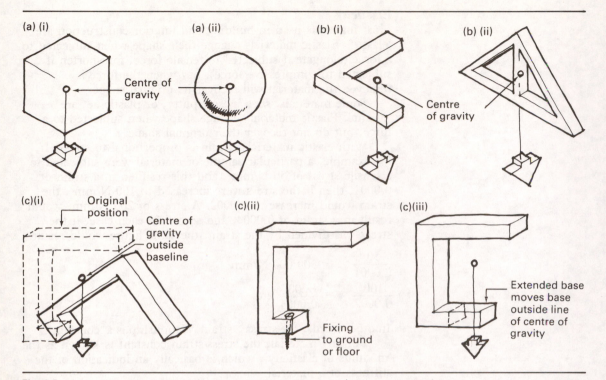

Fig. 1.5
(a) *Centre of gravity of simple solid shapes (i) cube (ii) sphere*
(b) *Centre of gravity of simple hollow shapes*
(c)
  (i) *Centre of gravity and stability*
 (ii) *Object stabilized by being held down*
(iii) *Object stabilized by increased size of base*

Centre of gravity

As already mentioned, gravity has the effect of pulling objects downwards. Although gravity has a pulling effect on all the particles of an object, it is sometimes useful for engineers to regard gravity as though it was pulling the entire weight of an object from one point. In any object, this point is known as the 'centre of gravity'.

In the case of simple symmetrical objects, it is easy to imagine where the centre of gravity lies, that is, in the centre of the object (Figs. 1.5(a)(i) and (ii)). In the case of hollow, framed and certain asymmetrical objects, the centre of gravity might lie outside the material of the object, as is the case in Fig. 1.5(b)(i), or inside the material (Fig.1.5(b)(ii)).

The centre of gravity is a useful concept in considering the stability of certain objects. If a vertical line is drawn down from the centre of gravity, and this line falls outside the effective base of the object, then the object will overturn (Fig. 1.5(c)(i)). To remedy such a problem, the object could be either fixed down to the floor or ground (Fig. 1.5(c)(ii)) or the base could be increased in size such that the centre of gravity falls within the base (Fig. 1.5(c)(iii)).

Moments

Sometimes the application of force to a structural member can cause it to rotate. Consider, for example, the see-saw illustrated in Fig. 1.6(a). The force exerted by child A, on the right-hand side of the see-saw, causes a clockwise rotation to occur. When child B of similar weight, sits on the left-hand side, the clockwise rotational effect is counterbalanced by the anticlockwise effect of child B (Fig. 1.6(b)). This rotational or 'turning effect' of a force is known as the 'moment' of a force. The moment of a force may be increased by:

- increasing the size of the force; and/or
- by increasing the distance between the force and the turning-point.

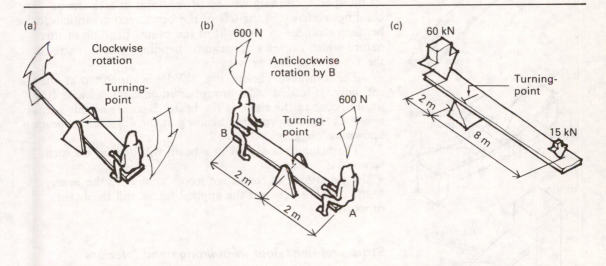

Fig. 1.6
(a) *Rotation in a see-saw*
(b) *Equilibrium achieved in a see-saw*
(c) *Examples of asymmetrical equilibrium*

The moment of a force is measured by multiplying the size of the force by the distance of the force from the turning-point. Thus in Fig. 1.6(b), each child exerts a force of 600 N and both children are sitting at a distance of 2 m from the turning-point of the see-saw. The moment of the force exerted by each child is therefore 600 N × 2 m = 1200 Nm.

Where the moments on both sides of the turning-point are the same, the structural member is considered to be in 'equilibrium' (or balance). In Fig. 1.6(b) equilibrium is achieved by a completely symmetrical arrangement of the forces, whilst in Fig. 1.6(c), equilibrium is achieved with an asymmetrical arrangement of forces, where the left-hand moment is 60 kN × 2 m = 120 kNm and the moment on the right-hand side is 15 kN × 8 m = 120 kNm.

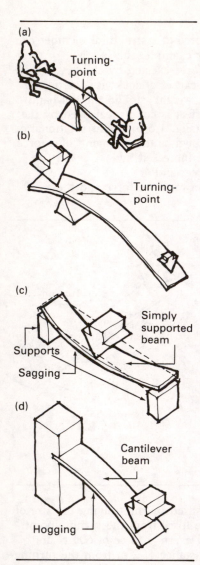

(a)

Turning-point

(b)

Turning-point

(c)

Simply supported beam

Supports

Sagging

(d)

Cantilever beam

Hogging

Fig. 1.7
(a) and (b) *Bending occurring instead of rotation*
(c) *Bending occurring in simply-supported beam ('sagging')*
(d) *Bending occurring in cantilevered beam ('hogging')*

Bending

The illustrations in Fig. 1.6 all rely on structural members which are subjected to moments of forces resulting in rotation. If the members were insufficiently rigid, and the forces were sufficiently large, bending would occur in the beam (Figs. 1.7(a) and (b)). A moment of a force which results in bending is known as a 'bending moment' (BM). Formulas are available whereby an engineer may calculate the size of BM that might occur in a particular beam. The engineer may then calculate a beam of suitable size, shape and material which will be capable of resisting such BMs.

Figure 1.7(c) illustrates a 'simply-supported beam' (a beam which rests on supports at each end). If the beam is loaded at its mid-point, bending will occur resulting in a clockwise bending tendency to the left of the beam, and an anticlockwise bending moment to the right of the beam. Bending of this nature which causes a downwards bending, in the centre of the beam, is known as 'sagging'.

The 'cantilever' beam (Fig. 1.7(d)) is supported at one end only. If loaded at its unsupported end, a clockwise BM would occur to the right of the beam. Bending of this nature which causes a downwards bending at the edge of the beam is known as 'hogging'.

The magnitude of BM in a beam will depend on such factors as:
● the magnitude of the active forces applied to the beam;
● the distance between the applied forces and the beam supports.

Structural behaviour in buildings and interiors

The structural elements of a building or interior must be designed to resist excessive deformation, or collapse, that would result from active forces. This section looks at some of the effects that active forces might have on: (*a*) vertical members; (*b*) horizontal members; (*c*) framed members.

Vertical members

Vertical members include:
● 'post'-type members such as columns of buildings, chair and table legs and newel posts of staircases;
● 'wall'-type members, such as the external walls of buildings, internal load bearing partitions, and vertical boards forming the sides of cabinets.

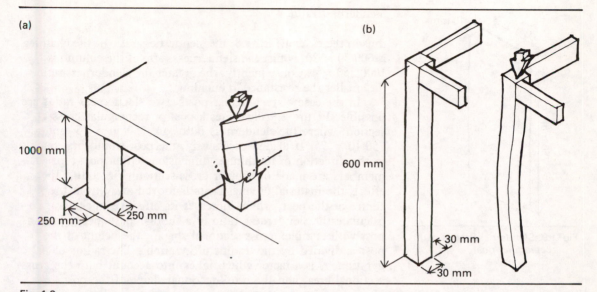

Fig. 1.8
(a) *Crushing in a column*
(b) *Buckling in a chair leg*

Post-type members Post-type members could fail due to:
- crushing, where an applied force crushes the material of the post (Fig. 1.8(a));
- 'buckling', where an applied force causes the post to bend (Fig. 1.8(b)). The Figures illustrate respectively the crushing of a column in a building and the buckling of a chair leg.

One of the main differences between these examples is that whereas the chair leg is 'tall' (high in relation to its thickness), the column is 'short' (short in relation to its thickness). If a short and a tall post (of similar thickness and made of similar material) were gradually loaded, the tall post would fail due to buckling before the short post failed due to crushing.

Basically, the ability of a post-type member to resist crushing depends on:
- the cross-sectional area of the post (the greater the area, the greater the resistance to crushing);
- the nature of the material from which the post is made (the higher the compressive strength of the material, the greater the resistance to crushing).

The ability of a post-type member to resist buckling depends on its 'slenderness ratio'.

Slenderness ratio In simple terms, the slenderness ratio of a post is the ratio of the height of the post to the thickness of the post:

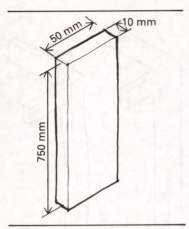

Fig. 1.9 *Slenderness ratios of rectangular post*

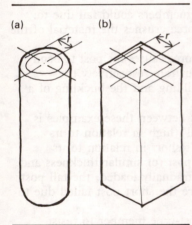

Fig. 1.10 *Tubular posts*

Slenderness ratio $\dfrac{h}{t}$

thus in the case of Fig. 1.8, the slenderness ratio of the chair leg is 600/30 = 20, whilst the slenderness ratio of the column is 1000/250 = 4. Consequently, the greater the slenderness ratio, the smaller the resistance to buckling.

In the case of rectangular posts, two slenderness ratios are possible. Figure 1.9 illustrates a post of rectangular cross-section, where the slenderness ratios are 750/50 = 15, and 750/10 = 75. If loaded excessively, the post would first buckle in the direction of its thinner dimension. Often post-type members are made of tubular cross-section (Fig. 1.10). In effect, the material of such posts is moved away from the centre of the post, which increases its effective thickness. To calculate the slenderness ratio of a tubular post (or, in fact, a post with complex cross-sectional shape), the height of the post is divided by the 'radius of gyration'. The radius of gyration (r) is a factor which takes into account both the cross-sectional area, and the cross-sectional shape of the post. Hence, for such posts, the slenderness ratio = h/r.

Formulae are available which enable engineers to calculate r for posts of given cross-sectional shapes and sizes.

Consider again the two posts illustrated in Figs. 1.8(a) and (b). In the case of the chair leg, the top of the leg is fixed by the joint with the seat structure, whilst the bottom is unfixed. The column of the building, however, is fixed both at its top and its bottom. The type of fixing at the top and bottom of a post affects the way in which buckling occurs. Figure 1.11(a,b,c) illustrates three conditions of fixing, and the type of buckling that could result from excessive loading. In: (a) the post is free of fixing both at the top and bottom, and consequently buckling could occur throughout its entire length; (b) one end is fixed (in similar fashion to the chair leg), and the effective height of the post is reduced; (c) both ends are fixed (in similar fashion to the column) and the effective height of the post is considerably reduced.

Therefore, the actual slenderness ratios of the chair leg and column will be rather less than the simple height divided by thickness figures previously mentioned. The actual slenderness ratio will be:

Effective height
——————————
Minimum thickness

Clearly, the quality of the fixings at the top and/or bottom of the post is important. In the case of the chair leg, for

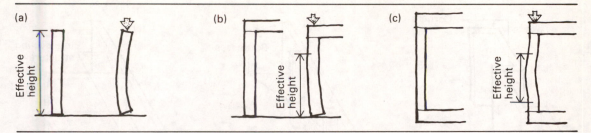

Fig. 1.11
(a) *Unfixed post*
(b) *Post fixed at one end*
(c) *Post fixed at both ends*

example, a non-rigid joint between the leg and seat frame would cause buckling to occur nearer to the top of the leg, thus the effective height would increase.

The effect of buckling could also be reduced by the introduction of other structural members which are joined to the post. Figure 1.12 illustrates the introduction of chair rails to the chair leg, which reduce the effective height of the leg. Such a measure, incidentally, would also limit the likely movement in the chair leg/seat frame joint.

Eccentric forces Sometimes a force acts directly on the centre-line of a post (Fig. 1.13(a)), which results in compressive stress in the material of the post. Sometimes, however, the force is applied in an 'eccentric' manner (off-centre). The degree of eccentricity is measured from the centre-line of the thickness of the post, to the point at which the force acts. If the degree of eccentricity exceeds one-sixth of the thickness(t) of the post, then the side of the post nearer the force will be in compression, whilst the side of the post further away from the force will be in tension (Fig. 1.13(b)).

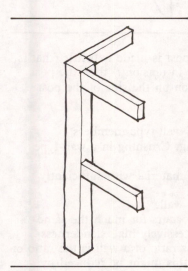

Fig. 1.12 *Chair leg with intersecting rails*

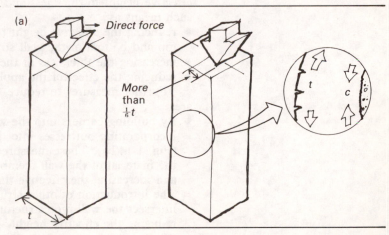

(a) Direct force

More than $\frac{1}{6}t$

t

c

t

Fig. 1.13
(a) *Direct force on post*
(b) *Eccentric force on post*

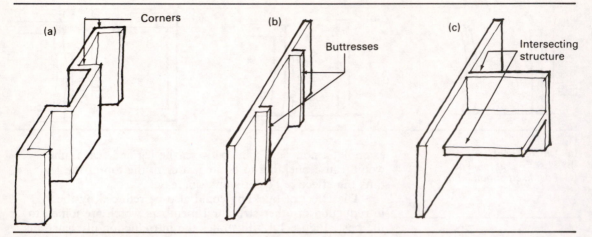

Fig. 1.14
(a) *Wall-type member with corners*
(b) *Wall-type member with buttresses*
(c) *Wall-type member with intersecting structural members*

If the material of which the post is made is brittle (that is, good in compression(c), but poor in tension(t) then the post could fail through excessive tension on the side of the post furthest from the force.

Wall-type members Like posts, wall-type members might fail by crushing or buckling. Crushing in a wall-type member could be avoided by:
- making the wall structure of a material with sufficiently high compressive strength;
- increasing the thickness of the wall.

Buckling in a wall-type member occurs for much the same reason as in a post: due to an excessively high slenderness ratio. Like a post, the slenderness ratio of a wall is the ratio of effective height to thickness and this might be reduced by such means as:
- reducing the effective height by ensuring good fixings at the top and bottom of the wall structure;
- increasing the thickness of the wall;
- reducing the effect of the applied force.

Further measures to reduce the buckling tendency in walls are:
- By building corners into the wall (Fig. 1.14(a)) or incorporating buttresses into the wall structure (Fig. 1.14(b)). These measures have the effect of moving the material of the wall member away from the centre-line, and increasing the effective thickness of the wall.
- The introduction of other structural members which intersect the wall-type structure (Fig. 1.14(c)) hence reducing the effective height. External and internal load

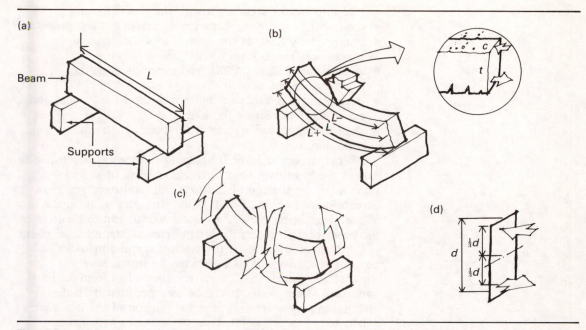

bearing walls may also suffer from 'settlement' (see 'Foundations', in Ch. 2, p. 23).

Horizontal members

Horizontal members include:

- 'beam'-type members, such as lintels for windows and doors and some of the rails used in chair and cabinet construction;
- 'slab'-type members which include roof and floor slabs, worktops for cabinets and shelves.

Beam-type members

Bending   Figure 1.15(a) illustrates a 'simply-supported' beam (a beam resting on a support at each end). When an active force is applied (Fig. 1.15(b)) bending will occur. As can be seen, bending results in the top surface of the beam being foreshortened (due to compressive stress in the material of the beam), and the bottom surface becoming elongated (due to tensile stress in the material of the beam). Notice that the middle of the beam remains the same in length, and the material of the beam in this area (known either as the 'neutral axis' in the cross-section of the beam, and the 'neutral layer' in the elevation of the beam) is neither in compression nor tension.

Taking a cross-section through the loaded beam:

- the material at the top experiences maximum compression;
- progressively towards the neutral axis the material experiences less compression;
- at the neutral axis, the material experiences neither compression nor tension;
- progressively towards the bottom of the beam, the material experiences increasing tension;
- the material at the bottom of the beam experiences maximum tension.

When an active force is applied to a beam, the left-hand side of the beam will tend to rotate or bend in a clockwise fashion, whilst the right-hand side will tend to rotate in an anticlockwise fashion (Fig. 1.15(c)). In order to avoid the top of the beam being crushed by compression and the bottom of the beam being stretched apart by tension, the material of the beam itself needs to resist the clockwise and anticlockwise bending tendencies. This resistance is illustrated in Fig. 1.15(d), which shows a vertical slice taken from the left-hand side of the beam. It can be assumed that if all the compressive force acting in the top portion of the beam were concentrated in one point, then the distance between this point and the neutral axis would be one-third the total depth of the beam. Similarly, if the total tensile force were concentrated in the bottom portion of a beam, it would act at a point one-third the depth of the beam below the neutral axis. As can be seen from the illustration, the combination of these forces cause an anticlockwise turning effect (or 'anticlockwise moment') which resists the clockwise turning effect (or 'clockwise moment') which results from bending in the left-hand side of the beam. A similar slice taken from the right-hand side of the beam would illustrate the material in the beam producing clockwise moments to resist the anticlockwise moment caused by bending in the right-hand side of the beam. The greater the depth of a beam member, the greater its ability to resist bending, because the moment of resistance is increased due to the distance $\frac{1}{3}d$ being increased. For this reason, beams are made deep in relation to their breadth. Notice, for example, that the joists illustrated in Figs. 3.7(a), (b) and (c) (Ch. 3), are deep in relation to their breadth. Since the greatest resistance to stress occurs towards the top and bottom of a beam, beam members are sometimes designed so that:

- they are deep in relation to their breadth;
- most of the material of the beam is concentrated in the

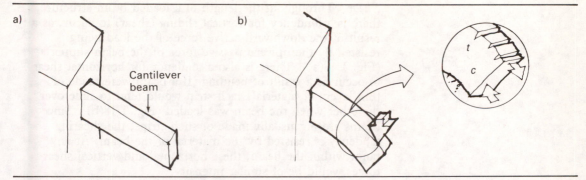

Fig. 1.16
(a) *Cantilever beam, unloaded*
(b) *Cantilever beam, loaded*

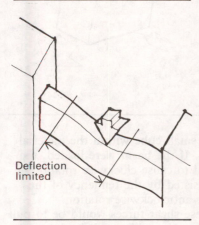

Fig. 1.17 *Built-in beam*

stress-resisting areas at the top and bottom of the beam.
See, for example, the illustrations of cross beams in
Figs. 3.7(e), (f) and (g) (Ch. 3).

In the case of a cantilever beam member, the effect of bending
due to an applied force results in elongation in the top of the
beam and foreshortening in the bottom of the beam. Thus the
distribution of stress is opposite to that of a simply-supported
beam; in a cantilever, tensile stress is experienced by the
material at the top of the beam, whilst compressive stress is
experienced by the material at the bottom (see Figs. 1.16(a)
and (b)).

Deflection 'Deflection' is a measure of the degree to which a
beam member will 'sag' (see Fig. 1.7(c)). Whilst a beam
could be designed to resist bending without collapse, it
could be unsuitable if excessive deflection took place.
Excessive deflection in a beam could look unsightly, and
could cause damage to materials fixed to the beam, such as
cracking of plaster, etc.

The amount of deflection can be calculated by use of a
formula which takes into account:

- the loads on the beam and the span of the beam (increased
  load and increased span result in increased deflection);
- the stiffness of the material (E), and the stiffness of the
  cross-sectional shape of the beam (an increase in these
  factors will reduce deflection).

Deflection is also affected by the nature of the beam supports.
In the case of a simply-supported beam, deflection is able to
occur throughout the beam's length. In reality, such beams are
unusual. Most beams are 'built in' or fixed at their ends, in
which case deflection only takes place in a limited part of the
length (Fig. 1.17).

Shear Throughout the length of a loaded beam structure, there is a tendency for vertical sliding (shear) to occur, as a result of the downward active force of the load being resisted by the upward passive force of the beam supports (Fig. 1.18(a)). There is also a tendency for horizontal shear to occur as a result of bending. If a beam were made of thin layers of material, each strip would tend to slide over the other when the beam was loaded (Fig. 1.18(b)). Since beams are not usually made of such strips, the shearing tendency is resisted by the material of the beam. At any point within the beam, these horizontal and vertical shear forces would be of similar intensity.

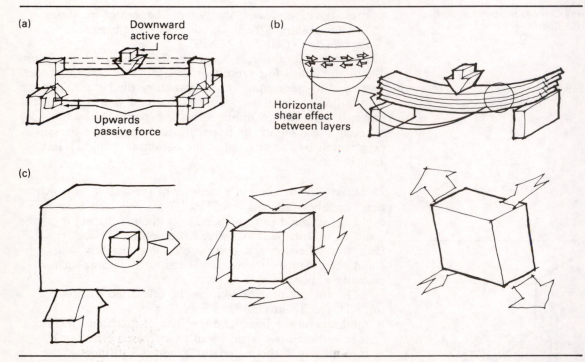

Fig. 1.18
(a) *Vertical shear in a beam*
(b) *Horizontal shear in a beam*
(c) *Diagonal compression and tension in a beam*

Figure 1.18(c) illustrates a small cube within the material of the left-hand side of a beam. If the beam were loaded, the vertical shear forces would tend to cause clockwise rotation of the cube, and this would be resisted by the tendency of the horizontal shear forces to cause anticlockwise rotation.

The combined effect of these shear forces would be to pull out two opposite corners of the cube and to push in the two other opposite corners, thus resulting in diagonal tension and compression. The effect of this is particularly important at the

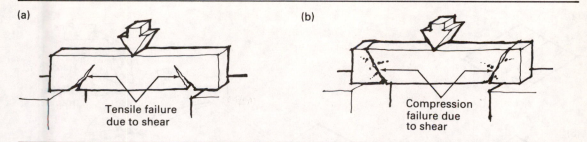

Tensile failure
due to shear

Compression
failure due
to shear

Fig. 1.19
(a) *Shear failure in beam weak in tension*
(b) *Shear failure in beam weak in compression*

ends of a beam, since a beam made of material weak in tension could fail in the manner illustrated in Fig. 1.19(a), whilst a beam made of material weak in compression could fail in the manner illustrated in Fig. 1.19(b).

Slab-type structural members A slab might be likened to a very broad, shallow beam, and in some respects certain slabs behave in a very similar manner.

Slab-type structural members might:
- span between two supports, when the slab is known as a 'one-way' slab (Fig. 1.20(a));
- span between two sets of supports arranged in a square or rectangular fashion, when the slab is known as a 'two-way' slab (Fig. 1.20(b)).

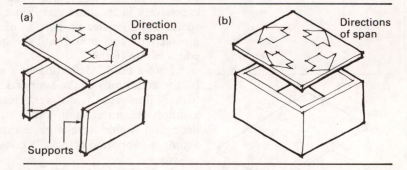

(a) Direction of span

(b) Directions of span

Supports

Fig. 1.20
(a) *One-way slab*
(b) *Two-way slab*

One-way slabs behave in a very similar fashion to beams, in fact some slabs are constructed of series of wide beams laid adjacent to each other (see for example the floor construction illustrated in Figs. 3.30(a) and (b), Ch. 3). When an active force is applied to a one-way slab, compression will occur in the top, and tension in the bottom of the slab, due to the effect of bending.

In the case of a two-way slab, the bending will take place in both axes of the slab. In the case of a square slab, the

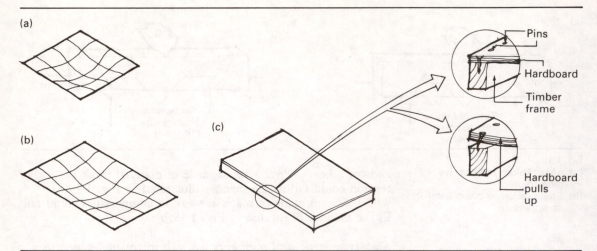

(a)

(b)

(c)

Pins

Hardboard

Timber frame

Hardboard pulls up

Fig. 1.21
(a) *Bending in square slab*
(b) *Bending in rectangular slab*
(c) *Bending in hardboard rectangular slab*

bending will be the same in both directions (Fig. 1.21(a)), whereas the effect of bending in a rectangular slab will be different (Fig. 1.21(b)). With a rectangular two-way slab, the short axis will sustain a greater proportion of the load. This can be demonstrated by considering a rectangular sheet of hardboard fixed to timber frame with pins (Fig. 1.21(c)). If an active force was gradually applied to the hardboard, failure would eventually occur due to the hardboard bending excessively in its short axis. Since the short axis would be sustaining a greater proportion of load, bending would be greater, and this would pull the hardboard up along the long edges of the frame.

Like a beam, deflection in a slab structure will be affected by the conditions of the supports. Figures 3.29(a) and (b) (Ch. 3) illustrate the different degree of deflection occurring in a simply-supported slab, and a built-in slab. Whereas deflection is able to take place throughout the entire length of a simply-supported slab, it is considerably restricted in a built-in slab.

Deflection is also affected by such factors as:
- size of span and size of applied force (the greater the span and force, the greater the deflection). The span could be reduced by provision of intermediate supports (Fig. 1.22);
- stiffness of material (the greater the E value, the smaller the deflection);
- the profile of the slab: by folding or shaping the slab, material is being positioned at a greater distance from the neutral axis which improves the resistance to bending of the beam (Fig. 1.23).

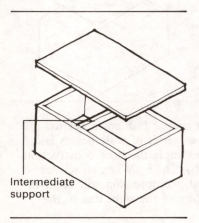

Intermediate support

Fig. 1.22 *Bending reduced by intermediate supports*

20

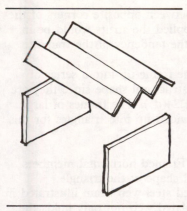

Fig.1.23 *Folded slab*

Framed members

Sometimes the weight of a structure made of solid elements can cause problems. For example:

- in the case of a heavy external wall, the weight of the wall could cause it to sink into the ground (see 'Foundations' in Ch. 2, p.23) and Figs. 2.1(a) and (b),
- in the case of a horizontal one-way slab structure, the weight of the slab itself could cause it to bend, or alternatively, it could cause the supports to crush, buckle or overturn.

In many cases it is necessary to reduce the weight of structural elements, and this is often achieved by use of a 'frame'. A frame is a structural element composed of at least three members, which are joined together.

Framed vertical members Figure 1.24(a) illustrates a simple frame for a wall-type structure, which would be suitable for supporting the end or side of a cabinet. If an excessive active force was applied to the frame, the corner joints could act like hinges, and the frame would deform to a parallelogram shape. (This effect is sometimes referred to as 'racking'.) This tendency could be overcome by such means as stiffening the joints, or by bracing the frame. If the joints were stiffened, for example, by the use of gussets (Fig. 1.24(b)) the joints would be better equipped to resist movement, although an excessive force could cause bending in the vertical frame members. Bracing of the frame could be achieved by either fixing a sheet of stiff material securely to the face of the frame, or by using

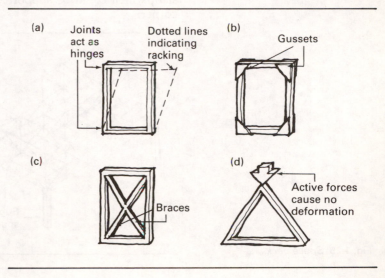

Fig. 1.24
(a) *Simple square frame*
(b) *Square frame with rigid joints*
(c) *Braced frame*
(d) *Force imposed on triangular frame*

diagonal braces (Fig. 1.24(c)) between opposite corners of the frame. If an active force was applied the struts would be in compression, and would resist the tendency of the frame to rack.

The use of struts divides the frame up into a series of triangles. Triangles are a good structural shape since they are very resistant to racking (Fig. 1.24(d)). The frames of large structural elements are often divided up into triangles for this reason.

Framed horizontal members   Framed horizontal members often use the efficient structural shape of the triangle. Consider for example the latticed steel-web beam illustrated in Fig. 3.7(f) (Ch. 3). The triangulated web not only holds the flange material apart in order to increase resistance to bending, but also the diagonal direction of the web resists the diagonal tension and compression due to shear (see Fig. 1.18(c)). Consider also the roof trusses illustrated in Figs. 3.16–3.19 (Ch. 3). Here, the basic triangular shape of the truss is subdivided into a series of smaller triangles by use of 'struts' (compression members) and 'ties' (tension members).

An example of triangulation in a horizontal slab-type member is the 'space deck', which is used for roof structures (Fig. 1.25). The structure consists of a series of inverted pyramid-shaped frames which are linked at the top by fixing together adjacent sides, and at the bottom, by linking members.

The structure provides an extremely lightweight two-way slab, which could be supported by as little as one post at each corner.

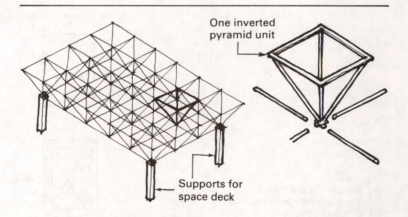

One inverted pyramid unit

Supports for space deck

Fig. 1.25 *Space deck.*

CHAPTER 2

External structure: foundations and walls

FOUNDATIONS

Introduction

With most traditional buildings, the walls are the major loadbearing element of the structure. The walls might bear:

- the combined loads of the building itself including roof, walls, floors;
- 'live' loads, such as people, furniture, equipment;
- wind loads.

Large-scale buildings, such as factories or multi-storey offices, usually employ a framed structure, where columns and beams are the major load-carrying elements. Here, the primary function of the walls is to provide weather protection. This chapter is mainly concerned with loadbearing external walls.

To prevent 'settlement' (sinking) of the building into the ground, the soil is required to provide an upward pressure equivalent to the downward pressure exerted by the building.

A foundation at the base of the walls is necessary to spread the load of the building over the soil (Fig. 2.1), thus reducing the pressure on the soil.

The nature of soil

For building purposes, soil consists of:

- 'topsoil', which is the soil immediately visible to the eye and usually consists of loose, organic material of 150–250 mm depth;
- 'subsoil', which lies between the topsoil and the earth's crust.

Subsoil is the material in which the foundation of a building is laid, and it may be classified as either:

(*a*) *Rock*: such as sandstone, hard chalk and limestone.

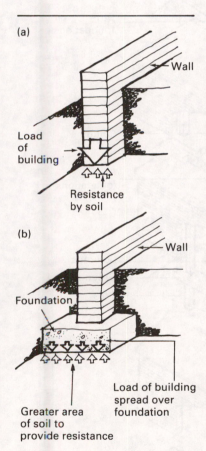

Fig. 2.1
(a) *Wall with no foundation*
(b) *Wall with foundation*

23

Table 2.1 Choice of type of foundation in accordance with soil and site conditions

Soil and site conditions	Possible type of foundation
Rock or solid chalk, sand and gravels	Shallow strips or pads
Firm, stiff clays with little vegetation liable to cause shrinkage or swellings	Strips (minimum 1 m below ground level) or piles with ground beam
Firm, stiff clays with trees close to the building	Piles and ground beam
Firm, stiff clays where trees have been recently felled and the ground is still absorbing moisture	Reinforced piles or thin reinforced rafts in conjunction with flexible building structure
Soft clays or soft, silty clays	Wide strips: up to 1 m wide or rafts
Peat or sites consisting partly of 'filled-in' soil	Piles driven down to a firm strata of subsoil
Where subsidence might be expected (e.g. mining districts)	Thin reinforced rafts

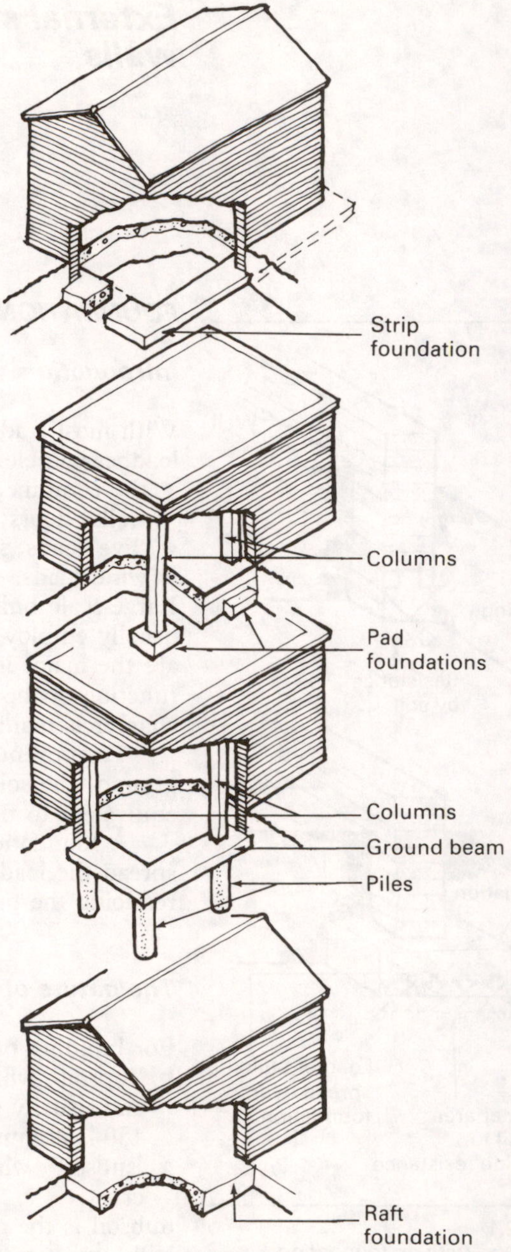

Strip foundation

Columns

Pad foundations

Columns

Ground beam

Piles

Raft foundation

(*b*) *Non-cohesive soils*: such as sand and gravel. Non-cohesive soils consist of large, closely packed, cohesionless particles which when loaded by a foundation, compress rapidly, but comparatively little. Variation of moisture levels in these soils result in little variation of volume ('shrinkage and swelling') of the soil.

(*c*) *Cohesive soils*: such as silts and clays which consist of fine particles separated by films of water which cause cohesion in the soil. When loaded by a foundation, the films of water are pushed out into areas of surrounding subsoil, resulting in compression of the subsoil. When the moisture level is increased in cohesive soils, the water films thicken out, and the soil swells. Conversely, when the moisture level is reduced, the water films are reduced, and the soil shrinks.

Site investigation

Before a foundation is designed, the nature of the soil must be established by a site investigation. This will involve careful observation of the site, to decide whether conditions such as fast-growing trees are likely to dry out the soil. Further investigation might involve:

- checking with local authorities to discover old drain lines, or areas of felled trees which could affect the future characteristics of the soil;
- digging inspection holes in order either to hand test the subsoil, or to provide samples of subsoil for laboratory analysis.

Types of foundation

Today, concrete is the material most commonly used for foundation construction. In the past, materials such as stone or brick were used. Table 2.1 illustrates some of the commonly used types of concrete foundation, and indicates the type of soil and site conditions that might influence the type of foundation to be used.

Strip foundations

Strip foundations are literally strips of concrete which are laid in trenches dug in the subsoil. Loadbearing walls may then be built off the top surface of the strip, and the remainder of the trench is filled in.

The design of a strip foundation must take into account the mode of excavation used for digging the trench. Manual excavation cannot be carried out unless the trench is more than 600 mm in width, since it would be impossible to lay the lower courses of bricks in a narrower trench. If a deep strip foundation is required (Fig. 2.2(a)) a trench of 300 mm or less could be excavated by a mechanical excavator. Here, the trench is almost filled with concrete, and the bricklayer has no difficulty in laying the first few courses of bricks. Strip foundations may be either 'shallow', 'deep', 'wide' or 'stepped' depending on the nature of the site and soil conditions.

Shallow strips (Fig. 2.1(b))   On firm, non-shrinkable soils, the depth of a strip foundation for a lightly loaded building may be as little as 600 mm. At depths less than this, however, the foundations could be disturbed by the action of frost causing the subsoil to swell.

Deep strips (Fig. 2.2(a))   Where the subsoil is of stiff or firm clay, and the building is lightly loaded, a deep strip foundation may be used. In shrinkable subsoils, such as clay, the depth of the foundation should be at least 1 m, since above this depth, differences in climatic conditions could cause shrinkage and swelling in the subsoil.

Buildings exerting heavier loads might require wider foundations (sometimes referred to simply as 'strip foundations' – Fig. 2.2(b))

Fig. 2.2
(a) *Deep strip foundation*
(b) *Strip foundation*
(c) *Steel-reinforced wide strip foundation*
(d) *Shear failure in wide foundation*
(e) *Stepped foundation*
(f) *Foundation supporting chimney breast*

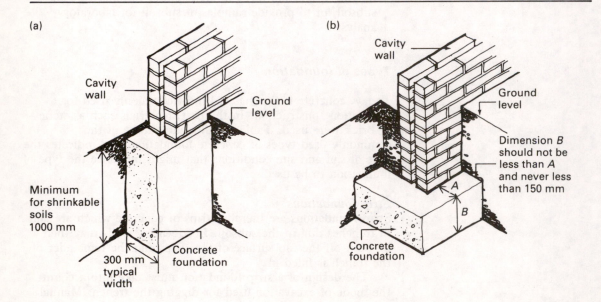

(a)

Cavity wall

Ground level

Minimum for shrinkable soils 1000 mm

300 mm typical width

Concrete foundation

(b)

Cavity wall

Ground level

Dimension B should not be less than A and never less than 150 mm

A

B

Concrete foundation

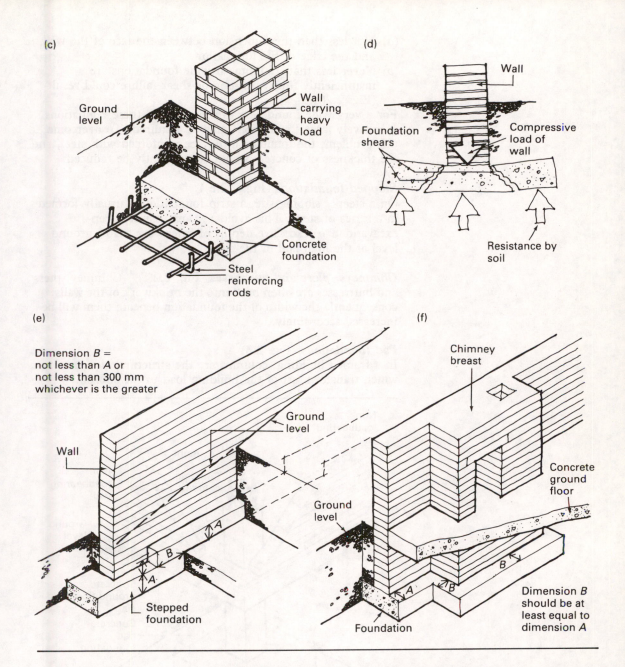

(c)

Ground level

Wall carrying heavy load

Concrete foundation

Steel reinforcing rods

(d)

Wall

Foundation shears

Compressive load of wall

Resistance by soil

(e)

Dimension B = not less than A or not less than 300 mm whichever is the greater

Wall

Ground level

A

B

A

Stepped foundation

(f)

Chimney breast

Ground level

Concrete ground floor

Ground level

A

B

B

Foundation

Dimension B should be at least equal to dimension A

Wide strips (Fig. 2.2(c)) Buildings exerting heavy loads on soft clay might require wide strip foundations. The Building Regulations require that the thickness of foundations should be:

(*a*) Not less than the dimension between the face of the wall and the edge of the foundation.

(*b*) Never less than 150 mm. If wide foundations are insufficiently thick, bending or shear failure could result (Fig. 2.2(d)).

For a very wide foundation to comply with these regulations, excessively large foundations may be required. To overcome this problem, the foundation may be reinforced with steel, and the thickness of concrete would consequently be reduced.

Stepped foundations (Fig. 2.2(e))

On a steeply sloping site, a strip foundation is usually formed in a series of steps. This avoids: (*a*) excessively deep excavation; (*b*) excessive depths of brickwork below ground level at the high end of the site.

Chimneys, piers and buttresses (Fig. 2.2(f))

Chimney, piers and buttresses are often built into the brickwork of the walls, consequently the width of the foundation beneath them will be increased accordingly.

Pad foundations (Fig. 2.3)

In a framed building, columns are the structural elements which transfer most of the building load to the subsoil.

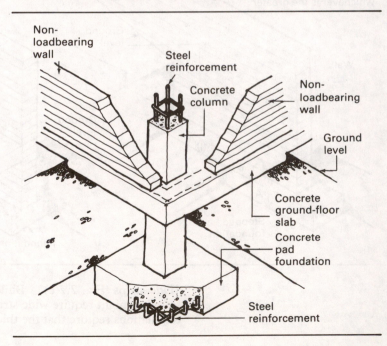

Fig. 2.3 *Pad foundation*

Usually a separate foundation will be made for each column, and this may take the form of a concrete 'pad'.

Pad foundations may also be used for isolated columns of buildings with loadbearing walls.

Raft foundations (Fig. 2.4)

Raft foundations are flat slabs of concrete formed near to ground level. The slab covers the entire ground-floor area of the building. The raft may project slightly beyond the outer face of the external walls. A raft foundation, therefore, combines the functions of both the foundation and ground floor of the building.

Fig. 2.4
(a) Effect of load distribution on raft foundation
(b) Raft foundation

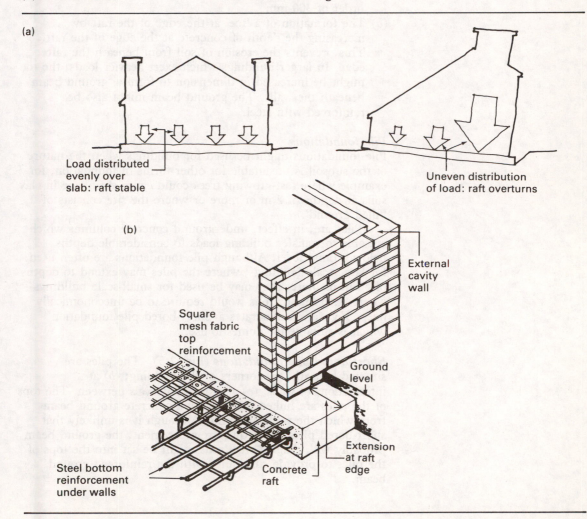

(a)

Load distributed evenly over slab: raft stable

Uneven distribution of load: raft overturns

(b)

External cavity wall

Square mesh fabric top reinforcement

Ground level

Steel bottom reinforcement under walls

Concrete raft

Extension at raft edge

Figure 2.4(a) illustrates the need to spread the loads of the building symmetrically on the raft, otherwise there would be a tendency for the raft to overturn.

Raft foundations for small buildings are usually in the order of 150 mm thick and are reinforced with square mesh steel fabric placed near the top surface of the slab. Where walls rest on the raft, reinforcement may also be placed towards the bottom of the raft beneath the walls. Features which may be incorporated into the raft slab include:

(a) An extension of the edge of the raft beyond the outer face of the external walls, which spreads the load of the wall over the soil. Such an extension will typically be in the order of 300 mm.

(b) The formation of a 'toe' at the edge of the raft by increasing the depth of concrete at the edge of the raft. This prevents the erosion of soil from beneath the raft edge. In larger buildings which exert heavier loads, the toe might be increased in dimension to form a 'ground beam' beneath the walls. The ground beam might also be reinforced with steel.

Pile foundations

Pile foundations might be used for buildings when the nature of the subsoil is unsuitable for other forms of foundation, for example where fast-growing trees could cause shrinkage in clay soils at depths of 2 m or more or where the site consists of 'filled ground'.

Piles are, in effect, underground concrete columns which are able to transfer building loads to considerable depths below ground level. Although pile foundations are often used for large-scale buildings, where the piles may extend to depths of 50 m or more, they may be used for small-scale buildings where strip foundations would require to be uneconomically deep. Figure 2.5 illustrates a 'short-bored pile foundation' which is suitable for domestic use.

Short-bored pile foundations (Fig. 2.5)

The piles are situated beneath: (a) corners of the building; (b) at intersections of walls; (c) at regular intervals between. The tops of the piles are linked together by a concrete ground beam from which the walls are built. Although it is unlikely that short-bored piles will require reinforcement, the ground beam probably will. Angled steel rods might be set into the tops of the piles to provide continuity with the reinforced ground beam.

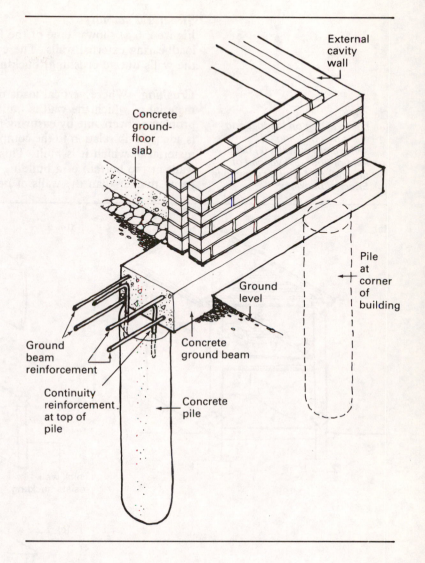

Labels on figure:
- External cavity wall
- Concrete ground-floor slab
- Pile at corner of building
- Ground level
- Ground beam reinforcement
- Concrete ground beam
- Continuity reinforcement at top of pile
- Concrete pile

Fig. 2.5 *Short-bored pile foundation*

EXTERNAL WALLS

Functional requirements

Some of the principal functional requirements of external walls are: (*a*) structural stability; (*b*) resistance to moisture; (*c*) resistance to fire; (*d*) sound insulation; (*e*) thermal insulation; (*f*) security.

Structural stability

Figure 2.6(a) shows some of the loads that might be applied to loadbearing external walls. These loads could cause failure in the walls due to crushing, buckling or overturning.

Crushing   Where vertical loads on a wall are excessive, the material of which the wall is built could be crushed. The problem is overcome by ensuring that the thickness of the wall is adequate in relation to the compressive strength of the material of which it is built. This could necessitate loadbearing walls at the bottom of a building (where the loading is greater) being thicker than the walls of upper storeys.

Fig. 2.6
(a) *Loads acting on loadbearing walls of a building (1) self-weight of walls (2) dead load of floor (3) live load of floor (4) roof load (5) wind load*
(b) *Effect of buckling*

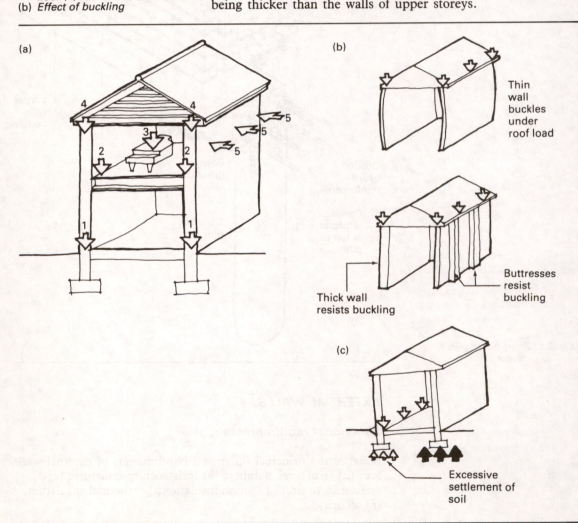

(a)

(b) Thin wall buckles under roof load

Thick wall resists buckling

Buttresses resist buckling

(c) Excessive settlement of soil

Buckling   Where a wall is tall in relation to its thickness (i.e. when the slenderness ratio is high), buckling could result. Buckling may be avoided by either increasing the thickness of the wall, or by stiffening the wall by the use of corners in the wall, junctions with internal walls or by buttresses (Fig. 2.6(b)).

Overturning   Excessive settlement of the soil on one side of a building, could cause an overturning tendency (Fig. 2.6(c)) due to the soil being overstressed. The problem could be overcome by increasing the width of the foundation.

Resistance to moisture

Moisture may pass through an external wall and cause dampness to occur on the internal face of the wall. Two major causes of dampness are: (*a*) rain penetration; (*b*) damp rising from the ground.

Rain penetration   Most external walls are built of materials which are, to an extent, porous. Consequently, rain may pass through the pores to the internal surface of the wall. Measures to prevent this include:
- use of relatively impervious materials, either in the construction of the wall itself (e.g. dense brickwork) or as a surface cladding (such as tiles or boards nailed to battens fixed over a sheet of impervious material (Fig. 2.7(a));
- building the wall of sufficient thickness to ensure rain is incapable of penetrating through its entire thickness;
- use of 'cavity wall' construction where the wall is built in two separate 'skins' or 'leaves' (Fig. 2.7(b)). Although rain could penetrate through the outer skin, it would be prevented, by the cavity, from passing through to the inner skin.

Damp rising from the ground   Moisture in the soil could rise up through the walls due to 'capillary attraction' (the ability of liquids to be drawn through minute passageways such as the pores in a brickwork wall). To prevent the damp rising to the inner surfaces of the building, a 'damp-proof course' (d.p.c.) is installed.

Resistance to fire

External walls should hinder the spread of fire from one building to neighbouring buildings. This may be achieved by:
- use of suitable materials and forms of construction;
- suitable distribution of 'unprotected area'.

Table 2.2 Fire resistance of loadbearing walls

Wall construction		Minimum thickness of wall (excluding plaster) for fire resistance of:				
		4 hr	2 hr	1½ hr	1 hr	½ hr
Solid wall Clay concrete or sand-lime bricks. Minimum thickness referred to in table	(a) No plaster	200	100	100	100	100
	(b) 12.5 mm thick gypsum/sand plaster	200	100	100	100	100
	(c) 12.5 mm thick vermiculite/sand plaster	100	100	100	100	100
Solid wall Concrete blocks of 'Class I aggregate' (lightweight aggregate such as pumice or foamed slag) Minimum thickness referred to in table	(a) No plaster	150	100	100	100	100
	(b) 12.5 mm thick gypsum/sand plaster	150	100	100	100	100
	(c) 12.5 mm thick vermiculite/sand plaster	100	100	100	100	100
Cavity wall Outer leaf of clay, concrete or sand-lime bricks of minimum 100 mm thickness. Minimum thickness referred to in table	(a) Internal leaf of clay, concrete or sand-lime bricks	100	100	100	100	100
	(b) Internal leaf of solid or hollow concrete blocks of Class 1 aggregate	100	100	100	100	100

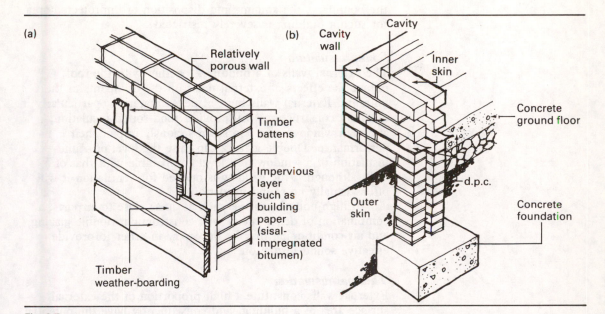

(a)

Relatively porous wall

Timber battens

Impervious layer such as building paper (sisal-impregnated bitumen)

Timber weather-boarding

(b) Cavity wall

Cavity

Inner skin

Concrete ground floor

d.p.c.

Outer skin

Concrete foundation

Fig. 2.7
(a) *Cladding on external wall*
(b) *Damp-proof course in cavity wall*

Materials and construction   Building materials may be classified as either 'combustible' or 'non-combustible' depending on their behaviour under standard fire tests. Examples of combustible materials include timbers, plastics and bitumen felt, examples of non-combustible materials include asbestos cement, concrete, metals, brick, stones, fibreglass insulation and gypsum and vermiculite plasters.

The Building Regulations stipulate the degree of 'fire resistance' required for various elements of a structure. Fire resistance relates to the length of time that a building element (such as a door, wall, floor) will survive under standard test conditions. Table 2.2 shows examples of external wall constructions, and the degree of fire resistance offered by them.

Unprotected area   Whilst external walls may be constructed mainly of non-combustible material, elements such as doors, windows and combustible claddings could reduce the ability of the wall to hinder the spread of fire from one building to another. Such elements are referred to in the Building Regulations as 'unprotected areas'. The regulations specify limits for the amount and the disposition of unprotected area in relation to the distance of the external wall from the boundary of the property. Since the risk of spread of fire from one building to another is increased if the building is close to the boundary, the amount and disposition of unprotected area for such a building is severely restricted.

Sound insulation

The external walls of a building, together with the roof, reduce the effects of external noise on the occupants of the building. External walls built of dense materials, particularly if of cavity construction, will provide good sound insulation, although windows and doors can seriously impair their performance. Double glazing can raise the level of sound insulation of a window to a level comparable with that of a wall, although if the window is opened for ventilation it will offer virtually no sound insulation.

In high-noise localities, such as areas close to airports, a combination of dense cavity wall construction, double glazing and air-conditioning may be necessary in order to provide effective sound insulation.

Thermal insulation

External walls constitute a high proportion of the external surface area of a building, and consequently have the potential for losing a lot of heat from a building. Thin dense materials, such as glass, offer little resistance to the passage of heat, whereas comparatively lightweight and thick materials, such as aerated concrete, offer much better resistance. The provision of airspaces such as the cavities in double glazing or cavity walling can greatly improve thermal insulation.

Table 2.3 shows the U values of various types of wall construction. The U value of a structure is the amount of heat that will pass through 1 square metre (m^2) of the structure, when the difference of temperature between the inner and outer surfaces of the structure is 1 °C. The figures take into account:

- the 'thermal conductivity' of each material in the construction (i.e. the ability of the material to resist the passage of heat);
- the thickness of each material;
- the resistance to the passage of heat offered by cavities in the structure;
- an allowance for the outer and inner surfaces of the structure to take account of such factors as the effect of cooling winds, the reduced effectiveness of corrugated surfaces, etc.

It is important to note that the lower the U value, the better is the thermal insulation offered by the structure. The Building Regulations 1991, in approved document L require a U value of 0.45 for external walls.

Security

The majority of burglaries in buildings take place via the windows and doors of the ground floor. The back of a building is particularly vulnerable since it is less overlooked or exposed to street lighting. Measures to combat burglaries include:

- strengthening door frames;
- provision of security locks and bolts for windows and doors;
- improving the construction used for doors and windows;
- installing alarm systems.

Table 2.3 Examples of *U* values for external wall constructions

Wall construction	*U* value (W/m K)
Cavity walls	
102.5 mm brickwork outer leaf 50 mm cavity 102.5 mm brickwork inner leaf 16 mm dense plaster	1.48
As above, but 75 mm cavity filled with mineral wool slabs	0.39
102.5 mm brickwork outer leaf 75 mm cavity 102.5 mm concrete block inner leaf 16 mm dense plaster	1.24
As above, but 50 mm mineral fibre batts in cavity fitted against inner leaf	0.44
102.5 mm brickwork outer leaf 50 mm cavity 100 mm aerated concrete block inner leaf 16 mm dense plaster	0.99
As above, cavity filled with glass mineral fibre slabs	0.45
Solid walls	
As above, only dense plaster replaced by 60 mm mineral fibre slabs fitted between timber battens covered by vapour barrier and 12.5 mm plasterboard	0.45
As above, only 80 mm mineral fibre slabs	0.36

Table 2.4 Type and characteristics of bricks

Brick-manufacturing process	Typical examples	Colours	Weight (kg)	Compressive strength (N/mm²)
Clay Clay dug, formed into brick shapes, dried, then fired in a kiln 1. Pressed bricks. Clay pressed into moulds under high pressure. Bricks have frogs. Most bricks are made by this process 2. Wire-cut bricks. Clay extruded and cut into lengths by wires. No frogs, but perforation may be easily formed 3. Handmade bricks. Clay placed in moulds by hand. Provides attractive textured surface	**Commons** Economically made. Very regular sizes and sharp arrises. Rather unattractive appearance	Usually pink	1.82–2.27	3.5–70.0 and over
	Facings Designed for good appearance. Colour may be natural or applied (such as sand applied to the face of common bricks)	Very wide range, e.g. natural: Leicester reds and yellow London stocks; applied: brown, grey, red and yellow sand-faced Fletton bricks	2.32–3.10	5–45
	Engineering Heavy, smooth bricks of high compressive strength and low porosity	Accrington reds	3.0–3.3	Class A: 70+
		Staffordshire blues		Class B: 49+
Sand–limes Sand, lime and pigment are mixed, moulded under high pressure, then steamed in an autoclave Produces uniform, regular bricks with sharp arrises	Eight different classes with varying compressive strengths. Generally, the higher the class, the greater the durability and resistance of the bricks to severity and exposure of site conditions	Mainly pastel shades: whites, creams, pinks, blues, browns, lilacs	3.0–3.2	Class 7: 48 Class 5: 34 Class 4: 27 Class 3A: 20 Class 3B: 20 Class 2A: 14 Class 2B: 14 Class 1: 7
Concrete Concrete is compacted into moulds under high pressure producing bricks of regular size with sharp arrises	Special Class A Class B	Wide variety of pigments available Textures vary, in accordance with aggregate used	Wide variation in accordance with aggregate used	Special: 17 Class A: 12 Class B: 7

Bricks

England is rich in clay deposits suitable for making bricks, and consequently brickwork is the predominant material for the construction of external walls for small-scale buildings.

Although clay is the material from which most bricks are made, sand–lime and concrete bricks are also manufactured. Table 2.4 provides details of the manufacturing processes and characteristics of different bricks. The traditional imperial dimensions of a brick have proved to be ideal in size for ease of handling and laying. The bricks also had useful modular characteristics. Metric brick dimensions were adopted which retained similar handling and modular characteristics to the old imperial-sized bricks. Figure 2.8(a) shows the standard

Fig. 2.8
(a) *Size and modular characteristics of bricks*
(b) *Standard and cut bricks*
 (i) *Features of a standard brick*
 (ii) *Half bat*
 (iii) *Queen closer*
 (iv) *King closer*
 (v) *Queen closer used in termination of one-brick English bond wall*
 (vi) *King closers used in pier construction in one-brick Flemish bond wall*
 (vii) *Half bat used in one-and-a-half-brick column*

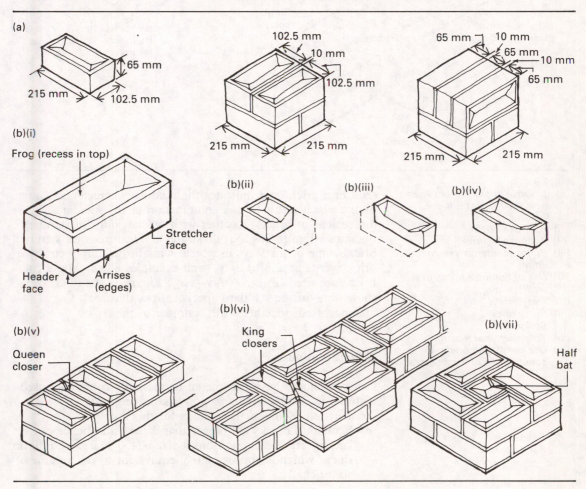

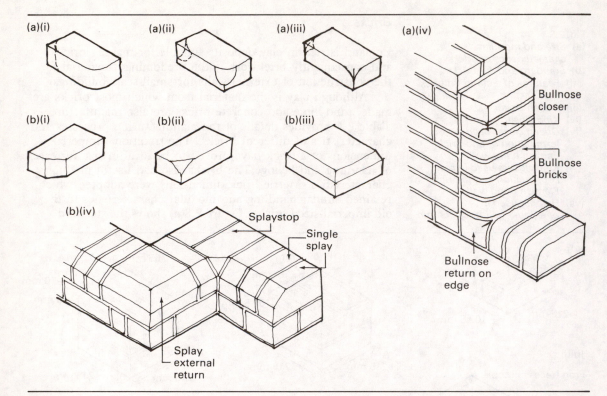

Fig. 2.9
(a) *Bullnose bricks (dotted lines
    indicate double bullnose
    feature)*
 (i) *Bullnose brick*
 (ii) *Bullnose return*
(iii) *Bullnose internal return on
     edge*
(iv) *Use of bullnose bricks in a
     wall opening*
(b) *Splay bricks*
 (i) *Splay brick*
 (ii) *Splay stop*
(iii) *Splay external return*
(iv) *Use of splayed bricks in
     construction of a wall recess*

sizes of a brick, and indicates the modular characteristics of bricks which result from a combination of the standard dimensions of bricks, together with mortar joints of standard thickness. Figure 2.8(b)(i) illustrates the features of a standard brick, some of the ways in which it might be simply cut by a bricklayer (Fig. 2.8(b)(ii–iv)) and examples of how these cut bricks are used (Fig. 2.8(b)(v–vii)). Figures 2.9(a) and (b) show some of the standard special bricks that are manufactured, together with examples of their use.

Concrete blocks

Concrete blocks have become an increasingly commonly used material for loadbearing and non-loadbearing wall construction. Some of the reasons for this are:
(a) Economy, blocks are bigger than bricks and hence can be laid faster. (A common block size is $450 \times 225 \times 100$ mm thick, which is approximately equivalent to the volume of six bricks.)

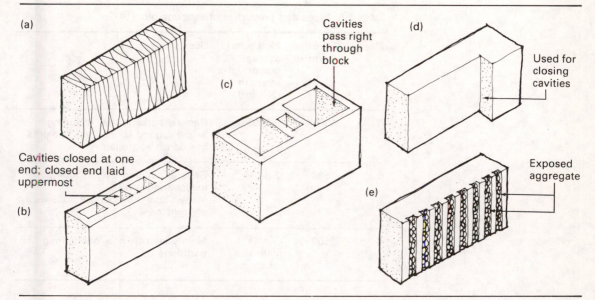

(a)

Cavities pass right through block

(d)

Used for closing cavities

(c)

Cavities closed at one end; closed end laid uppermost

(b)

(e)

Exposed aggregate

Fig. 2.10
(a) *Solid concrete block*
(b) *Cellular block*
(c) *Hollow block*
(d) *Reveal block*
(e) *Facing block*

(*b*) Most blocks provide better thermal insulation than most bricks.

(*c*) The lighter-density blocks can be very easily cut (or even sawn) and fixing may be easily made by nailing or drilling and screwing into the block.

Figure 2.10 illustrates some of the types of blocks used, typical sizes being 450 × 225 mm with thicknesses of 75, 100, 140 and 215 mm.

Concrete blocks are made of cement, aggregate (which may be lightweight, dense, natural or synthetic) and water. Additives may be included in the mix for purposes such as:

- reduction of colour variation in the blocks;
- to adjust the quality of the concrete mix;
- to generate gas within the mix for production of aerated concrete blocks.

Dense and lightweight aggregate concrete blocks are made by compacting the mixed concrete into a mould box, immediately removing the box, then curing the blocks. Aerated blocks are made by placing the concrete mix (including a chemical additive in order to aerate the mixture) in a large steel box. After a few hours, the large block of aerated concrete is cut into individual blocks by means of taut wires. The blocks are then cured in a high-pressure steam autoclave.

Concrete blocks are classified as either Class A, B or C. Table 2.5 indicates some of the properties and uses of each class of block.

Table 2.5 Types and properties of concrete blocks

Class	Density (kg/m³)	Minimum average compressive strength (N/mm²)	Uses
A	1500+	3.5–35.0	General building use, including below ground level. Class A blocks use dense aggregates
B	1500–	2.8–7.0	General building use, including loadbearing walls Many Class B blocks use lightweight aggregates
C	1500–	Varies with size of block	Mainly internal non-loadbearing partitions

Mortar

The principal function of mortar is to bind together individual bricks or blocks to form a homogeneous wall structure. Since bricks are laid rapidly, it is important that the mortar develops a sufficiently high initial strength to prevent the mortar from being squeezed out by the weight of courses of bricks laid above it.

To ensure consistent colour and strength, the constituents of the mortar are carefully measured in a gauge box prior to mixing.

Mortar mixes

Cement and sand mortar Cement and sand are mixed together dry, the water is added, producing an extremely hard but brittle mortar which develops rapid early strength. The disadvantage of this mortar is its inflexibility: movements in the wall due to moisture, heat or settlement could result in cracks developing in the mortar, which might extend through the bricks. Consequently, cement–sand mortar should only be used with engineering quality bricks where they are specified for locations such as work below ground level, or in the formation of sills, coping stones or parapet walls.

Cement, lime and sand mortars Although cement–lime–sand mortar is less strong than cement–sand

mortar, it is more pliable. The mortar may be made from
either:

- the dry materials which are delivered to the site, then
  mixed in the proportions of 1 : 1 : 6 for external brickwork
  or 1 : 2 : 9 for internal work;
- gauged mortar, where a wet mixed 1.5 or 2.6 lime–sand
  mortar is delivered to the site, then one part of cement is
  added. This method is more convenient than the above
  method and is consequently largely replacing it.

Cement and sand with plasticisers   Plasticisers are chemical
compositions which, when added to the mortar, improve its
workability by generating minute bubbles in the mix. The
resulting mortar has a reduced tendency to drying shrinkage,
and has increased frost resistance. A typical mix might be
1 : 6 (cement : sand) plus the plasticiser.

Mortar joints

In order to provide the modular characteristics of brickwork,
mortar joints are usually laid to a thickness of 10 mm. Figure
2.11 illustrates some of the different profiles used for mortar
joints. The recessed joint should only be used in conjunction
with dense bricks which are able to resist the effects of rain
laying on the exposed horizontal surface.

Fig. 2.11
(a) *Flush joint*
(b) *Struck weathered joint*
(c) *Square recessed joint*
(d) *Ironed joint*

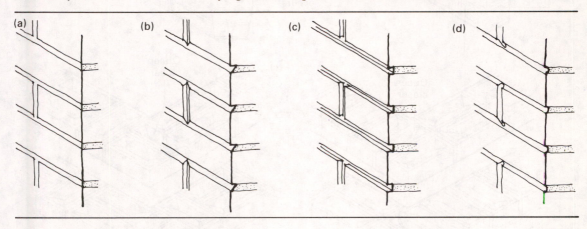

Wall construction

In small-scale buildings, loadbearing walls are usually
constructed of brickwork, concrete blockwork or a
combination of both, using either solid or cavity wall
construction.

Solid walls in brickwork and bonding of brickwork

Brick walls are built in 'courses'. A course of brickwork is a horizontal layer of bricks with vertical mortar joints between each brick. To ensure stability in a wall, it is important that the vertical joints of successive courses of brickwork should not coincide. This is achieved by 'bonding' which is an arrangement of brick courses with staggered vertical mortar joints.

Stretcher bond (Fig. 2.12(a)) Stretcher bond is used for walls

Fig. 2.12
(a) *Stretcher bond*
 (i) *Half-brick wall*
(ii) *Failure in a one-brick stretcher bond wall*
(b) *English bond*
(c) *Flemish bond*

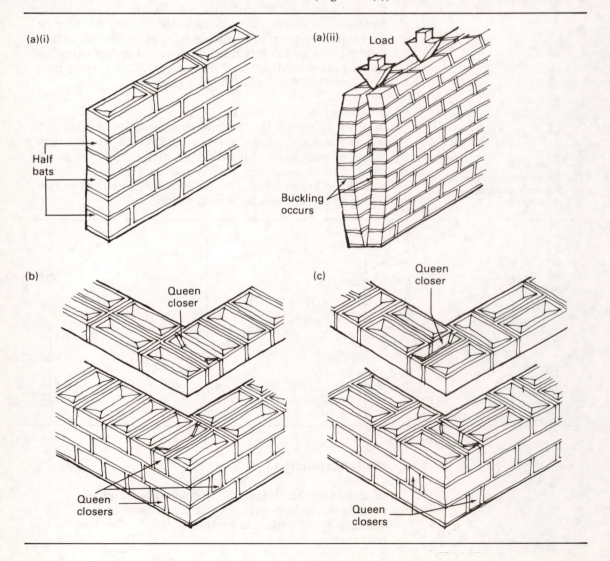

(a)(i) Half bats

(a)(ii) Load — Buckling occurs

(b) Queen closer — Queen closers

(c) Queen closer — Queen closers

which are the thickness of half a brick ('half-brick walls') and it is commonly used for the external skin of cavity walls. To terminate the wall, or to form openings, a half-bat brick is necessary for each alternate course.

Where a wall of one brick thickness ('one-brick wall') is required, stretcher bond is unsuitable, since it would result in a vertical joint midway along the entire wall thickness. If such a wall was loaded, the wall might split and buckle (Fig. 2.12(a)(ii)). A one-brick wall will require a bond such as English or Flemish bond.

English bond (Fig. 2.12(b)) English bond walls employ alternate courses of headers and stretchers. In order to maintain the bond at corners or terminate the wall, queen closers are incorporated.

Flemish bond (Fig. 2.12(c)) In Flemish bond walls, each course consists of alternate headers and stretchers. Again, queen closers will be incorporated to maintain the bond at openings, or to terminate the wall.

Cavity walls

Cavity walls consist of two separate parallel walls (referred to as 'skins' or 'leaves') with an air space ('cavity') between (Fig. 2.7(b)).

The outer skin is often of half-brick stretcher bond brickwork and the inner skin of lightweight concrete blockwork. The cavity is usually of 50–75 mm thickness completely or partially filled with insulation material.

To prevent the skins from buckling when loaded they are tied together with plastic or metal cavity-wall ties installed at 450 mm intervals vertically and 900 mm horizontally. The broad ends are bedded in the mortar of each skin whilst the drip prevents moisture in the outer skin travelling across the tie to the inner skin.

Wall ties with discs are used to hold insulation boards against the inner skin of partially filled cavities (Fig. 2.13(d)).

The cavity below ground is filled with fine concrete up to a level of 75 mm below d.p.c. level.

Damp-proof courses

A damp-proof course (d.p.c.) is a layer of impervious material which is sandwiched between the mortar joints of a wall to prevent the passage of moisture from one course of bricks to another.

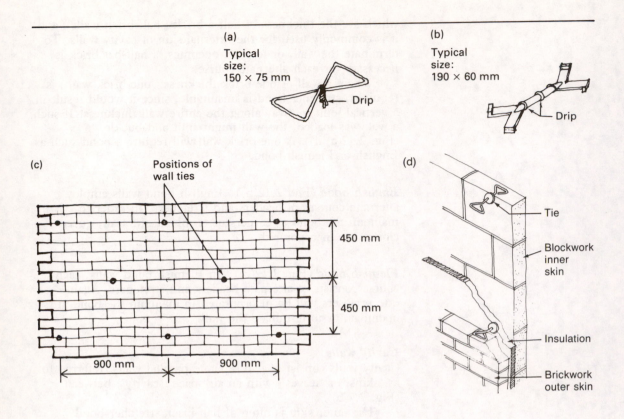

(a)
Typical
size:
150 × 75 mm
← Drip

(b)
Typical
size:
190 × 60 mm
← Drip

(c)
Positions of
wall ties

450 mm

450 mm

900 mm   900 mm

(d)
Tie

Blockwork
inner
skin

Insulation

Brickwork
outer skin

Fig. 2.13
(a) *Butterfly wall tie (galvanised
    steel wire)*
(b) *Polypropylene wall tie*
(c) *Spacing of wall ties*
(d) *Cavity wall insulation and
    ties*

Damp-proof courses at the bases of walls are installed at a level of at least 150 mm above the adjoining ground level. In cavity walls a separate d.p.c. is installed in each skin. As will be seen elsewhere in this book, d.p.c.s are also necessary around openings in cavity walls, in parapet walls and in chimney stacks.

The traditional material used for d.p.c.s was slate, which was installed in two layers with overlapping joints. Materials used more commonly today include:

- bitumenised felt, which sometimes incorporates materials such as asbestos and lead;
- black polythene.

Both of these materials are obtained in rolls, the widths of which are suitable for standard wall thicknesses of up to 450 mm and in lengths of typically 8 m. At corners, or at one end of a roll, the d.p.c. is overlapped by at least 100 mm.

OPENINGS IN EXTERNAL WALLS

Figure 2.14 illustrates some of the terms associated with openings in walls:

- *head*: the area of wall above an opening;
- *jambs*: the area of wall either side of an opening;
- *sill*: the area of wall below a window opening;
- *threshold*: the area of wall below a door opening;
- *soffit*: the horizontal area within the thickness of a wall below the head of an opening;
- *reveals*: the vertical areas within the thickness of a wall at the jambs of an opening;
- *lintel*: a beam which spans an opening and carries the load of the wall and other structure above the opening.

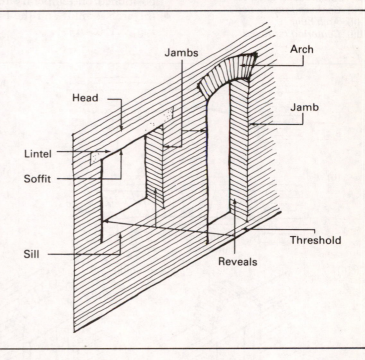

Fig. 2.14 *Terms used in wall openings*

Arches and lintels

Arches

An arch is a means of spanning an opening by use of small units such as bricks or stones. Figures 2.15(a) and (b) illustrate two common profiles of brick arch, respectively 'segmental' and 'flat'.

The 'rough ring' arch (Fig. 2.15(a)(ii)) is constructed of

uncut bricks, and the mortar joints are varied in thickness in order to form the curvature of the arch. The 'axed' arch (Fig. 2.15(a)(iii)) is constructed of bricks which are cut into wedge shapes (known as 'voussoirs') and are set with parallel 10 mm mortar joints. Voussoirs may also be made from bricks rubbed to a wedge shape, using soft bricks known as 'red rubbers' which are set in 1–2 mm thick joints of white lime putty. Flat arches tend to give an illusion of sagging, and to counteract this a small camber is introduced to the soffit (Fig. 2.15(b)).

Figure 2.15(c) shows the construction stages for forming an arch, using a temporary profiled timber support known as a centre:

- the jambs of the wall are built and the centering is positioned on temporary timber posts; Fig. 2.15(c)(i)
- the arch is built and the brickwork is carried over the arch; Fig. 2.15(c)(ii)

Fig. 2.15
(a) *Segmental arch*
 (i) *Profile of segmental arch*
 (ii) *Use of standard bricks to form a 'rough ring arch'*
 (iii) *Use of voussoirs to form an 'axed' or 'rubbed' arch*
(b) *Flat arch*
(c) *Construction of an arch*
 (i) *Centering positioned*
 (ii) *Arch built*
 (iii) *Centering removed*

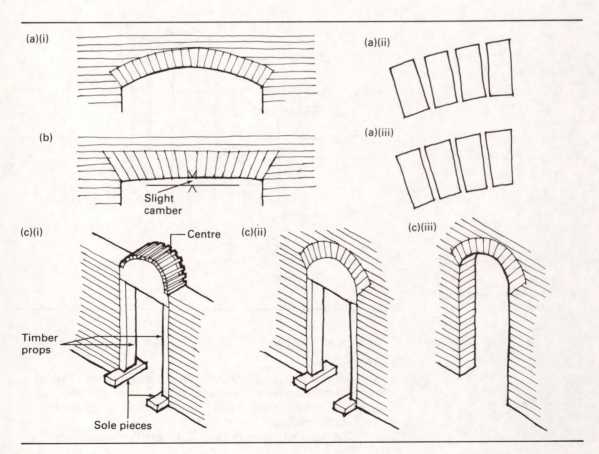

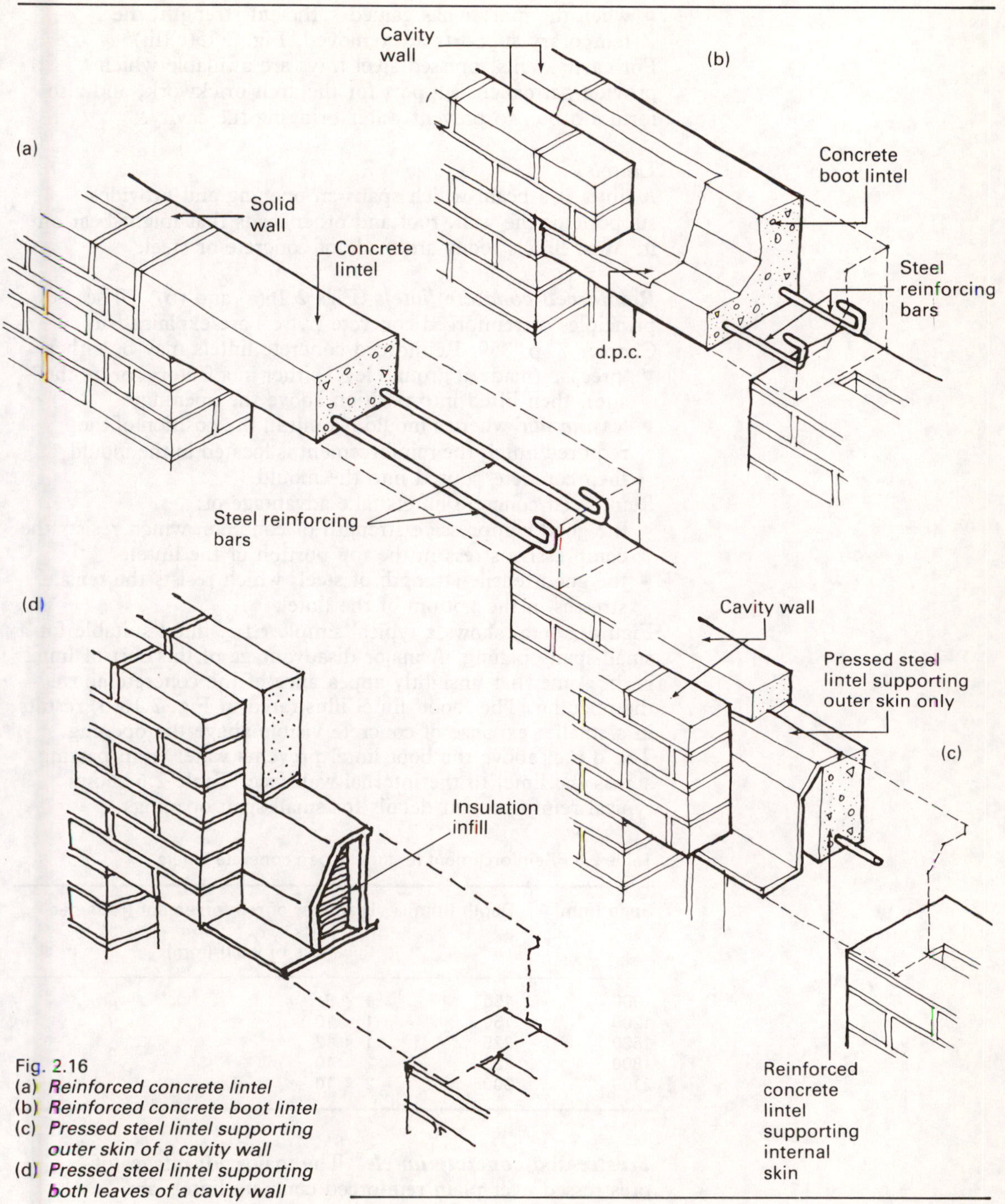

(a)
Solid
wall
Concrete
lintel
Steel reinforcing
bars
(b)
Cavity
wall
Concrete
boot lintel
Steel
reinforcing
bars
d.p.c.
(d)
Insulation
infill
(c)
Cavity wall
Pressed steel
lintel supporting
outer skin only
Reinforced
concrete
lintel
supporting
internal
skin

Fig. 2.16
(a) *Reinforced concrete lintel*
(b) *Reinforced concrete boot lintel*
(c) *Pressed steel lintel supporting outer skin of a cavity wall*
(d) *Pressed steel lintel supporting both leaves of a cavity wall*

- when the mortar has gained sufficient strength, the temporary supports are removed. Fig. 2.15(c)(iii)

For cavity walls, pressed steel trays are available which provide permanent support for the arch brickwork, and also form a d.p.c. to prevent water bridging the cavity.

Lintels

A lintel is a beam which spans an opening and provides support for the wall, roof and other loads that might bear on it. Most lintels today are made of concrete or steel.

Reinforced concrete lintels (Figs 2.16(a) and (b))   The principles of reinforced concrete (r.f.c.) are explained in Chapter 7, p. 359. Reinforced concrete lintels may be either:
- 'precast' (made at ground level either in a factory, or on the site), then lifted into position above the opening;
- 'cast *in situ*' where a mould is built in the position of the required lintel, the reinforcement is located in the mould, then concrete poured into the mould.

Reinforced concrete lintels take advantage of:
- the good compressive strength of concrete, which resists the compressive stress in the top portion of the lintel;
- the good tensile strength of steel, which resists the tensile stresses in the bottom of the lintel.

Figure 2.16(a) shows a typical simple r.f.c. lintel suitable for a small-span opening. A major disadvantage of this sort of lintel is the somewhat unsightly appearance of the concrete above the opening. The 'boot' lintel illustrated in Fig. 2.16(b) results in a smaller expanse of concrete visible above the opening. The d.p.c. above the boot lintel prevents water from passing across the lintel to the internal wall face. Table 2.6 shows typical reinforcement details for small-span openings.

Table 2.6   Reinforcement for small-span concrete lintels

Span (mm)	Depth (mm)	Diameter of reinforcement (rods per half-brick thickness of wall) (mm)
900	150	1 × 8
1200	150	1 × 10
1500	225	1 × 12
1800	225	2 × 10
2100	300	2 × 10

Prestressed concrete lintels   The major advantages of prestressed over plain reinforced concrete lintels are:

- they are light in weight and easily handled;
- they are shallow, and therefore less unsightly than the r.f.c. lintel illustrated in Fig. 2.16(a): a span of between 1.8 and 2.1 m would probably require a lintel of only 75 mm depth.

Pressed steel lintels (Figs 2.16(c) and (d))   Figure 2.16(c) illustrates a pressed steel lintel which supports the outer skin of a cavity wall. The inner skin is supported by a conventional r.f.c. lintel. Figure 2.16(d) illustrates a similar lintel which supports both the outer and inner skins of the wall.

The advantages of these lintels include:
- their lightness in weight, which makes them easy to handle;
- a separate d.p.c. is unnecessary due to the shape of the lintel and because steel is impervious to water;
- they are scarcely visible, once installed.

Typical lintel sizes are:

Supporting both skins: span 900–2100 mm; depth 145 mm
span 2100–4200 mm; depth 220 mm
Supporting outer skin: span 900–1500 mm; depth 145 mm
span 1500–3600 mm; depth 220 mm

Jambs, sills and thresholds

Jambs

Figure 2.17(a) shows a corner detail of a window opening in a solid 1½-brick wall. The jamb is 'rebated', that is, the outside face of the wall projects slightly into the window opening, as was often the case when 'sliding sash windows' were installed.

In cavity wall construction, the cavity has to be 'closed' at the jamb. This may be done by means such as:
- returning the outer skin across the cavity to abut against the inner skin (Figs 2.17(b) and (d)) and setting the window or door frame towards the inner face of the reveal;
- returning the inner skin across the cavity to abut against the outer skin (Figs 2.17(c) and (e)), and setting the window or door frame towards the outer face of the reveal.

In both cases, a vertical d.p.c. is installed between the abutment of the inner and outer skins to prevent the passage of moisture to the internal face of the wall.

Sills and thresholds

Sills and thresholds are required to prevent the ingress of water at the bottom of a window or door opening. Since windows and doors are designed to be impervious to water, the quantity of rain-water driven to the sill can be

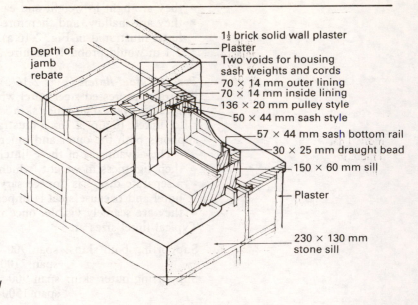

1½ brick solid wall plaster
Plaster
Two voids for housing sash weights and cords
70 × 14 mm outer lining
70 × 14 mm inside lining
136 × 20 mm pulley style
50 × 44 mm sash style
57 × 44 mm sash bottom rail
30 × 25 mm draught bead
150 × 60 mm sill
Plaster
230 × 130 mm stone sill

Depth of jamb rebate

Fig. 2.17
(a) *Sliding sash window installed in wall with rebated reveals*

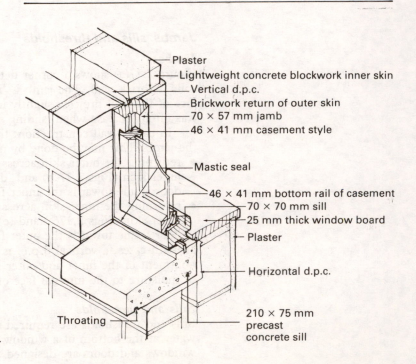

Plaster
Lightweight concrete blockwork inner skin
Vertical d.p.c.
Brickwork return of outer skin
70 × 57 mm jamb
46 × 41 mm casement style
Mastic seal
46 × 41 mm bottom rail of casement
70 × 70 mm sill
25 mm thick window board
Plaster
Horizontal d.p.c.
210 × 75 mm precast concrete sill

Throating

Fig. 2.17
(b) *Timber casement window installed in cavity wall – outer leaf closes the cavity*

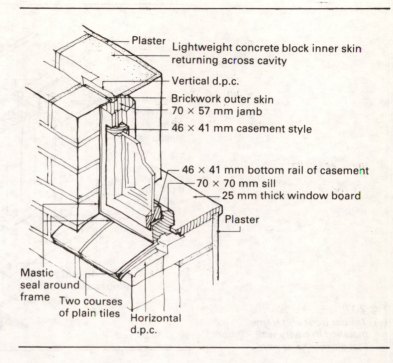

Fig. 2.17
(c) *Timber casement window installed in cavity wall – inner leaf closes the cavity*

Labels:
- Plaster
- Lightweight concrete block inner skin returning across cavity
- Vertical d.p.c.
- Brickwork outer skin
- 70 × 57 mm jamb
- 46 × 41 mm casement style
- 46 × 41 mm bottom rail of casement
- 70 × 70 mm sill
- 25 mm thick window board
- Plaster
- Mastic seal around frame
- Two courses of plain tiles
- Horizontal d.p.c.

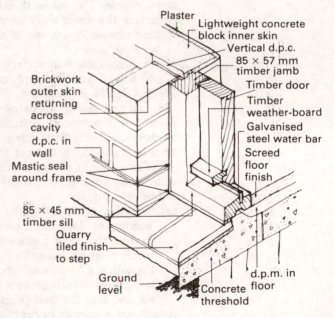

Fig. 2.17
(d) *Timber door and frame installed in cavity wall – outer leaf closes the cavity*

Labels:
- Plaster
- Lightweight concrete block inner skin
- Vertical d.p.c.
- 85 × 57 mm timber jamb
- Timber door
- Timber weather-board
- Galvanised steel water bar
- Screed floor finish
- Brickwork outer skin returning across cavity
- d.p.c. in wall
- Mastic seal around frame
- 85 × 45 mm timber sill
- Quarry tiled finish to step
- Ground level
- Concrete threshold
- d.p.m. in floor

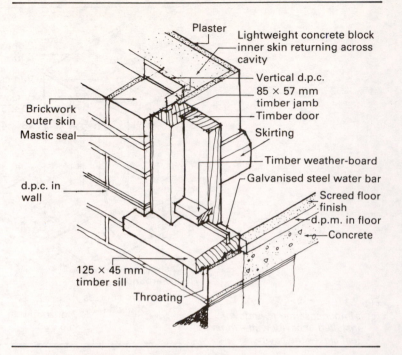

Plaster

Lightweight concrete block inner skin returning across cavity

Vertical d.p.c.

85 × 57 mm timber jamb

Timber door

Skirting

Brickwork outer skin

Mastic seal

Timber weather-board

Galvanised steel water bar

d.p.c. in wall

Screed floor finish

d.p.m. in floor

Concrete

125 × 45 mm timber sill

Throating

Fig. 2.17
(e) *Timber door and frame installed in cavity wall – inner leaf closes the cavity*

considerable. The sill or threshold should be capable of deflecting the water away from:

- the vulnerable junction of the window frame and sill of the opening; and
- away from the wall surface below the sill.

Sills and thresholds are made of impervious material. Figures 2.17(a–e) illustrate sills of stone, concrete and clay tile, and thresholds of quarry tile and timber (which would be painted if of softwood, or treated with a clear finish if of hardwood). These materials prevent water from becoming absorbed into the bottom of the opening.

The shape of the sill or threshold is designed to deflect water away from the opening. This may be achieved by the following methods:

(a) 'Weathering' (sloping) the top surface away from the window or door frame and projecting the sill or threshold beyond the outer surface of the wall. This throws the water clear of the wall surface.

(b) Providing a 'throating' underneath the sill. The throating is a groove which prevents water being driven under the sill and into the wall (Figs 2.17(b) and (e)).

Additional measures to prevent the ingress of moisture might include:

- providing a galvanised steel 'water bar' which prevents water penetration between the sill and the window or door frame (Figs 2.17(a), (b) and (e));
- sealing the junction between the sill and window or door frame with 'mastic' (a flexible material often based on synthetic rubbers (Figs. 2.17(b), (c) and (d));
- preventing moisture rising up from the wall into the sill by installing a d.p.c. (Figs 2.17(b) and (c)).

Thresholds have the additional functional requirement of being capable of withstanding the weight of people. Where a door frame is set towards the inside of an opening, the threshold might form a door step (Fig. 2.17(d)).

Window frames

Functional requirements

Daylight   The amount and quality of daylight entering a room will depend largely on:
- the size and shape of the windows;
- the size and distribution of the members of the window frame;
- the position of the window in relation to the room;
- the effects of internal and external light reflection.

Where the effect of daylight in a room is important, calculation may be made to ensure an adequate level of daylight in various parts of the room.

Ventilation   Ventilation may be achieved by use of openable portions of the window and, in some cases, by doors. The Building Regulations 1991, approved document F require habitable rooms to be ventilated by an area of openable window/door equivalent to at least one-twentieth of the room floor area. Kitchens, bathrooms and sanitary accommodation are required to have a mixture of mechanical and/or natural ventilation.

Thermal and sound insulation   Windows are the weak link in the ability of an external wall to provide good thermal and sound insulation.

The use of two separate sheets of glass, with an air space between, will improve the thermal and sound insulation of the window. An air gap of 150–200 mm is usually necessary for improving sound insulation, whereas an air gap of 20 mm will usually provide a suitable improvement in thermal insulation (see 'Double windows and double glazing', p. 68 below).

Providing a good seal between the openable portions and the fixed portions of the window frame will reduce the heat lost through the window and improve sound insulation. This may be achieved by fitting metal or plastic sealing strips around the rebates of the frame.

Weather resistance (Fig. 2.18)   Window frames are designed to prevent the passage of water to the inside of the frame. Design features which are often incorporated in a frame include:

(a) '*Weathering*': where rain-water would lie on horizontal surfaces, the top surface of a sill is sloped (or 'weathered') to spill water off the surface.

(b) '*Anti-capillary grooves*': where rain might be driven between the frame and the window opening, or between fixed and openable portions of the window frame, 'anti-capillary grooves' will be necessary. These grooves prevent the passage of rain-water.

(c) '*Throatings*': grooves which are made in the underside of projecting portions of the frame to prevent water from being driven beneath them.

Security   All openable portions of windows should be fitted with suitable latches, catches or stays which prevent them from being opened from outside. Where security is particularly important, special security fittings may be installed.

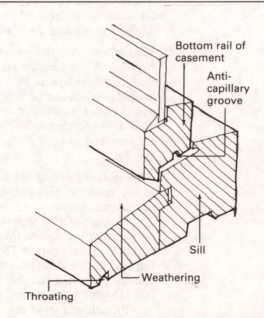

Fig. 2.18 *Features of a timber window frame designed to prevent rain penetration*

Cleaning The means of opening a window will affect the ease with which it may be cleaned. The outside surfaces of windows (see (c), (e) and (f) in Fig. 2.20) may be cleaned relatively easily from the inside of the building.

Also, the location of a window within a room is important. A window located in a high position above a stairwell might be difficult to clean without the use of a stepladder.

Types of window frame

Figure 2.19 illustrates some of the terminology associated with window frames. A basic window frame consists of a head, sill and two jambs. Such a frame might be divided by vertical members known as 'mullions' and/or horizontal members known as 'transoms'.

An area of glass fitted in a frame is known as a 'light'. Thus, Fig. 2.19 illustrates a three-light window consisting of one fixed light and two openable lights. Openable lights are referred to as 'casements' or 'sashes'.

Figure 2.20 illustrates some of the different methods of opening windows. Figures 2.20 (a–d) are based on hinged opening lights, whilst Figs. 2.20(e) and (f) are based on sliding opening lights.

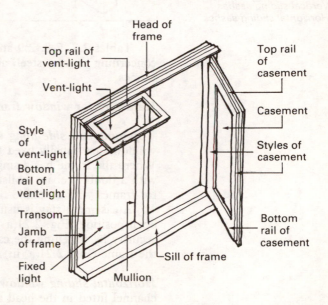

Fig. 2.19 *Window terminology*

57

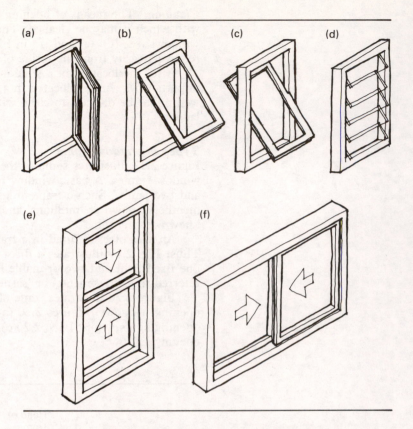

Fig. 2.20
(a) *Side-hung casement*
(b) *Top-hung casement*
(c) *Pivot-hung casement*
(d) *Louvred window*
(e) *Vertical sliding sashes*
(f) *Horizontal sliding sashes*

Tables 2.7, 2.8, 2.9 and 2.10 provide information concerning timber, steel, aluminium and plastic window frames.

Hardware for window frames

Top-hung and side-hung windows Top-hung and side-hung windows are usually fixed to the frame with a pair of hinges. Where the frame has a single rebate, flat hinges will be used with a double rebate whilst cranked hinges will be necessary for frames with a double rebate (Fig. 2.21(a)(i)).

A casement stay is usually fitted to hold the window in an open position (Fig. 2.21(a)(ii)). A casement fastener provides security by fastening the casement sash against the mullion of the frame (Fig. 2.21(a)(iii)).

Horizontal sliding windows Each sash slides in a horizontal channel fitted in the head and sill of the frame. Figure 2.21(b) illustrates an aluminium window where the sashes are fitted

Table 2.7 Timber window frames

Material	Typical section	Typical range
The top illustration shows a traditional window frame, usually of softwood, primed in the factory, then 'touched in' and painted on-site. Casements are usually top or side-hung and opened outwards. Typical frame sizes (mm) are: Head 90 × 57 Transoms and mullions 70 × 57 Styles and rails of casements 46 × 41 Frame jambs The lower illustration shows a high-performance window frame of softwood or hardwood. The timber is pressure impregnated with preservative in the factory (see p. 159), then usually stained. The cross-section is smaller and more sophisticated than traditional timber frames and incorporates (i) a large groove accommodating the hinge and stay mechanism and (ii) a small groove into which weather stripping is installed. Casements are available for inwards opening with 'tilt and turn' facility (bottom and side hung). Typical frame sizes (mm) are: head: 78 × 56 transoms and mullions: 92 × 56 styles and rails of casements: 78 × 56 frame jambs: 78 × 56	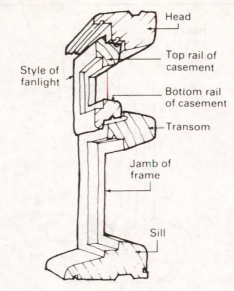Style of fanlight — Head — Top rail of casement — Bottom rail of casement — Transom — Jamb of frame — Sill 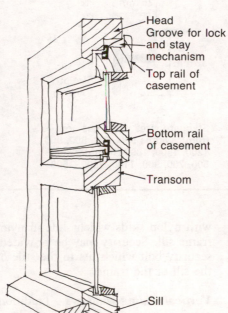Head — Groove for lock and stay mechanism — Top rail of casement — Bottom rail of casement — Transom — Sill	The illustrations below indicate some of the many configurations of timber windows. Tilt and turn apply to high-performance windows only. Sizes (mm) standard traditional frames: width: 600, 900, 1200, 1800, 2400 height: 600, 900, 1050, 1200, 1500 High-performance: often purpose made but maximum casement sizes of: width: 1600 height: 2350 Traditional 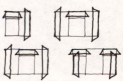 Top hung 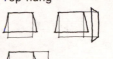 Pivot hung 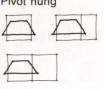 Tilt and turn

Table 2.8 Steel window frames

Material	Typical section	Typical range
The frame members of steel windows are all made from the same basic 'Z' section of steel framing. The framing is produced by passing white-hot metal through rollers, which gradually reduce the metal to the correct profile. The steel is then cut to length and welded together to form the frame. The steel is normally galvanised to prevent rust, by dipping the complete frame in molten zinc. The frames still require priming and regular painting to prevent rust from occurring. Sometimes the frames are fixed directly to the opening and sometimes they are set in separate timber surrounds (as illustrated)	Head of timber surround / Steel framing / Sill of surround — 25 mm / 32 mm — Size of framing. Standard sizes of frames are (mm): widths 500, 600, 800, 900, 1200, 1500, 1800; heights 200, 500, 700, 900, 1100, 1300, 1500	Some typical steel windows: dark line indicates openable portion of window. Opening lights. Pivot hung. Steel transom and mullion coupling pieces are also available to link together steel windows

with nylon skids which slide in nylon channels fitted to the frame sill. Security may be provided by such means as a security bolt which fits to the stile of the sash, and locks into the sill of the frame.

Vertical sliding windows Traditional vertical sliding windows required a system of cords, weights and pulleys which were housed in a timber box frame fixed to the jambs of the reveal, such as the window illustrated in Fig. 2.17(a). Modern timber

Table 2.9 Aluminium window frames

Material	Typical section	Typical range
Aluminium window frames are made by extrusion, where heated aluminium is forced through a die of the required profile of frame. The frames tend to be box-like in section, which provides stiffness to the frame. The frames can be accurately made and are ideal for sliding windows. The extrusions can be easily made to incorporate features such as weather strips. Aluminium windows may be left unfinished (mill finish) since aluminium resists corrosion in all but the most corrosive industrial atmospheres. Alternatively, the aluminium may be 'anodised', an electrochemical process which increases the corrosion resistance of the aluminium and also may incorporate a dye to provide colours such as gold or brown	Section of vertical sliding window 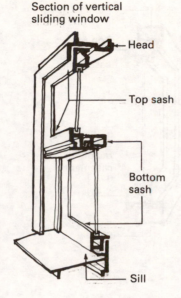 Sizes and profiles of section vary from manufacturer to manufacturer	Some typical aluminium windows: dark line indicating openable portion of window 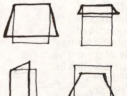 Most aluminium frames are made to modular sizes, the module being 100 mm. Typical sizes (mm): widths 600, 900, 1200, 1500, 1800, 2100, 2400; heights 300, 600, 700, 900, 1050, 1200, 1500

Labels in section diagram: Head, Top sash, Bottom sash, Sill

versions use a spiral spring balance mechanism which fits into a groove in the back of the outer sash style. Aluminium vertical sliding windows use spring balances which are held in extruded plastic guides fitted to the back of each sash style. A sash fastener (Fig. 2.21(c)) is fitted between the top rail of the inner sash, and the bottom rail of the outer sash.

Pivot windows   The window is hinged at the centre of each style with a pivot friction hinge (Fig. 2.21(d)(i)) which enables the windows to be held open without the use of stays. A sash

Table 2.10 Plastic window frames

Material	Typical section	Typical range
Plastic window frames are manufactured from either g.r.p. (glass-reinforced plastic) which is moulded, or p.v.c. (polyvinyl chloride) which is extruded. Most plastic frames are made of p.v.c. The frames are box-like in cross-section to provide rigidity. Sometimes steel sections are placed within the frame to act as a rigid core. Features such as clip-on seals, weather strips and glazing beads are often incorporated in the frame. Plastic windows require no further decoration, and only require periodic cleaning with water and detergent. The colours usually available are white, grey and black	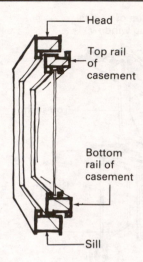 Head Top rail of casement Bottom rail of casement Sill	Some typical plastic windows: dark line indicates openable portion Windows are largely 'custom built', consequently standard size ranges are rarely quoted by manufacturers 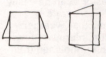 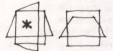 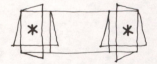

✳ 'Tilt and turn' windows capable of top and side-hung opening

Fig. 2.21
(a) *Hardware for top-hung and side-hung casements*
 (i) *Flat and cranked hinges*
(ii) *Casement stay*
(iii) *Casement fastener*
(b) *Hardware for aluminium horizontal sliding window*
(c) *Sash fastener*
(d) *Hardware for pivot windows*
 (i) *Friction hinge*
(ii) *Sash fastener*

(a)(i)

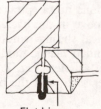
Flat hinge

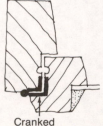
Cranked hinge

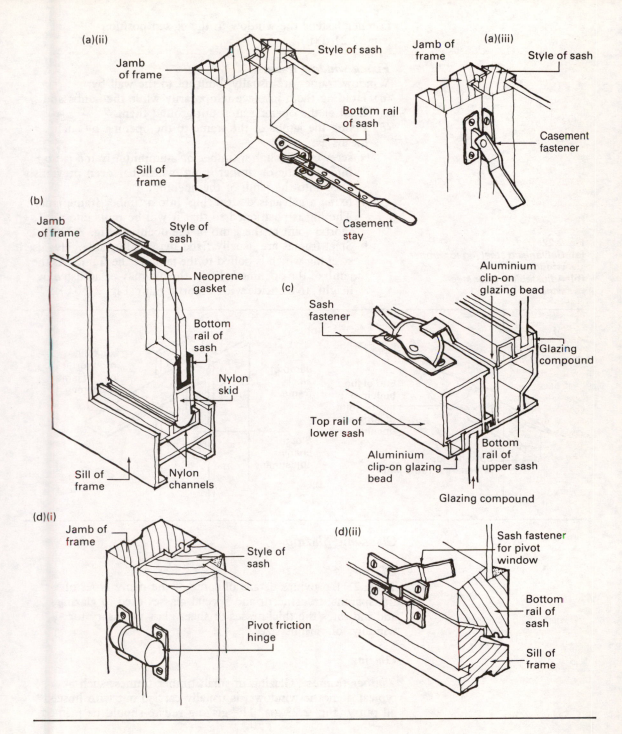

(a)(ii)

Jamb of frame

Style of sash

Bottom rail of sash

Sill of frame

Casement stay

(a)(iii)

Jamb of frame

Style of sash

Casement fastener

(b)

Jamb of frame

Style of sash

Neoprene gasket

Bottom rail of sash

Nylon skid

Sill of frame

Nylon channels

(c)

Sash fastener

Aluminium clip-on glazing bead

Glazing compound

Top rail of lower sash

Aluminium clip-on glazing bead

Bottom rail of upper sash

Glazing compound

(d)(i)

Jamb of frame

Style of sash

Pivot friction hinge

(d)(ii)

Sash fastener for pivot window

Bottom rail of sash

Sill of frame

fastener fastens the window in the closed position
(Fig. 2.21(d)(ii)).

Fixing window frames

Window frames are usually 'built-in' to the wall by:

(*a*) Holding them in place temporarily while the jambs and head of the opening are built around them.

(*b*) Fixing the jambs of the frame to the opening by such means as:

- screwing through a timber or aluminium frame jamb, into a plastic or timber 'plug' which has been previously fixed into the jamb of the opening;
- fixing a galvanised steel 'lug' into a timber frame jamb. The lug is positioned so that it will be built into a mortar joint at the jamb of the opening (Fig. 2.22(a)).
- Steel frames are usually fixed by an adjustable galvanised steel lug which is bolted to the jamb of the frame. As the jamb of the opening is built, the lug may be adjusted in height, to coincide with a mortar joint (Fig. 2.22(b)).

Fig. 2.22
(a) *Galvanised steel lug for timber windows*
(b) *Adjustable lug for steel windows*

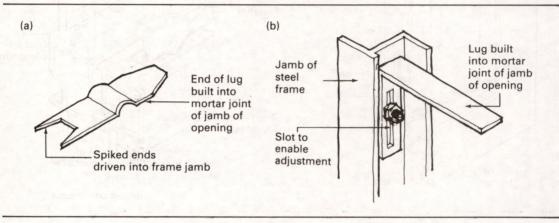

Glass and glazing

Glass

Table 2.11 provides details of some of the many types of transparent, translucent and special glasses used in glazing, and indicates the thicknesses of sheet glass necessary for windows of various sizes.

Glazing

Timber frames   Glazing in small timber frames, such as typical domestic windows, is usually carried out with linseed oil putty (Fig. 2.23(a)). The glazing rebate should be primed, prior to puttying, otherwise the oil from the putty will become absorbed into the frame. Before applying the putty, glazing sprigs (small headless nails) are tapped into the rebate to hold the glass.

Larger timber frames usually employ glazing beads which are small section lengths of timber screwed to the glazing rebate. The glass is bedded in non-setting glazing compound, and the beads are fixed with screws and cups to facilitate easy removal (Fig. 2.23(b)).

Fig. 2.23
(a) *Glazing timber casement using putty*
(b) *Glazing timber casement by use of glazing beads*
(c) *Glazing steel casement*

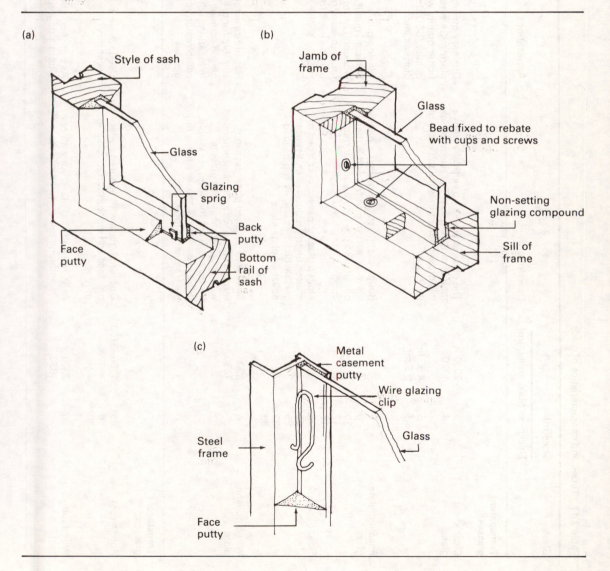

(a)

Style of sash

Glass

Glazing sprig

Back putty

Face putty

Bottom rail of sash

(b)

Jamb of frame

Glass

Bead fixed to rebate with cups and screws

Non-setting glazing compound

Sill of frame

(c)

Metal casement putty

Wire glazing clip

Glass

Steel frame

Face putty

Table 2.11 Types and characteristics of glass

Glass	Sizes(m)		Description	Applications
	Thickness	Max. area (mm)		
Transparent glasses				
Clear sheet glass	3 4 5 5.5 6	2030 × 1220 4.65 m² 9.3 m²	The glass is drawn up vertically from a tank of molten glass. The rate at which it is drawn determines the thickness of the glass	Ordinary quality (OQ) for general glazing. Selected quality (SQ) for better quality work. Selected special quality (SSQ) for doors of cupboards, pictures, etc.
Polished plate glass	5 6 10 12 15 19 22 25 32 38	Max. width 2640	Rough-cast glass (see below) is passed between sets of grinders and is then polished to produce an undistorted finish	Glazing quality (GG) for general glazing. Selected quality (SG) for high-class work and mirrors. Silvering quality (SQ) for high-class mirrors. Thick glasses may be used for loadbearing purposes such as shelves, table-tops, etc.
Float glass			Molten glass is passed over a bed of molten tin to produce an undistorted glass which has largely superseded polished plate glass	
Wired polished plate	6	3300 × 1830	Similar to polished plate glass, but formed with a wire mesh incorporated within its thickness. The glass is transparent, but the wires are visible. If the glass breaks, the wire mesh holds the fragments together	Has fire-resisting qualities, and is suitable for the glazing of fire-resisting glazed screens and doors. May be used in locations where damage or injury could result from breakage of other glasses
Translucent glasses				
Rough cast	5 6 10	1900–3700 × 1280	Obscured glass with textured surface on one face and smooth surface on the other	Used where a degree of privacy is necessary
Wired rough cast	6	3300 × 1830	As wired polished glass, but with one smooth and one textured surface	As polished plate wired glass, but where a degree of privacy is necessary

Patterned glasses

Made with one smooth surface and one surface with a patterned texture

Arctic cross-reeded Large Flemish Flemish small	Arctic / Cross reeded / Large Flemish / Flemish small / Broad reeded / Narrow reeded	3 / 1540–2140 × 1280	Glasses are classified in accordance with their degree of diffusion (A high, E low) and their degree of obscuration (a high, e low)
Broad reeded Narrow reeded		3 / 1540–2140 × 1280	The classification for the glasses described are Arctic C–b and D–c Cross-reeded D–c Large Flemish B–a Flemish small A–a Broad reeded A–a Narrow reeded A–a Pacific B–c Spotlyte B–c Stippolyte C–c Patterned glasses are used either for the decorative effect of the pattern and/or for the degree of privacy they offer
Pacific Spotlyte Stippolyte	Pacific / Spotlyte / Stippolyte	5 / 1840–2140 × 1320	

Special glasses

Toughened glass		5 6 10 12 15 19 / 1525 × 914 2540 × 1524 3950 × 1524 and 3100 × 2410	The glass is heated then suddenly cooled, resulting in a more flexible and much more impact-resistant glass	Used where other glasses could cause damage or injury due to heavy loading or impact, such as doors, balustrades, etc.
One-way glass			Provides mirror effect on one side and transparent effect on the other	For security supervision in shops, offices, etc.
Prismatic glass		6 / 1220 × 2540	One smooth surface and one surface covered with rows of prisms which deflect light	Can be used to reduce glare or to throw light to the back of rooms
Solar control glass		3–12	The glass, which may be tinted, reduces the effect of heat from the sun by either absorbing, or by absorbing and reflecting heat	Used to provide a comfortable environment in offices and other buildings
Security glass (laminated)		33–56 / 2133 × 1220	A laminated glass incorporating layers of plastic designed to withstand impact such as small-arms fire	Security screens in banks, offices, etc.

Steel frames   Steel frames may be glazed by using either:

- metal casement putty; the glass is bedded in metal casement putty, and held in place by wire clips prior to applying the face putty (Fig. 2.23(c)); or
- by use of hollow steel glazing beads which either screw to the frame, or are held in place by being clipped to studs in the frame.

Aluminium frames   Aluminium windows are usually glazed by such means as:

- neoprene or p.v.c. gaskets (see Fig. 2.21(b));
- by use of glazing compound and clip-on aluminium glazing beads.

Plastic frames   Plastic windows are usually glazed by use of neoprene gaskets and a p.v.c. clip-on glazing bead (see Fig. 2.24(b)(ii)).

Double windows and double glazing

Double windows are separate glazed frames which are fixed or hinged to the existing window frames. Double glazing consists of two sheets of glass which are fixed in the same glazing rebate of a window frame.

Double windows   Double-window systems might involve the installation of a second glazed casement, which is either:

- hinged on the inside of the existing window frame (Fig. 2.24(a)(i));
- fixed or hinged to the inside of the existing casement or light (Fig. 2.24(a)(ii)).

An alternative approach is to fix a separate sheet of glass on the inside of the existing window frame. This might be achieved by either:

- fixing panes of glass (which are edged with a plastic or aluminium sleeving) to the window frame with clips;
- installing a set of sliding aluminium windows within the reveal of the opening, on the inside of the existing window frame.

An advantage of double windows over double glazing, is that it is relatively easy to achieve a sufficiently large air gap between the sheets of glass for purposes of sound insulation. Compared with double glazing, however, double windows are cumbersome, and could be rather unsightly.

Fig. 2.24
(a) *Double windows*
 (i) *Hinged casement of each side of frame*
 (ii) *Second casement fixed or hinged to first casement*
(iii) *Separate sheets of glass clipped to inside of frame*
(b) *Double glazing*
 (i) *Double glazing unit*
 (ii) *Hermetically sealed unit*

Double glazing Figure 2.24(b)(i) illustrates a system of double glazing which consists of two separate sheets of glass, sealed at the edges with glass. Such sheets are made in standard window-frame sizes. The system illustrated in Fig. 2.24(b)(ii) comprises two sheets of glass, sealed at the edges with a metal spacer. This system, known as 'hermetically sealed' double glazing, can be made to order to suit the required size of window.

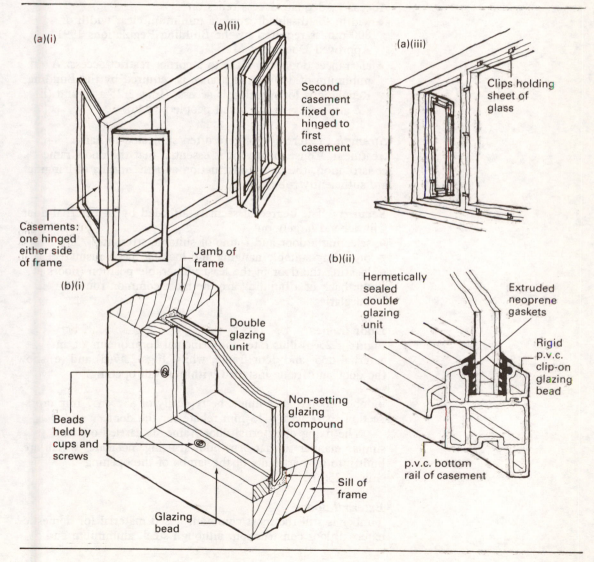

(a)(i)

(a)(ii)

Second casement fixed or hinged to first casement

(a)(iii)

Clips holding sheet of glass

Casements: one hinged either side of frame

Jamb of frame

(b)(i)

Double glazing unit

Non-setting glazing compound

Beads held by cups and screws

Glazing bead

Sill of frame

(b)(ii)

Hermetically sealed double glazing unit

Extruded neoprene gaskets

Rigid p.v.c. clip-on glazing bead

p.v.c. bottom rail of casement

External doors and door frames

Functional requirements of doors
The principal functions of an external door and door frame are access, strength and security. Other requirements may include daylight, ventilation, thermal and sound insulation and weather resistance (see 'Functional requirements of windows', p. 55 above) and fire resistance and means of escape (see 'Resistance to fire', pp. 209–11).

Access Adequate access relies on:
- width; for disabled people a minimum clear width of 800 mm is required by the Building Regulations 1991, Approved Document M;
- clearance; doors too close to a corner restrict access. A minimum of 300 mm clearance is required by the Building Regulations for wheelchair access. (Figs 2.25(a)(i) and (ii) see also access for disabled people p. 246.)

Strength External doors are often subjected to harsh treatment. Consequently, it is essential that the door-frame construction, the door construction and the means of hanging are sufficiently robust.

Security The degree of security afforded by an external door will depend largely on:
- selecting a door and frame of suitable strength;
- providing suitable hanging and locking mechanisms;
- locating the door in the least vulnerable position (doors at the back of a building are the most common route for burglaries).

Door frames
Figure 2.25(b) illustrates the principal components of an external door and door frame, whilst Figs 2.17(d) and (e) show the door and frame installed within a wall opening.

Timber door frames might be typically of 85×57 mm cross-sectional size with a 13 mm rebate for the door.

The fixing of external door frames is carried out in a similar manner to that of windows, using metal ties which are built into mortar joints of the jambs of the opening (Fig. 2.25(c)).

External doors
Timber is still the most commonly used material for domestic external door construction, although steel, aluminium and

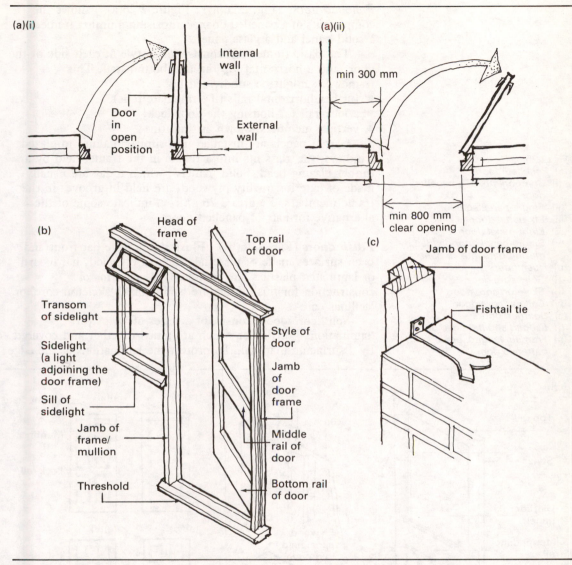

Fig. 2.25
(a) *Ease of access through external door*
(i) *Door near corner wall*
(ii) *Door away from corner walls*
(b) *Components of an external door frame*
(c) *Fixing of external door frame*

plastic are also used. Metal doors are frequently used for buildings such as shops and offices. Prestigious buildings may employ metals such as stainless steel or bronze for external doors and door frames. Timber doors may be categorised as: (*a*) panelled doors; (*b*) flush doors; (*c*) boarded doors. Standard sizes available are 1981 mm high × 762 or 838 mm in width.

Panelled doors (Fig. 2.26(a)) Figure 2.26(a)(i) shows the components of a panelled door, comprising a timber frame, a solid panel and a glass panel.

The basic framework consists of a style at each side of the door, and a horizontal top rail and bottom rail. This framework might be subdivided by:
- further horizontal rails (intermediate rails);
- a lock rail for housing the door lock;
- vertical members known as 'muntins'.

Glazed panels are installed in a similar manner to glazed windows; the glass fits into a rebate in the frame, held by a timber glazing bead. Solid 'panels', which today are usually made of exterior quality plywood, are held in grooves in the frame members. Figures 2.26(a)(ii–v) indicate some of the alternative formats of panelled doors.

Flush doors (Fig. 2.26(b)) Flush doors have flat front and back surfaces made of materials such as plywood, hardboard or laminated plastic. The commonly used forms of construction for flush doors are 'solid core', 'skeleton core' or 'cellular core'.

Solid core doors consist of a series of strips (or 'laminations') of timber which are glued together and covered by a surfacing material. To protect the laminations at the edge

Fig. 2.26
(a) *Panelled doors*
 (i) *Components of a panelled door*
 (ii) *Single-panelled door*
 (iii) *Six-panel door*
 (iv) *Eight-panel door*
 (v) *Single-panel door divided by glazing bars into 15 panels*
(b) *Flush doors*
 (i) *Solid core door*
 (ii) *Skeleton core door*
 (iii) *Cellular core door*
(c) *Boarded doors*
 (i) *Ledged and battened door*
 (ii) *Framed, ledged, braced and battened door*

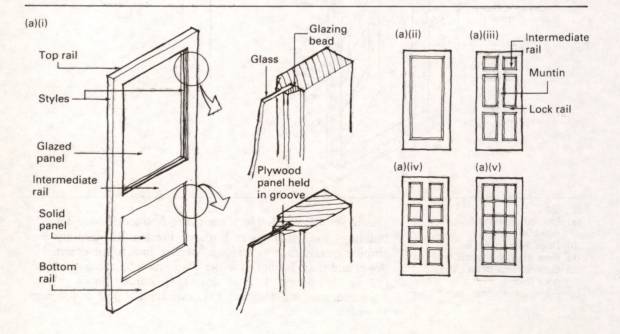

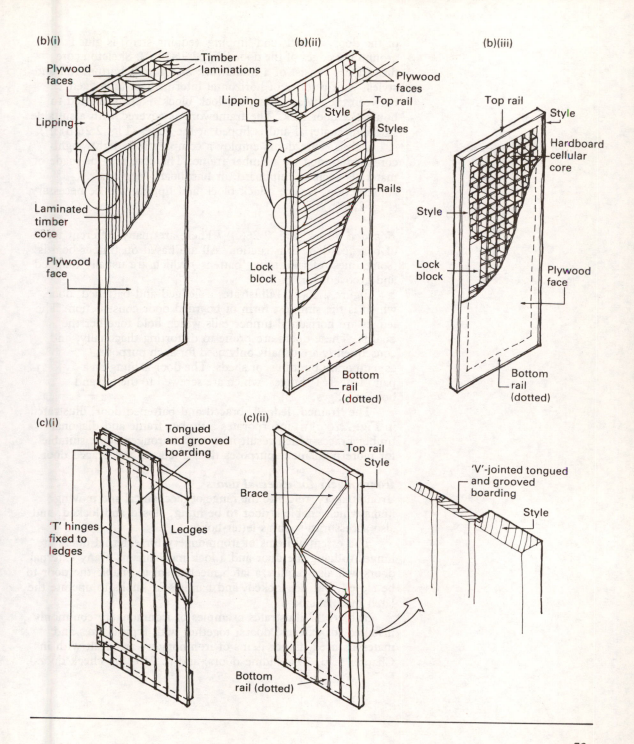

(b)(i)

Timber laminations

Plywood faces

Lipping

Laminated timber core

Plywood face

(b)(ii)

Plywood faces

Lipping

Style

Top rail

Styles

Rails

Lock block

Bottom rail (dotted)

(b)(iii)

Top rail

Style

Hardboard cellular core

Style

Lock block

Plywood face

Bottom rail (dotted)

(c)(i)

Tongued and grooved boarding

'T' hinges fixed to ledges

Ledges

(c)(ii)

Top rail

Style

Brace

Bottom rail (dotted)

'V'-jointed tongued and grooved boarding

Style

of the door, a hardwood 'lipping' (edging strip) is glued around the edges of the door (Fig. 2.26(b)(i)). Skeleton core flush doors consist of a framework of top rail, bottom rail and styles, with a series of horizontal intermediate rails fixed between the styles. A timber 'lock block' is incorporated to house the door lock. The framework is covered with a suitable surfacing material and is lipped at the edges (Fig. 2.26(b)(ii)).

Cellular core doors employ a comparatively lightweight core surrounded by a timber frame. The core may be made of materials such as chipboard, or hardboard 'egg-box' construction. Again, a lock block and lipping will be necessary (Fig. 2.26(b)(iii)).

Boarded doors (Fig. 2.26(c))   There are many different forms of boarded door construction. All are based on timber boards (sometimes referred to as 'battens') which are usually tongued and grooved.

Figure 2.26(c)(i) illustrates a 'ledged and battened' door, which is the simplest form of boarded door construction. The ledges are horizontal timber rails which hold together the boards. These doors are prone to distorting diagonally, and consequently are usually only used for such purposes as temporary buildings or sheds. The door is hung on a pair of steel 'T' hinges, which are screwed to the top and bottom ledges.

The 'framed, ledged, braced and battened door' illustrated in Fig. 2.26(c)(ii) incorporates a timber frame and diagonal timber braces which result in a much stronger door, suitable for more permanent purposes than a ledged and braced door.

Ironmongery for external doors

'Ironmongery' refers to the range of mechanical and moving items which enable a door to be hung, closed and locked, and also accessories such as letter-boxes.

The essential items of ironmongery for external doors are hinges to hang the door and a lock mechanism. Many external doors will also require a latch mechanism to enable the door to be closed, but not locked, and handles or knobs to operate the latch mechanism.

Table 2.12 illustrates examples of ironmongery commonly installed for external doors, together with typical sizes and materials used. Other items of ironmongery are dealt with in Chapter 5, p. 205 (sliding doors) and p. 210 (fire-check doors).

Table 2.12   Ironmongery for external doors

Item	Typical sizes (mm)	Typical materials and uses	Illustrations	
Hinges				
Butt hinge	75 or 100	Usually steel or brass. External doors usually hung on 3 hinges (1½ pairs of hinges)		Hinge leaves recessed into door style
Parliament hinge	100 × 100 or 100 × 150 wide	Brass or steel parliament hinges allow a door to be opened to 180°		Centre of hinge projects beyond face of door
'Tee' hinge	200–600	Steel – may be galvanised or 'japanned black' (a black lacquer). Used for ledged and battened doors		

Locks and latches

Latch and lock mechanisms are housed in a case which is designed to fit:
(a) to the surface of the door ('rim' latch or lock);
(b) fitted within the thickness of the door (mortise latch or lock).
Latches and locks operate a bolt which is activated by either handles, knobs or keys. In the closed position, the bolt is thrust into a plate ('striking plate') which is fitted to the jamb of the door frame.
Latches and locks may be fitted separately, or combined in one case known as a two-bolt set

Item	Typical sizes (mm)	Typical materials and uses	Illustrations
Rim lock and latch	150 × 75 150 × 100	Case: steel or aluminium Bolts: aluminium or brass usually used for sheds and outbuildings	Opening for key / Striking plate / Latch bolt / Lock bolt / Opening for spindle to which handles or knobs are attached

Table 2.12 continued

Item	Typical sizes (mm)	Typical materials and uses	Illustrations
Mortise lock	75 × 100 × 12 thick	Case: steel – fore end: steel, aluminium or brass. Bolts: aluminium or brass. Locks may be vertical for fitting to door style, or horizontal for fitting to lock rail. Used for external doors generally	Opening for spindle to which handles or knobs are attached. Striking plate screwed to frame. Opening for key. Latch bolt. Lock bolt.
Cylinder rim lock	90 × 65 × 32 thick	Case: often enamelled steel, brass or iron. Lock mounted on inside surface of door. Some locks have 'dead latch' which automatically locks door when it is closed. The handle may be turned from inside the building to operate the bolt, as though it were a latch	Staple screwed to frame jamb. Knob handle. Lock bolt. Thumb slide: when operated will hold bolt open or shut. Face of key cylinder.

Bolts

Barrel bolts	50–200	Brass or steel. The shoot slides in a barrel-shaped guide and locates in a metal keep. Usually used as an additional security measure	Back plate screwed to door. Keep screwed to jamb of frame. Shoot.

Item	Typical sizes (mm)	Typical material and uses	Illustrations
Panic bolts	For single doors up to 2440 × 1220 or double doors up to 2440 × 1830	Aluminium or steel. Can be adapted for exact door sizes. In cases of emergency, the horizontal bar is pushed, which releases a latch (for a single door) or a bolt at the top and bottom of the door (for double doors). Used for fire-escape doors. Pressure on the bars will activate the bolts or latches	

Double doors — Top bolt — Single door — Bottom bolt — Arrows indicate direction of pressure required to activate bolts

Lever handles and knobs

Lever handles	90 × 40 150 × 40	Aluminium, brass, bronze, chromium plate. Vast variety of styles available. For the bolt sets, the lever back plate may incorporate a key opening. Otherwise a separate 'escutcheon' may be used. Behind the handle the spindle is located which passes through an opening in the latch/lock case and into the back plate on the opposite side of the door	
Knobs	50	Aluminium, brass, bronze, chromium plate. Knobs operate in a similar manner to handles. Where a lock is positioned near to a projecting frame a knob could be awkward to use, since knuckles would be scraped against the frame	

Handle — Keyhole — Back plate — Rose — Handle — Escutcheon — Knobs on roses — Knob incorporated in back plate

Letter plates

Letter plates	250 × 75 300 × 75 200 × 45	Aluminium, brass, bronze, chromium plate. May be vertical, to fit into a style, or horizontal to fit into lock plate or bottom rail	

Vertical letter plate incorporating key opening and door-knocker — Horizontal letter plate

CHAPTER 3

Roofs and floors

ROOFS

Functional requirements

Figure 3.1 illustrates the functional requirements of a roof.

Weather resistance

Roofs are required to protect a building from the damaging effects of: (*a*) rain and snow: (*b*) wind.

The ingress of rain or snow into a building interior can lead to damage of decorations, and structural damage to timber and other materials due to attacks by fungus and insects. A roof is covered by a layer of impermeable material which prevents the passage of rain or snow into a building.

Figure 3.2(a) illustrates the way in which wind may be deflected around a building, resulting in the occurrence of suction on the windward slope of a roof. The problem could become acute for buildings on exposed sites, where the roof could be torn from the walls. This effect could be accentuated by the wind pushing up against overhanging eaves. Vulnerable flat or low-pitched roofs should have a roof finish which is firmly secured to the structure, and if necessary the roof structure should be anchored to the walls by means of steel straps (see Fig. 3.7(h)).

Structural stability

A roof structure is necessary in order to provide a means of support for the roof covering. Figure 3.2(b)(i) illustrates a very simple roof structure comprising of a series of 'joists' (or beams), the ends of which bear on the external wall of the building.

In designing a roof structure, an engineer must take into account the span of structural members. If the span of a joist or other structural member is too great, it will bend (see

Fig. 3.1 *Function requirements of roofs*

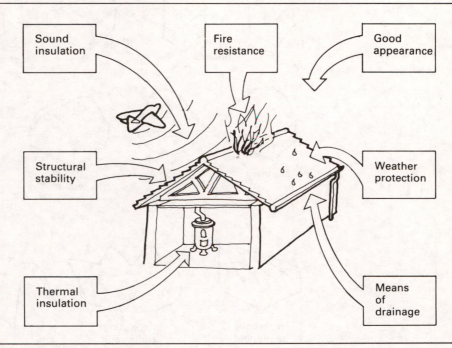

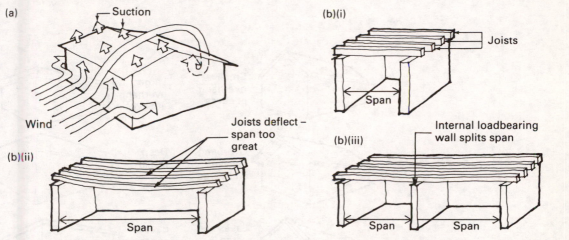

Fig. 3.2 *Roofs and structural
Stability*
(a) *Effect of wind*
(b) *Effect of span*

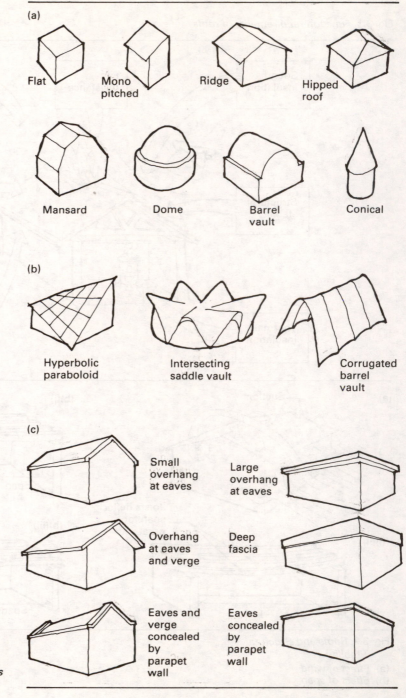

Fig. 3.3 *Roofs and appearance*
(a) *Traditional roof shapes*
(b) *Modern roof shapes*
(c) *Different treatments for eaves and verge*

Fig. 3.2(b)(ii)). This may be overcome by either increasing the depth of the joist, or by introducing a structural element which reduces the span (such as the internal loadbearing wall illustrated in Fig. 3.2(b)(iii)).

Provide good appearance

The roof might be a major visual element in the design of a building, particularly a low building since the roof is more visible to the eye. Figure 3.3(a) shows some traditional roof shapes, whilst Fig. 3.3(b) shows some of the roof shapes which have resulted from modern building materials and techniques.

The detailing of a roof can significantly affect the appearance of a building. Figure 3.3(c) illustrates:

- different treatments of 'eaves' detail (the junction between a flat roof and external wall, or the junction between the horizontal edge of a pitched roof and an external wall);
- different treatments of 'verge' detail (the junction between the angled edge of a roof and an external wall).

Provide thermal insulation

The roof constitutes a substantial proportion of the external surface area of a building and consequently has considerable potential for heat loss. Roof insulation consists of either:

- flexible materials (such as glass-fibre quilting);
- boards (such as expanded polystyrene);
- loose granules (such as perlite).

Pitched roofs are usually insulated between the ceiling joists (Fig. 3.4(b)). Alternatively insulation might be applied

Fig. 3.4 *Thermal insulation for pitched roofs*
(a) *Insulation above rafters*
(b) *Insulation between ceiling joists*

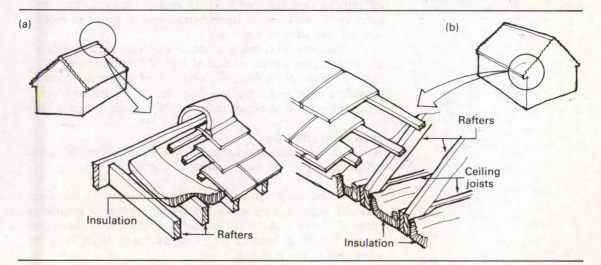

(a)

(b)

Rafters

Ceiling joists

Insulation

Rafters

Insulation

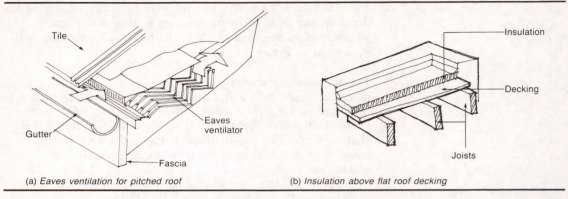

(a) Eaves ventilation for pitched roof   (b) Insulation above flat roof decking

Fig. 3.5   *Thermal insulation for roofs*
(a) *Eaves ventilation for pitched roof*
(b) *Insulation above decking of flat roof*

between the rafters (Fig. 3.4(a)). To avoid condensation occurring within an insulated pitched roof space, ventilation must be provided. Figure 3.5(a) illustrates a p.v.c. ventilator which allows fresh air through the eaves into the roof space.

Flat roof insulation is usually laid above the 'decking' (surface made of boards or slabs fixed over the joists) either below or above the roof covering (Fig. 3.5(a)).

Roof insulation laid above heated rooms could deteriorate due to absorption of moisture from the room below. Consequently the insulation is laid on an impermeable layer ('vapour barrier').

Provide sound insulation

The degree of sound insulation offered by roofs of domestic buildings is usually adequate to reduce external noise (such as aircraft) to an acceptable level. In areas which suffer from excessive levels of noise interference, special forms of roof construction may be necessary.

A concrete roof, being of dense material, will provide a noise reduction similar to that of typical brick or block external walls. Most domestic timber roof constructions will provide a noise reduction less than that of a brick or block wall, but somewhat better than that of a single-glazed window.

Protection from fire

A roof could contribute to the spread of fire either: (*a*) within a building; (*b*) from one building to another.

Within a building   Some factory roofs consist of a covering material, separated by an air space from a layer of insulation material. Fires within such buildings could spread through the air space causing burning pieces of insulation material to fall elsewhere in the building.

Measures designed to combat this problem include:

- using non-combustible insulation material;
- restricting the area of the air gap (see also 'safety in fire', in ch. 5, p. 231);
- fixing the insulation in such a way that no air gap exists;
- lining the insulation material with non-combustible material.

From one building to another Roof constructions are designated with two letters in accordance with British Standard test (BS 476: part 3: 1958). Each letter ranges from A (good) to D (poor). The first letter refers to resistance to external penetration from fire, and the second to the resistance to spread of flame over the outer surface. Natural slate roofs on a timber structure, for example, have an AA designation whilst bituminous felt pitched roofs laid on certain timber-based boards have designations varying from AB to CC.

Approved document B of the Building Regulations 1991 specifies certain minimum distances that buildings with particular roof designations may be situated from the boundary of the property.

The minimum distance for dwelling houses with roof designations BA, BB and BC is 6 m, whilst for DA, DB, DC and DD designations (other than for terraced houses) the minimum distance is 20 mm.

Provision of drainage

Pitched roofs During heavy storms, large quantities of water fall on a roof. Water falling down the surface of a pitched roof is collected at the lower end in a gutter. A vertical pipe connected to the underside of the gutter ('downpipe') carries the rain-water to ground level (Fig. 3.6(a)). The lower end of the downpipe discharges the rain-water into a 'gully' which is a clayware or plastic chamber installed below ground level (Fig. 3.6(b)). The downpipe is connected to the upper end of the gully, and the lower end of the gully connects to the drain. The gullies are 'trapped' (shaped with a bend which allows water to remain in the gully). The trap prevents the escape of any noxious gases from the drain, which otherwise could rise up the downpipe and into the open air.

Flat roofs Flat roofs are designed and built to slope slightly in one direction. This slope is known as the 'fall' of a roof. At the lower end of the fall, a gutter and downpipe might be fitted (Fig. 3.6(c)).

Fig. 3.6 *Roof drainage*
(a) *Pitched roof, gutter and downpipe*
(b) *Detail of gully*
(c) *Flat roof with external downpipe*
(d) *Flat roof with internal downpipe*

Alternatively, the edge of the roof could be raised, which in effect forms a gutter. In this case, the downpipe is installed inside the building (Fig. 3.6(d)).

Whilst it is usual to calculate the sizes of gutters and downpipes for roofs of large buildings, a typical domestic pitched roof would usually require 100 mm diameter gutters and 62 mm diameter downpipes.

(a) (b)

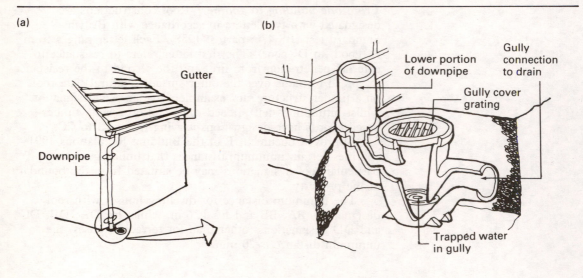

Gutter

Downpipe

Lower portion of downpipe

Gully connection to drain

Gully cover grating

Trapped water in gully

(c) (d)

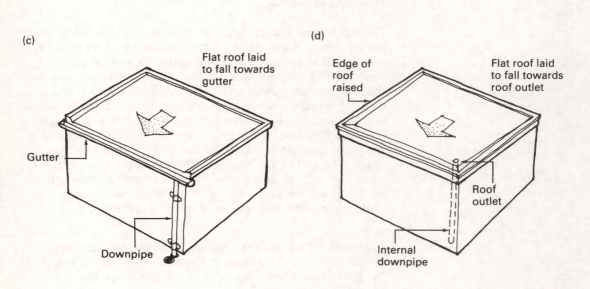

Flat roof laid to fall towards gutter

Gutter

Downpipe

Edge of roof raised

Flat roof laid to fall towards roof outlet

Roof outlet

Internal downpipe

Flat roof structures

The roof structure of most small flat-roofed buildings consists of a series of timber joists which span between the external walls of a building. The joists are covered by a 'decking' (some form of board, slab or rigid sheet material) which is designed to act as a suitable surface on which to apply the waterproof roof covering.

Small-span flat roofs

For small spans, timber is the most commonly used material for roof joists.

Tables in the Building Regulations 1991 (approved document A) list the maximum clear spans for joists in relation to their cross-sectional size, spacing and loading.

The figures in Table 3.1 are extracted from the regulations and relate to roof loadings of $0.50-0.75$ kN/m^2 using 'SC3' class timbers (basically general structural grade). Slightly greater spans are permitted for 'SC4' (basically 'special structural grade') timbers. Where joists are deep in relation to their thickness, buckling of the joists could occur. In order to prevent this, 'herringbone strutting' is fitted at intervals between the joists. Herringbone strutting consists of a series of diagonally crossing timbers or steel struts. Typical sizes of timber struts might be 50×32 mm fitted at intervals of 2.5 m (Fig. 3.7(a)). Whilst timber struts are made *in situ*, steel struts may be obtained from builders' merchants.

Table 3.1 Permissible spans for flat-roof joists

Joists for flat roofs with access only for the purpose of maintenance and repair

Size of joist (mm)	Spacing of joists (mm)		
	400	450	600
	Maximum clear span of joint (m)*		
38×97	1.67	1.64	1.58
50×97	1.89	1.86	1.78
50×122	2.53	2.49	2.37
50×147	3.19	3.13	2.97
50×195	4.48	4.36	3.97
63×195	4.86	4.69	4.28
75×195	5.13	4.95	4.53

* These spans relate to roofs with dead loads of $0.50-0.75$ kN/m^2, excluding the mass of the joist.

Source: From Building Regulations 1991, Approved Document B, Table A17.

A roof with overhanging eaves, or on an exposed site, may be susceptible to displacement by the wind. In such circumstances galvanised mild steel anchor straps are used to secure the joists to the wall. Anchor straps might typically be fitted to every fifth joist (Fig. 3.7(h)). The method used for fixing the joists to the external wall will depend on whether:

- the roof overhangs the wall;
- the wall is built up above the level of the roof joists to form a 'parapet wall'.

Roof overhanging wall (Fig. 3.7(b))   The wall is built up to the lower level of the joists, the joists are installed resting on the wall, then the wall is built up between the joists. A vertical board (known as a 'fascia') is fixed across the joist ends, and a horizontal board (known as a 'soffit') is fixed to the underside of the projecting joists. To provide ventilation of the roof space, a small gap might be left between the soffit and the wall. Materials commonly used for fascias and soffits include plywood, asbestos cement boards, timber boarding and plastic boarding.

When joists are built into a solid brick wall, they are usually fixed to a 'wall plate' (Figure 3.7(j)). A wall plate is a length of timber, typically of 100×75 mm section, bedded on top of the wall.

Fig. 3.7 *Flat roof structures*
(a) *Herringbone strutting*
(b) *Roof overhanging wall*
(c) *Parapet wall*
(d) *Cross beam*
(e) *Rolled steel joist cross beam*
(f) *Latticed steel cross beam*
(g) *Corrugated plywood web cross beam*
(h) *Anchor straps*
(j) *Wall plate*

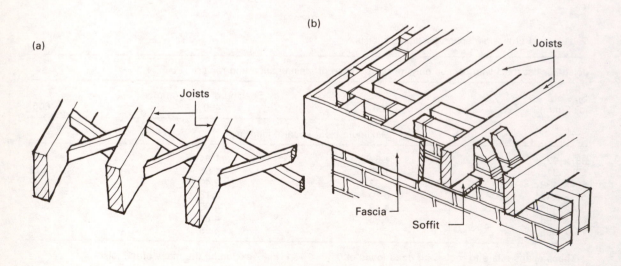

(a)

(b)

Joists

Joists

Fascia

Soffit

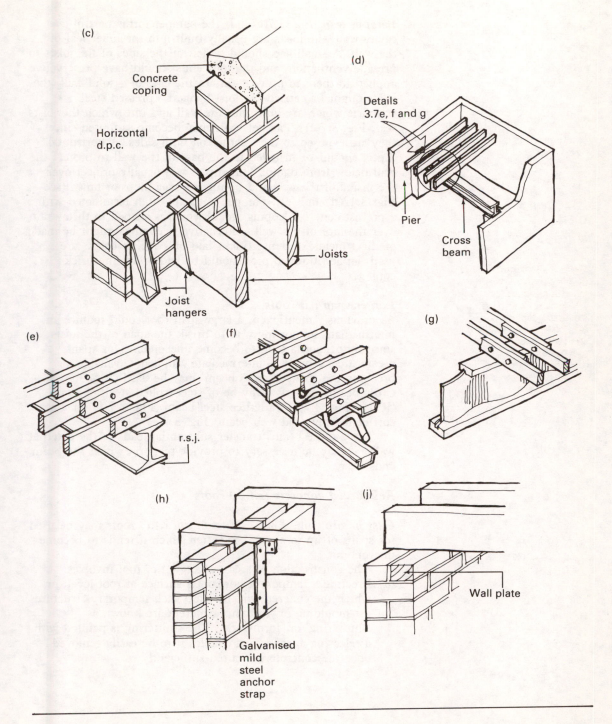

(c)

Concrete coping

Horizontal d.p.c.

Joists

Joist hangers

(d)

Details 3.7e, f and g

Pier

Cross beam

(e)

(f)

(g)

r.s.j.

(h)

Galvanised mild steel anchor strap

(j)

Wall plate

Parapet wall (Fig. 3.7(c)) If the parapet forms part of a cavity wall, the joists are usually built into the inner leaf of the wall. A small gap should be left at the sides of the joists to provide ventilation, and the joist ends should have preservative applied to them to prevent rot. In the case of a solid wall, the joists might be carried by 'joist hangers' (pressed steel supports which are built into the wall and into which the joists rest, Fig. 3.7(c)). Parapet walls may become saturated since they are exposed to the weather on both sides. A horizontal d.p.c. should be installed in the base of the wall to protect the wall below from damp. This d.p.c. will usually project over the inside of the wall and be turned over the roof finish (see Figs 3.12(d) and (e)). Parapet walls are often terminated with a precast concrete 'coping stone' which is shaped to throw rain over the face of the wall. Alternatively, copings might be made of sheet metal, or dense bricks laid on edge. If bricks are used, an additional d.p.c. should be laid under the brick coping.

Larger-span flat roofs

As previously mentioned, a large-span roof could require an intermediate supporting wall to break the span of the joists, since timber joists tend to become uneconomic for spans greater than 5 m. If an intermediate wall is not available for breaking the span, a 'cross beam' may be used (Fig. 3.7(d)). Cross beams may typically be of steel (rolled steel joist or 'RSJ') (Fig. 3.7(e)), a lattice steel beam (Fig. 3.7(f)) or a corrugated plywood web beam (Fig. 3.7(g)).

Since these beams transfer substantial loads to the external walls, it may be necessary to provide piers on which the beam ends bear.

Reinforced concrete (r.f.c.) roofs

Cast in situ (Fig. 3.8(a)) Cast *in situ* r.f.c. roofs may be used for spans of up to about 5 m, after which it tends to become uneconomic.

The construction of a cast *in situ* r.f.c. roof involves:
(a) Erecting a temporary horizontal surface at roof level, on which the concrete will be laid. Such temporary structures for moulding or 'forming' concrete are known as 'shuttering' or 'formwork'. The shuttering is painted with a releasing agent which enables it to be easily removed once the concrete has cured sufficiently.

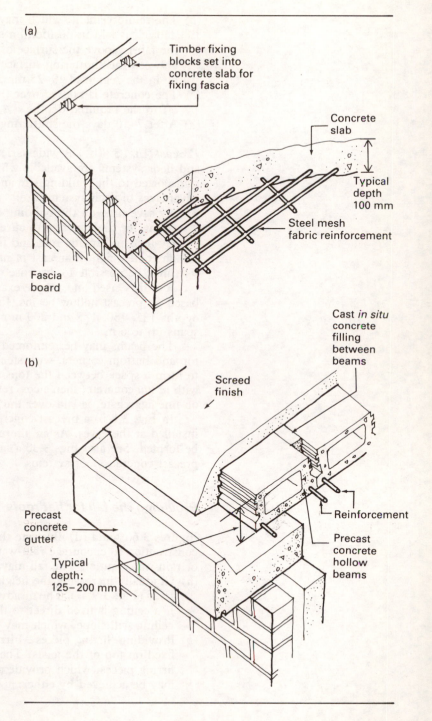

(a)

Timber fixing blocks set into concrete slab for fixing fascia

Concrete slab

Typical depth 100 mm

Steel mesh fabric reinforcement

Fascia board

(b)

Cast *in situ* concrete filling between beams

Screed finish

Precast concrete gutter

Typical depth: 125–200 mm

Reinforcement

Precast concrete hollow beams

Fig. 3.8 *Reinforced concrete roof*
(a) In situ *concrete*
(b) *Precast concrete*

(*b*) The reinforcement, which may consist of square mesh steel 'fabric', is laid in position on special blocks which support the fabric above the surface of the shuttering. The distance between the shuttering surface and the reinforcement will be in the region of 19–25 mm.

(*c*) The concrete is poured over the shuttering and is levelled off to the required depth.

(*d*) After 7–10 days the shuttering is removed.

Precast (Fig. 3.8(b)) A wide variety of precast concrete roof and floor systems are available which are cast in a factory, transported to the building site and then installed. Some of the advantages of such systems are:

- they save time on the building site since there is no delay waiting for the concrete to cure;
- they tend to provide roof and floor structures which are lighter in weight than cast *in situ* slabs;
- it may be possible to dispense with shuttering altogether.

A widely used and simple example of such a roof is that based on precast hollow beams (Fig. 3.8(b)). Typical sizes are: depths 125, 150, 175 and 200 mm; widths 300 and 350 mm; spans up to 6 m.

The beams may be reinforced in the bottom corners or top and bottom corners, with steel rods. The sides are profiled to form a space between the tops of the beams, which is filled with *in situ* concrete, then a concrete screed (concrete based on fine aggregate) is laid over the entire surface.

In Fig. 3.8(b) a precast concrete gutter unit has been installed at the eaves. As an alternative, a parapet wall could be formed. See also Figs 3.30(a) and (b) for examples of precast concrete floor systems.

Providing the fall for flat roofs

Figures 3.6(c) and (d) illustrate the 'fall' required in flat roof construction to encourage rain-water to flow towards the gutter or rain-water outlet. The fall may be achieved by either:

(*a*) Gradually increasing the height at which the roof joists are fixed (Fig. 3.9(a)). The disadvantage of this system is that if a ceiling is fixed direct to the underside of the joists, the ceiling will slope, which may be visually unsatisfactory.

(*b*) Providing 'firring pieces'. Firring pieces are strips of wood fixed on top of the joists. The decking is fixed to the firring pieces, which provide a surface with a fall. The fall may be achieved by either:

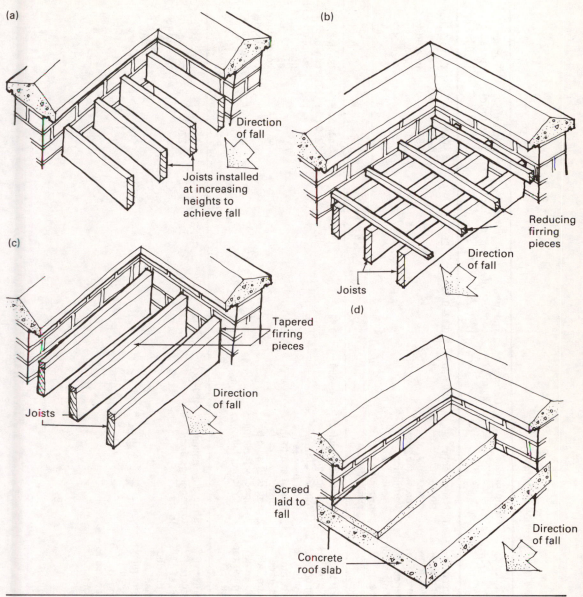

Fig. 3.9 *Providing the fall for flat roofs*
(a) *Varying height of roof joists*
(b) *Firring pieces across joists*
(c) *Tapered firring pieces*
(d) *Fall for concrete roof*

● fixing firring pieces of gradually decreasing thickness on top of the joists – the firring pieces may be fixed either perpendicular to the joists (Fig. 3.9(b)) or by fixing one firring piece on top of each joist;
● fixing tapered firring pieces to each joist (Fig. 3.9(c)).

Table 3.2 Materials used for roof deckings

Material	Description	Fixing details	Density (kg/m³)	Thermal conductivity (W/m K)*	Typical thickness (mm)	Typical sizes (mm)	Surface spread of flame
Timber boards	Best used in tongued and grooved form for better fire resistance and a smoother top surface	1. Nailed either at right angles or diagonally across the joists 2. Ventilation should be provided to avoid dry rot	400–500	0.12–0.15	25	Width of boards: 100, 150, 175	Class 3, but can be treated to provide Class 1
Chip board	Made from wood particles combined with resin glue and formed into sheets under pressure. Also available with roofing felt top surface	1. Should be nailed to 'noggings' (pieces of timber fixed between joists) and to the joists 2. Joints should be taped with strips of roofing felt	600–650	0.12	18, 22	2440 × 1220	Class 3, but can be supplied with finish (such as vermiculite) to provide Class 1
Plywood	Made from thin layers ('plies') of wood bonded together. Adjacent plies have grain running at right angles to each other	Should be nailed to noggings and joists	400–500	0.12	18, 21, 24	2440 × 1220	Class 3, but can be treated to provide Class 1

Material	Description	Application	Density	Thermal conductivity*	Thickness (mm)	Size (mm)	Fire class
Polystyrene board (insulation only)	Rigid vacuum extruded foam board	Laid on a waterproof membrane above the decking. Ballast or paving laid over the board to prevent dislodgement	27–35	0.035	50, 60, 75	1250 × 600	
Wood-wool slabs	Long wood fibres mixed with cement and compressed into a mould. Available as: 1. Unreinforced 2. Plain channel edged (channel reinforced) 3. Interlocking channel 4. Interlocking channel with toe	1. Nailed to joists 2. Joints between slabs filled with sand and cement 3. A sand and cement screed is laid on top: minimum thickness 12 mm	450–500	0.09	1. Unreinforced: 50,75, 100 2. Channel reinforced 50, 75, 100 3. Interlocking channel 75, 100 4. Interlocking with channel 50, with toe 150	All 600 mm wide Lengths: unreinforced 1600, 1800, 2000; channel reinforced 1800, 2000, 2100, 2400, 2700 Interlocking with channel 1800–4000 Interlocking with toe 4500–6300	Class1
Fibre board	Made from wood fibres or sugar-cane fibres which are rolled to the required thickness	In conditions where the board might become damp, due to condensation, a vapour barrier must be laid underneath	Up to 400	0.06	10 13 16 19	2440 × 1220	Class 4, but can be faced with asbestos paper or aluminium foil to provide Class 1

*Thermal conductivity: basically a measure of the ability of a material to conduct heat through itself. A low figure such as wood-wool slabs represents better thermal insulation properties than a high figure.

93

The advantage of fixing firring pieces on top of each joist is that their thickness may be reduced to a very small dimension (minimum 12 mm). Firring pieces fixed perpendicular to the joists will have to be sufficiently thick to span from joist to joist, and would therefore need to be a minimum of 40 mm in depth.

(c) Using screeds. Some flat roofs, such as r.f.c. roofs, may be provided with a fall by laying a screed which diminishes in depth towards the gutter or rain-water outlet (Fig. 3.9(d)).

Deckings for flat roofs

The 'decking' or 'substrata' of a flat roof is the material which spans the joists to provide a flat surface on which the waterproof roof covering (or 'roof finish') may be laid. For small-scale roofs, deckings tend to be in the form of boards or slabs or rigid sheets. Timber boarding is the traditional material used, although today other decking materials are also commonly used, some of which are detailed in Table 3.2. Figures 3.5(a) and (b) show typical slab-type deckings fixed to a timber-joisted roof.

For larger buildings, particularly of framed construction, composite deckings are frequently used. These deckings (which may also be used in industrial pitched roof construction) are usually based on aluminium or steel corrugated sheeting, with insulation board and roofing felt fixed on top (Fig. 3.10). Sometimes a 'vapour barrier' is laid

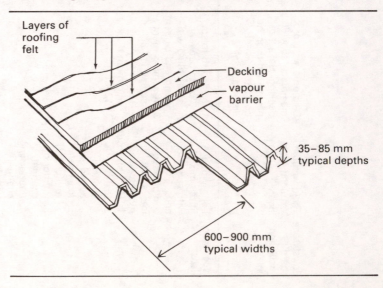

Fig. 3.10 *Corrugated metal roof decking.*

between the metal and insulation board to prevent condensation destroying the insulation material.

These deckings might typically span between 4 and 5 m.

Flat roof finishes

The finish of a flat roof is the material which covers the decking to provide protection from the weather. The roof finish may also contribute to the thermal insulation and resistance to fire and condensation of the roof.

Joints in the roof finish and junctions between the roof and walls are particularly vulnerable to rain penetration and care must be taken in the design of these details.

Flat roof finishes may be of either:

- single-sheet material: metal or proprietary material;
- multiple-layer material, usually based on bitumen felt;
- monolithic material which is applied as a liquid, and dries to form a solid covering.

Single-sheet materials

Metals   Metals are usually chosen because of their durability. Lead and copper roof finishes have been used for centuries, but are very expensive today; zinc and aluminium provide less expensive alternatives. Some of the characteristics of these metals are compared in Table 3.3.

Since metals are liable to movement due to temperature changes, metal roof finishes are laid on a layer of felt ('underlay') which reduces abrasion caused by movement of the metal over the decking. The underlay may also protect the metal finish from the corrosive effects of certain timbers and metals which may form part of the decking. The basic process of laying and jointing sheet metal roof finishes is similar whether lead, copper, zinc or aluminium are used. Figure 3.11(a) shows a flat roof with zinc roof finish.

Joints   Joints made parallel to the fall of a roof may be formed by either 'wooden rolls' or standing seams. Wooden rolls consist of timber battens nailed to the underlay and decking, fixed parallel to the direction of the fall. The sheet metal is laid between these rolls. In the case of lead roofing, one sheet is dressed over the roll and nailed to it, whilst the adjacent sheet is dressed over the lower sheet and the roll (Fig. 3.11(c)). In the case of copper, zinc and aluminium, adjacent sheets are dressed up the sides of the roll, and a separate metal roll cap is dressed over the roll

Table 3.3 Metal roof coverings

Metal	Colour	Density (kg/m³)	Thickness (mm)	Sheet size (mm)	Reaction with other materials	Comments
Lead	Dull grey	11 350	2.24 2.50 3.15	Length 2000 –2700 Max area 2250	Corrosion can be caused by contact with certain timbers, particularly oak and teak	Very malleable, and hence suitable for forming into complex shapes. Extremely expensive but rarely used for new roofs today
Copper	Initially yellow/red but develops a light green patina	8940	0.56	Length 1200–2400 Width 600–1200	Direct contact with aluminium, zinc and steel should be avoided since the copper could cause them to corrode	
Zinc	Dull light grey	7140	0.8 1.0	Lengths up to 2400 Width up to 1000	Liable to corrosion if in contact with copper or damp timbers, particularly oak and western red cedar	May be slowly attacked in a polluted atmosphere
Aluminium	Whitish grey	2700	0.7 0.9	Lengths up to 10 000 Widths 450 and 500	Liable to corrosion if in contact with copper, iron or steel, and timbers such as oak and western red cedar	The lightest of metal roof coverings

(a)

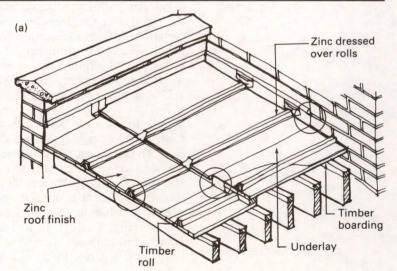

Fig. 3.11 *Metal roof coverings*
(a) *Flat roof with zinc roof covering*
(b) *Wood roll for zinc roof*
  (i) *roll fixed to decking*
  (ii) *zinc dressed up sides of roll*
  (iii) *zinc cap fitted over roll*
(c) *Wood roll*
  (i) *roll fixed to decking*
  (ii) *first sheet of lead dressed over roll*
  (iii) *second sheet dressed over roll and first sheet*
(d) *Metal drip*
(e) *Junction between metal roof and wall covering*
(f) *Twin-rib roof junction*

Zinc dressed over rolls

Timber boarding

Underlay

Zinc roof finish

Timber roll

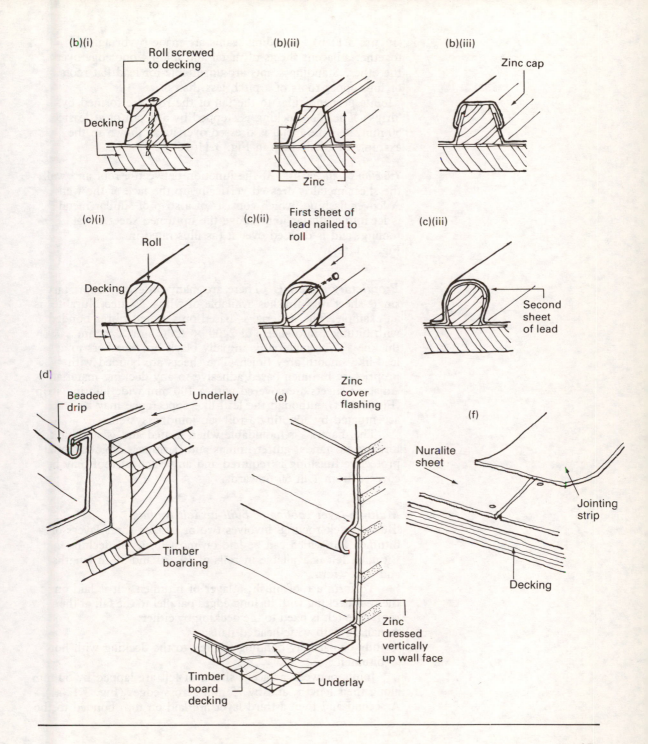

(b)(i)
Roll screwed to decking
Decking

(b)(ii)
Zinc

(b)(iii)
Zinc cap

(c)(i)
Roll
Decking

(c)(ii)
First sheet of lead nailed to roll

(c)(iii)
Second sheet of lead

(d)
Beaded drip
Underlay
Timber boarding

(e)
Zinc cover flashing
Zinc dressed vertically up wall face
Timber board decking
Underlay

(f)
Nuralite sheet
Jointing strip
Decking

(Figure 3.11(b)). Standing seams are made by bringing together adjacent sheets of metal and folding one edge over the other. Standing seams are unsuitable for lead flat roofs, or for copper roofs of a pitch less than 6°.

Joints perpendicular to the fall of the roof are formed by 'drips'. The roof decking is stepped by a depth of at least 50 mm, and the metal is dressed over it, as shown in the example of a zinc drip in Fig. 3.11(d).

Junctions with walls At the junction between a roof and wall, the sheet metal is dressed vertically up the face of the wall. A 'cover flashing' which consists of a strip of similar metal is let into a mortar joint above the upturned sheet metal roofing, and is dressed over it (as illustrated in Fig. 3.11(e)).

Proprietary materials There are many different proprietary single-sheet roof finishes available of which 'Nutec Nuralite' is an example. The material is based on selected fibres bonded with bitumen into sheets of 2400 × 1200 mm × 2 mm thickness. The material is initially black but weathers to a lead-like smooth grey finish. The sheets are bonded with proprietary bitumen-based adhesive to any decking material. Adjacent sheets are covered with a 100 mm wide jointing strip (Fig. 3.11(f)) although the lead-like appearance may be accentuated by adopting a roll-cap jointing system.

The material is mouldable when heated and is suitable for forming upstands, gutter linings and flashings. No additional protective finishing is required and any pitch of roof may be covered from 1 in 60 upwards.

Multiple-layer roofing – 'built-up felt roofing'
'Built-up felt roofing' involves two or usually three layers of bitumen felt which can be laid on most forms of decking. Bitumen felt is available in rolls of 10, 20 and 30 m lengths and 1 m width.

To form a roof finish, a layer of bitumen felt is laid on the roof decking with its long edges parallel to the fall of the roof. The felt is fixed to the decking by either:
- nailing with wide-headed nails;
- fully or partially bonding the felt to the decking with hot bitumen.

Joints between adjacent sheets of felt are lapped by 50 mm along their length, and by 75 mm across edges (Fig. 3.12(a)). A second and then a third layer are laid on top, bonded to the layers below with hot bitumen. The layers are laid to 'break joint' (with no adjacent layer coinciding joints). The fall of the roof should be not less than 1.7 in 100.

A wide variety of bitumen felts are available and comprise a fibre core impregnated or coated with bitumen. British Standard 747 1977 (1986) classifies bitumen felts by core material: Class 1 felt, Class 2 asbestos, Class 3 glass-fibre and Class 5 polyester. Each class is then further subdivided by surface characteristics and weight.

Bitumen felts may be applied by heating with a gas torch, hot applied bitumen or by a self-adhesive surface incorporated in one surface of the felt.

Polyester felts coated with polymer modified bitumen and high-performance characteristics are expensive but require only two layers for a roof finish.

When felt roofing is applied to a screeded concrete roof deck, condensation can be a problem, causing the felt to blister. To help alleviate this problem, a vapour barrier is installed underneath the insulating material on which the felt is laid (Fig. 3.12(b)). Sometimes a vented base layer of felt is used which enables trapped moisture to escape.

Sunlight tends to damage the top surface of built-up felt roofing, and usually a protective layer of stone chippings is applied, bedded in bitumen. On vertical and angled faces of the roof, where the chippings could become dislodged, 'mineral-surfaced' felt is used.

Eaves details   Figures 3.12(b) and (c) show alternative treatments for the eaves. In Fig. 3.12(b) rain-water from the roof is discharged into an external gutter fixed to the fascia (see also Fig. 3.6(c)).

In Fig. 3.12(c) an 'upstand' is formed around the perimeter of the roof, and rain-water is discharged into a rain-water outlet and internal downpipe (see also Fig. 3.6(d)). The upstand is formed by an angled piece of timber (kerb) over which the felt is dressed. the felt is then dressed into an eaves trim which is screwed into the kerb. Eaves trims are available in several sizes and profiles, and may be made of aluminium or glass-reinforced plastic.

Junction with parapet wall   Figures 3.12(d) and (e) show alternative methods of forming a junction between a bitumen felt roof and a parapet wall.

In Fig. 3.12(d) an 'angle fillet' (strip of triangular section timber) is installed at the junction of the decking and wall to reduce the severity of the junction angle.

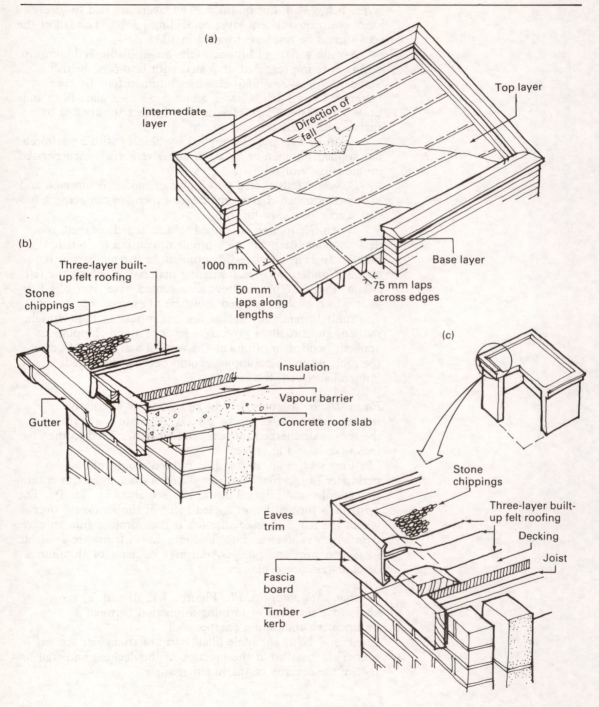

(a)

Top layer

Intermediate layer

Direction of fall

Base layer

1000 mm

50 mm laps along lengths

75 mm laps across edges

(b)

Three-layer built-up felt roofing

Stone chippings

Insulation

Vapour barrier

Gutter

Concrete roof slab

(c)

Stone chippings

Three-layer built-up felt roofing

Eaves trim

Decking

Joist

Fascia board

Timber kerb

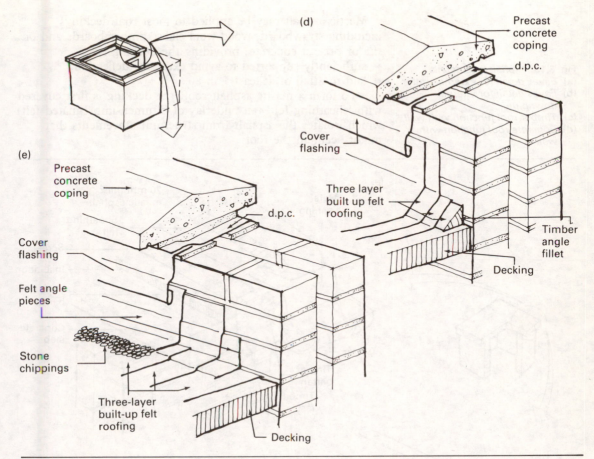

Fig. 3.12 *Built-up felt roofing*
(a) *Joints and laps*
(b) *Eaves detail for external gutter*
(c) *Eaves detail for internal
    downpipe*
(d) *Parapet wall detail with angle
    fillet*
(e) *Parapet wall detail without
    angle fillet*

In Fig. 3.12(e) a right-angled bend has been formed, but an angled piece of bitumen felt reinforces the junction. In both details a sheet metal cover flashing has been used to protect the upturned roofing felt. Bitumen felt cover flashings could be installed as an alternative.

Monolithic roof finishes – mastic asphalt
Mastic asphalt is a mixture of:
- asphaltic cement, which is either 'bitumen' (a black water-resisting material which softens when heated, and is derived either naturally or from the distillation of petroleum) or a blend of bitumen and 'lake asphalt' (a naturally occurring mixture of bitumen with inert minerals);
- aggregate which may be either 'asphalt rock' (a limestone naturally impregnated with bitumen) or limestone.

Fig. 3.13 *Asphalt roof details*
(a) *Eaves detail for external gutter*
(b) *Eaves detail for internal downpipe*
(c) *Timber roof junction with wall*
(d) *Concrete roof junction with wall*

Mastic asphalt may be applied to most roof deckings including strawboard, wood-wool slabs, timber boards and *in situ* or precast concrete, providing the decking is:

- sufficiently supported to avoid excessive deflection;
- laid to a fall of at least 1 in 80.

To form a mastic asphalt roof, the decking is first covered with 'sheathing felt' (an underlay of bitumen-impregnated felt) which isolates the asphalt from structural movements that might occur in the roof.

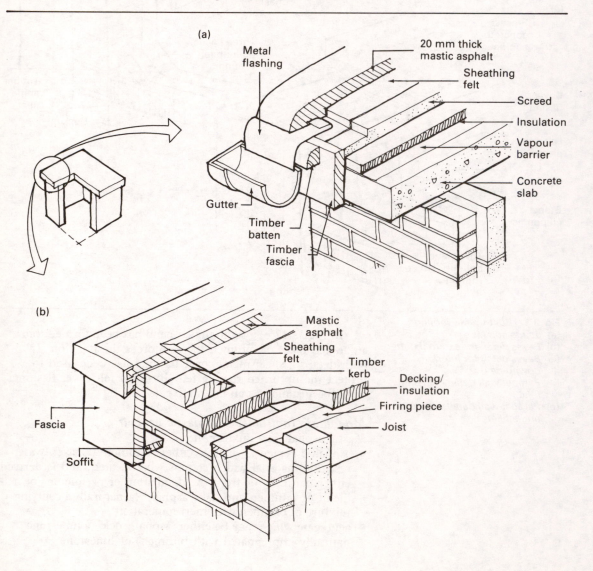

(a)

Metal flashing

20 mm thick mastic asphalt

Sheathing felt

Screed

Insulation

Vapour barrier

Concrete slab

Gutter

Timber batten

Timber fascia

(b)

Mastic asphalt

Sheathing felt

Timber kerb

Decking/ insulation

Firring piece

Joist

Fascia

Soffit

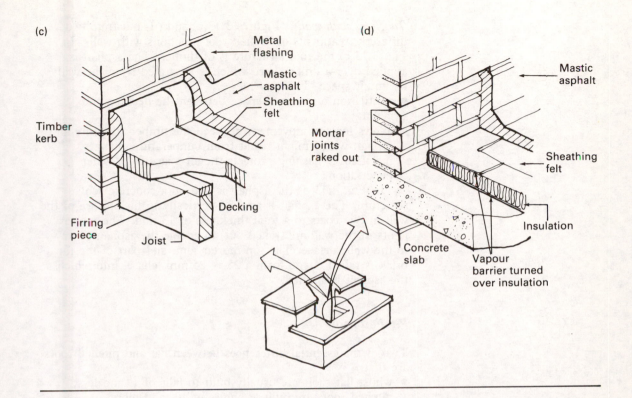

(c)

Metal flashing

Mastic asphalt

Sheathing felt

Timber kerb

Firring piece

Joist

Decking

(d)

Mastic asphalt

Mortar joints raked out

Sheathing felt

Insulation

Concrete slab

Vapour barrier turned over insulation

The mastic asphalt is heated to such a temperature that it becomes soft enough to be spread over the sheathing felt by use of a 'wood float' (a wooden trowel). The asphalt is applied in two coats, each 10 mm thick. Since asphalt tends eventually to break down due to the result of sunlight, a surface of protective stone chippings is applied on top of the asphalt.

When a concrete roof structure is used, a vapour barrier should be provided below the insulation. This will prevent water vapour damaging the insulation or the asphalt roof finish (Figs 3.13(a) and (d)). When a timber roof structure is used, the space between the decking and ceiling should be ventilated to avoid water vapour condensing within the roof space and causing deterioration of the roof timbers (Figs 3.13(b) and (c)).

Eaves details Figures 3.13(a) and (b) show alternative eaves treatments for mastic asphalt roofs. Figure 3.13(a) shows a projecting concrete roof with the asphalt laid over a metal flashing which discharges rain-water into an external gutter.

Figure 3.13(b) shows a timber roof structure with an upstand, where the asphalt is applied over the upstand and into an eaves trim.

Junctions with walls   Figures 3.13(c) and (d) illustrate two different treatments of asphalt roof junctions with walls. In Fig. 3.13(c) the roof structure is of timber. The asphalt is applied over a timber kerb which is separated from the wall by an air space. This provides:

- ventilation of the roof space between the decking and ceiling;
- allows for any movement between the timber structure and asphalt which might result from temperature changes. A metal cover flashing protects the air space from water penetration.

In Fig. 3.13(d) the asphalt is laid on a concrete roof structure. The asphalt is applied vertically to the surface of the wall in two coats to a total thickness of 13 mm. The mortar joints in the wall are 'raked out' to enable the asphalt to key to the wall surface. The top mortar joint and part of the top brick is raked out to form a 25 × 25 mm 'chase' into which the asphalt is applied.

Pitched roofs

Two of the essential differences between flat and pitched roofs are that:

- whilst flat roofs are usually built to falls of 1° and less, pitched roofs are built at angles of 10° and more;
- whereas materials used for flat roof finishes may also be used for pitched roofs, roof finishes consisting of small units such as tiles or slates may only be used on pitched roofs.

Pitched roof structures
For small-scale and domestic buildings, timber is the predominantly used material for pitched roof construction. For larger and framed buildings, steel and concrete may be used.

Monopitched roofs   The monopitched roof is the simplest form of pitched roof construction and resembles a timber flat roof structure, in that the structural timbers span from wall to wall. However, whilst flat roof joists are laid horizontally and bear on the horizontal top surface of the wall, pitched roof joists (known as 'rafters') are sloped and therefore could have a tendency to slide down off the wall. In order to reduce this tendency, the rafters are jointed over the wall plate with a 'birdsmouth' joint (Fig. 3.14(a)). This joint provides an area of horizontal bearing surface on the wall plate.

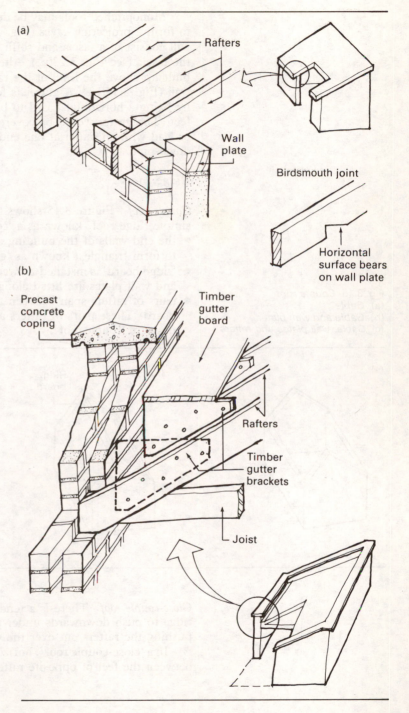

(a)

Rafters

Wall plate

Birdsmouth joint

Horizontal surface bears on wall plate

(b)

Precast concrete coping

Timber gutter board

Rafters

Timber gutter brackets

Joist

Fig. 3.14 *Monopitched roof eaves*
(a) *Overhanging eaves*
(b) *Parapet walls*

Monopitched roofs may be designed to overhang the wall to form a projecting eaves (Fig. 3.1(a)), where the eaves detail will consist of a fascia and soffit board, and a gutter fixed to the fascia (see Fig. 3.23(c)). Alternatively, the walls could be built up above the level of the rafter feet, to form a parapet wall (Fig. 3.14(b)). A gutter is formed behind the parapet wall by fixing a horizontal board to brackets attached to the rafter feet. The gutter would be covered in sheet metal, such as zinc or lead with an outlet at one end discharging into an internal down-pipe.

Ridge roofs

Couple roof Figure 3.15 shows the principal components of a simple ridge roof, known as a 'couple roof' where:
- the end walls of the building are built up above eaves level to form triangles, known as 'gables';
- 'ridge board' is installed between the apexes of the gables and wall plates are fitted along the side walls;
- pairs of rafters span between the wall plates and ridge board. These pairs of rafters are known as 'couples', hence the term 'couple roof'.

Fig. 3.15 *Couple roof*
(a) *Gable*
(b) *Gable and wall plates*
(c) *Gable, wall plates and rafters*

(a)

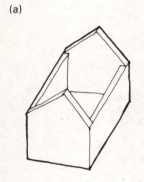

(b)

Ridge board

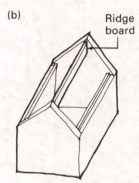

(c)

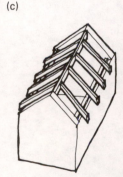

Close-couple roof There is a tendency in couple roofs for the ridge to push downwards under the weight of the roof, thus pushing the rafters out over the walls (Fig 3.16(a)).

In a 'close-couple roof', horizontal timbers are attached between the feet of opposite rafters, which resist the tendency

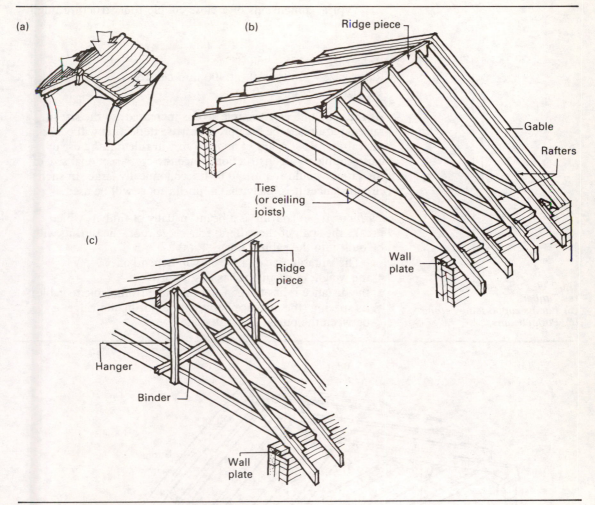

Fig. 3.16 *Close-couple roof*
(a) *Outward thrust of roof on walls*
(b) *Timbers of close-couple roof*
(c) *Close-couple roof with hangers and binder*

of the rafters to be pushed outwards. These timbers are known as 'ties' or (if the ceiling is fixed to the underside of them) 'ceiling joists' (Fig. 3.16(b)).

If the ties are to act as ceiling joists or if the ties are of considerable span, it may be necessary to provide the ties with additional support, otherwise deflection could occur in them.

This may be achieved by fixing 'hangers' between the ridge piece and the ceiling joist, thereby halving their span (Fig. 3.16(c)). Rather than fitting a hanger to each joist, it is usual to fix a 'binder' on top of the joists which runs parallel to the ridge piece. Hangers are then attached from the ridge to the binder at intervals of every third or fourth joist.

Typical dimensions of a close-couple roof structure might be:

Span	4 m
Ridge size	175 × 38
Rafter size	100 × 50 at 400 mm centres
Tie size	100 × 50

As the span of a close-couple roof is increased, so the size of rafters requires to be greater, otherwise deflection will occur in the rafters. Roof spans of above 6 m will rule out the use of close-couple roof construction, since the cross-sectional size of the rafters would require to be uneconomically large. In such circumstances it is likely that a 'purlin roof' will be used.

Purlin roofs A 'purlin' is a beam, usually of timber, which breaks the span of the rafters. In a ridge roof, the purlins will be built into the gables (Fig. 3.17(a)).

The suitable size of a purlin will depend on:
- the weight of roofing it has to support;
- the distance it has to span (for example from gable to gable);
- its spacing (the distance between the purlin and ridge or between the purlin and wall plate).

Fig. 3.17 *Purlin roof*
(a) *Timbers*
(b) *Purlins supported by struts*
(c) *Purlin beams*

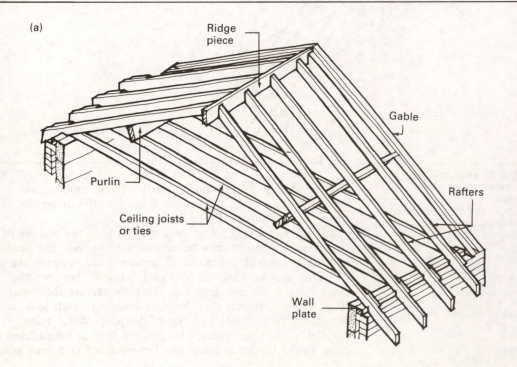

(a)

Ridge piece

Gable

Purlin

Rafters

Ceiling joists or ties

Wall plate

(b)

Rafters

Hanger

Ceiling
joists

Struts
to every
fourth
pair of
rafters

Purlin

Supporting
wall

Binder

(c)

Top
flange

Rafters

End of purlin
beam built into
loadbearing
wall

web

Bottom
flange

Purlin beams

Ceiling
joists

N.B. In the case of building
employing cross-wall construction,
these walls would be of lightweight
non-loadbearing construction

It is likely that the distance between the gable walls of a building would be too great for a purlin to span without providing it with additional support, otherwise the purlin would need to be of excessive cross-sectional size.

The usual method of supporting purlins is by means of 'struts'. Struts are pieces of timber which transfer some of the weight of the purlin to another structural support (typically an internal wall). It is usual for struts to be arranged in pairs (one from each purlin), with their feet bearing against each other (Fig. 3.17(b)).

A typical purlin size for a domestic roof where the span and spacing are in the order of 2.5 m might be 50×225 mm or 75×200 mm.

If the span of the ceiling joists is excessive, hangers may be attached from the purlins to a binder, which runs parallel to the purlin and is fixed to the top of the ceiling joists (Fig. 3.17(b)).

If the distance between the gable walls of a building is too great for an unsupported purlin (purlin without struts), and no internal walls are available from which the purlin might be supported, then alternatives which may be considered include purlin beams, roof trusses and trussed rafters.

Purlin beams (Fig. 3.17(c))   Purlin beams are deep beams consisting of timber top and bottom members (known as 'flanges') separated by a plywood 'web'. The rafters are birdsmouthed over the top flange of the beam. The bottom flange of the beam usually lies on top of the ceiling joists, in which case the span of the ceiling joists may be split by nailing them to the flange of the purlin beam. An alternative type of purlin beam is the corrugated plywood web beam, similar to that illustrated in Fig. 3.7(g).

Purlin beams are often used for buildings employing 'cross wall construction' (a form of construction in which the loads of the building are carried principally on the internal walls, which are built at right angles to the length of the building).

Roof trusses   A roof truss is a supporting framework of structural members. A truss supporting a ridge roof is triangular in shape, following the lines of a pair of rafters and the ceiling joist. Within this triangle, a series of 'struts' (members resisting compression) and 'ties' (members resisting tension) divide the triangular space into a series of smaller triangles (Fig. 3.18(a)). The triangle is a useful structural shape since it resists deformation, whereas a square structural shape could deform into a parallelogram shape. The trusses, which in a domestic building might be placed at 2 m intervals, provide support for the purlins.

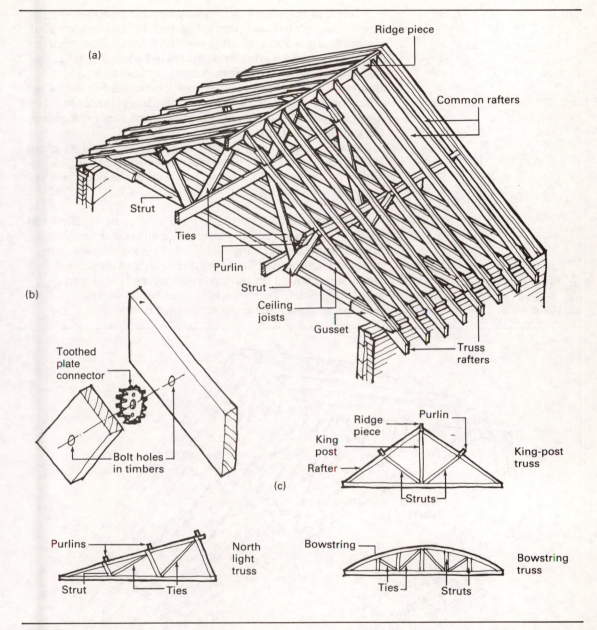

Fig. 3.18 *Roof trusses*
(a) *Roof trusses for small roof*
(b) *Toothed plate connector*
(c) *Some other truss designs*

In the small-scale roof illustrated in Fig. 3.18(a), trusses are positioned so that the truss rafters lie in the same plane as the 'common' rafters. In larger-span roofs, the trusses are usually positioned so that the truss rafters lie beneath the purlins.

Joints in truss construction may be fixed by either glue, nails or bolts and timber connectors. Figure 3.18(b) shows a typical connector known as a 'toothed plate connector'. The connector is placed between the pieces of timber to be joined, and as the bolt (which passes through the centre of the connector) is tightened, the connector teeth become embedded in the timber. The junction between the ties and rafter feet is made with 'gussets' (pieces of timber which bridge, and connect together, the adjoining timber members).

Many forms of truss design are possible, some of which are shown in Fig. 3.18(c). The apexes (or 'nodes') of triangles within the truss should occur at loadbearing positions such as at junctions with purlins.

Trussed rafters   In a trussed rafter roof, each pair of rafters and linking ceiling joist is triangulated with struts and ties, to form a 'trussed rafter' (Fig. 3.19(a)). Trussed rafters are, in effect, self-supporting lightweight trusses, and their use eliminates the need for purlins, ridge pieces or loadbearing supporting walls. Trussed rafters have become very widespread for domestic and small-span roofs.

Fig. 3.19 *Trussed rafters*
(a) *Trussed rafters for small roof*
(b) *Truss plate*

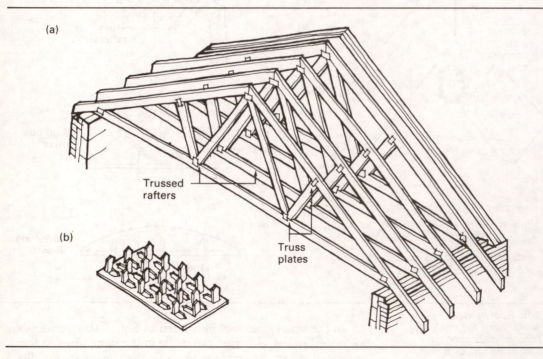

(a)

Trussed rafters

(b)

Truss plates

The trussed rafters are usually prefabricated in a factory, then transported to the building site, and fixed in position on the wall plates. This considerably reduces the amount of site work compared with a traditional purlin roof built on the site by carpenters. The volume of timber used in the roof construction is also less than that of a traditional roof.

The timbers used in a trussed rafter are all of the same thickness, although their depths will vary – a typical rafter size might be 100×38 mm. The trussed rafters are usually spaced at intervals of 600 mm, compared with the usual spacing of 400 mm for rafters and ceiling joists of traditional roof structures.

The trussed rafter roof illustrated is jointed by 'truss plates' (sometimes referred to as 'gang nail' connectors) (Fig. 3.19(b)). One connector is placed each side of the joint, hydraulic pressure is applied and the teeth of the plate are driven into the timber.

Junctions and openings in pitched roofs

Hipped roof A 'hip' is an inclined triangle of roof (Fig. 3.20(a)). The structure of a hip is formed by two hip rafters which span from the intersections of the wall plates at the corners of the walls to the end of the ridge piece. A series

Fig. 3.20 *Hipped roof*
(a) *General view*
(b) *Timbers*

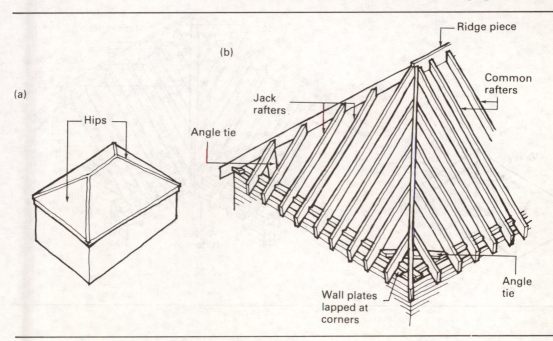

of 'jack' rafters of diminishing length, span from the wall plates to the hip rafter (Figure 3.20(b)). Hip rafters are of substantial cross-sectional size since they carry the weight of the jack rafter ends. A typical hip rafter size might be 225 × 50 mm.

There is a tendency for the thrust of the hip rafter to push out against the corners of the walls of the building. This thrust is resisted by a timber 'angle tie' which joints into, and ties together, the wall plates at the corner of the walls. Sometimes, as an additional measure, a 'dragon beam' is installed. This is a timber member which joints into, and ties together the intersection of the wall plates, and the midway position of the angle tie.

Valleys When an intersection between two pitched roofs occurs, a 'valley' is formed (Fig. 3.21(a)). Like hip construction, a valley requires jack rafters which span from the ridge to a 'valley rafter', which is of similar size to a hip rafter (Fig. 3.21b).

A 'valley board' is sometimes fixed above the valley rafter which forms the base of a sheet-metal-lined 'valley gutter' (see Fig. 3.23(j)).

Fig. 3.21 *Valley*
(a) *General view*
(b) *Timbers*

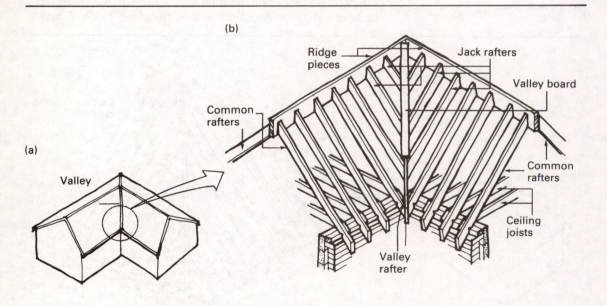

Fig. 3.22 *Dormers*
(a) *Flat-roofed dormer*
(b) *Trimming of dormer opening*
(c) *Timbers*
(d) *Different styles of dormer* (i) *partial dormer* (ii) *recessed dormer* (iii) *gable-ended dormer*

Dormers Dormer windows are vertical windows formed within the roof area, a simple example of which is shown in Fig. 3.22(a).

The construction of a dormer requires an opening to be made in the roof (Fig. 3.22(b)). This is achieved by use of 'trimmer joists', which are timber members fixed perpendicular to the direction of the rafters, and into which the 'trimmed rafters' are jointed.

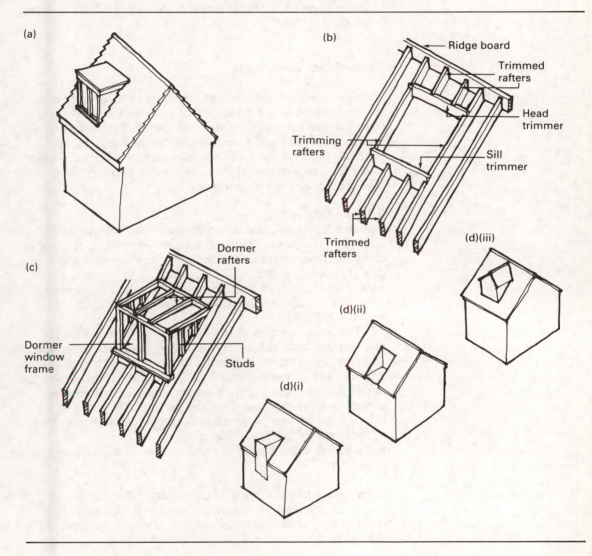

(a)

(b)

Ridge board

Trimmed rafters

Head trimmer

Trimming rafters

Sill trimmer

Trimmed rafters

(c)

Dormer rafters

Dormer window frame

Studs

(d)(i)

(d)(ii)

(d)(iii)

The dormer itself is formed by:

- a window frame which is fixed to the lower (or 'sill') trimmer;
- dormer rafters, which are fixed between the upper (or 'head') trimmer and the head of the window frame;
- vertical timbers ('studs') which are fixed between the dormer rafters and the common rafters below to form the sides ('cheeks') of the dormer (Fig. 3.22(c)). Figure 3.22(d) shows some of the other forms of dormer windows that are possible.

Pitched roof coverings

The functional requirements of a pitched roof covering are similar to those of a flat roof, namely, to provide protection from the weather and to contribute towards fire protection, thermal insulation and sound insulation. Although many materials used for flat roof coverings may be used for pitched roofs, the coverings usually chosen consist of materials of overlapping sheet or tile form.

Tiles and slates

Tile and slate roof coverings comprise of units small enough to be held in the hand. Before laying tiles or slates, it is usual to cover the rafters with an 'underslating' or 'sarking', which is a thin sheeting of either:

- bitumen felt which may incorporate an aluminium foil surface to improve thermal and vapour insulation;
- polythene sheet.

The main function of the underslating is to protect the roof timbers from any water which may penetrate through the joints in the tiling or slating. Timber battens with a cross-sectional size in the order of 25 × 50 mm are nailed across the rafters and over the underslating and the tiles or slates are hooked over, and/or nailed to, the battens. The distance between the battens (known as the 'gauge') is determined by:

- the length of the tile or slate;
- the distance by which the tiles or slates overlap (known as the 'lap').

Tiles Roof tiles are made of either clay (which is the traditional material) or concrete. Clay is used comparatively rarely today, apart from repair works. Table 3.4 compares the characteristics of some of the tiles available.

Plain tiles   Plain tiles are the simplest form of roof tile, and are laid with no overlap at their sides. To prevent rain penetration, the side joints of successive courses of tiles are staggered. The overlap of one course over another is large, which results in there being three thicknesses of tiles at any lap (Fig. 3.23(a)).

The tiles are cambered in profile to prevent water creeping up between the courses of tiles due to capillary attraction.

At the back of the tile are two projecting nibs for hooking the tile over the batten, and a hole for nailing the tile to the batten. Nailing of tiles is carried out:

- usually to every fourth course of tiles;
- to tiles laid at the verge;
- to tiles laid at the junction with walls.

Single-lap tiles   Single-lap tiles overlap at their sides, which prevents the passage of water between them; consequently, each course of tiles requires only one small overlap to provide a water-resistant roof finish. The side overlap may be achieved by either the side of one tile overlapping the side of the adjacent tile (such as a clay pantile) or by interlocking, where a series of ribs in the side of one tile locates into a series of grooves in the adjacent tile.

Since single-lap tiles resist water penetration through their side joints:

- it is not necessary for the joints of successive courses to be staggered:
- the tiles can be used on lower pitched roofs.

Tiled ridge   A ridge tile is usually of half round or angled profile. It fits over the ridge piece, and is bedded in mortar which is laid on the top courses of tiles (Fig. 3.23(b)). Joints between ridge tiles are jointed with mortar.

The illustration shows a plain tiled roof, where a course of short 'top tiles' are necessary at the ridge.

Tiled eaves   Figure 3.23(c) shows a typical eaves detail with an interlocking tiled roof. An eaves formed with plain tiles will require a course of short eaves tiles.

Tiled verge   The verge is formed with an 'undercloak' which consists of a layer of slate or asbestos cement laid on the top surface of the gable and wall. The underslating battens and tiles are laid across the undercloak, and the edges of the verge tiles are bedded in mortar (Fig. 3.23(d)).

Table 3.4 Characteristics of roof tiles

Clay

Tile	Min. pitch	Size (mm)	Typical gauge (mm)
Plain tile (reverse side)	40°	265 × 165	100
Pantile	35°	340 × 240	265
Double Roman	35°	420 × 355	345

Tiled hips The hips are tiled in a similar manner to the ridge by using either:
- ridge tiles which are bedded in mortar either side of the hip, and a wrought iron or steel hook is screwed to the foot of the hip rafter to support the hip tiles (Fig. 3.23(e));
- 'bonnet tiles' (used in conjunction with plain tiling) where the head of the bonnet tile is nailed, whilst the tail is bedded in mortar (Fig. 3.23(f)).

Tiled valleys Tiled valleys may be formed by using either:
- 'valley tiles' which may be used to form a plain tiled valley (Fig. 3.23(g));

Table 3.4 continued

Concrete

Tile	Min. pitch	Size (mm)	Typical gauge (mm)
 Plain tile (reverse side)	35°	265 × 165	100
Interlocking Pantile	30°	380 × 230	345
Double Roman interlocking	30°	420 × 330	345
Interlocking slate	17½°	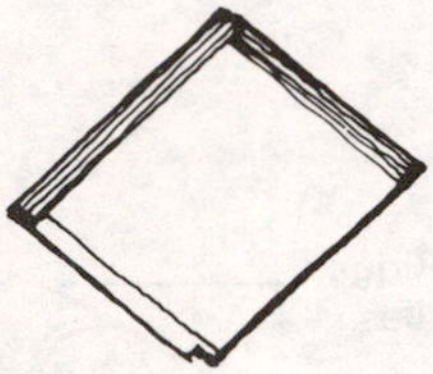 430 × 380	355

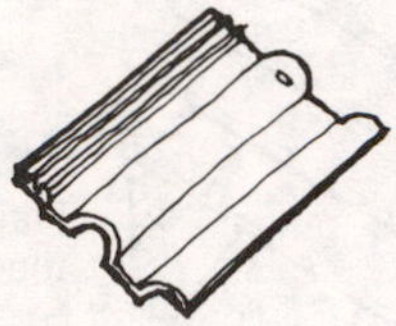

- 'trough valley tiles' which may be used to form a valley with interlocking tiles. The trough valley tiles form a gutter, and the roof tiling is cut to fit the valley. The roof tiling overlaps the trough valley (Fig. 3.23(h));
- forming a sheet metal, or vinyl valley gutter. In the case of a sheet metal valley gutter, the metal may be dressed over the valley board, and up over triangular timber fillets fixed either side of the valley board (Fig. 3.23(j)).

Junctions with walls and chimneys Junctions between tiled roofs and walls or chimneys are made with sheet metal 'flashings'. Figure 3.23(k) shows the four flashings installed around a chimney:

- a 'stepped flashing' to each sloping side;
- an 'apron flashing' at the front end;
- a 'gutter flashing' at the back of the chimney.

The stepped and apron flashings are dressed over the tiles, turned up the face of the brickwork and tucked into a mortar joint. The gutter flashing is dressed under a layer of tiles above the chimney, over a horizontal 'gutter board', then vertically up the face of the brickwork and tucked into a mortar joint.

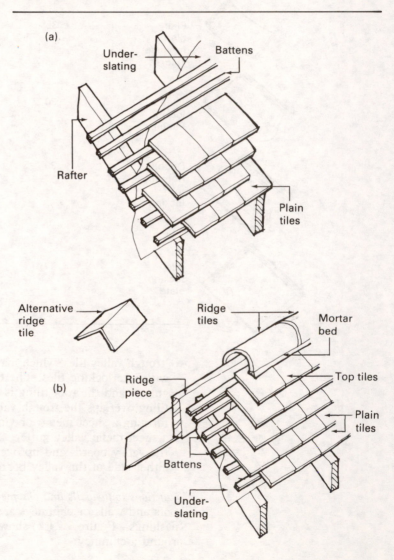

Fig. 3.23 *Roof tiling*
(a) *Plain tiling*
(b) *Plain tiling at ridge*
(c) *Interlocking tiling at eaves*
(d) *Interlocking tiling at verge*
(e) *Tiling at hip using ridge tiles*
(f) *Tiling at hip using bonnet tiles*
(g) *Tiling at valley using valley tiles*
(h) *Tiling at valley using trough valley tiles*
(j) *Tiling detail for valley gutter*
(k) *Flashings around chimney*

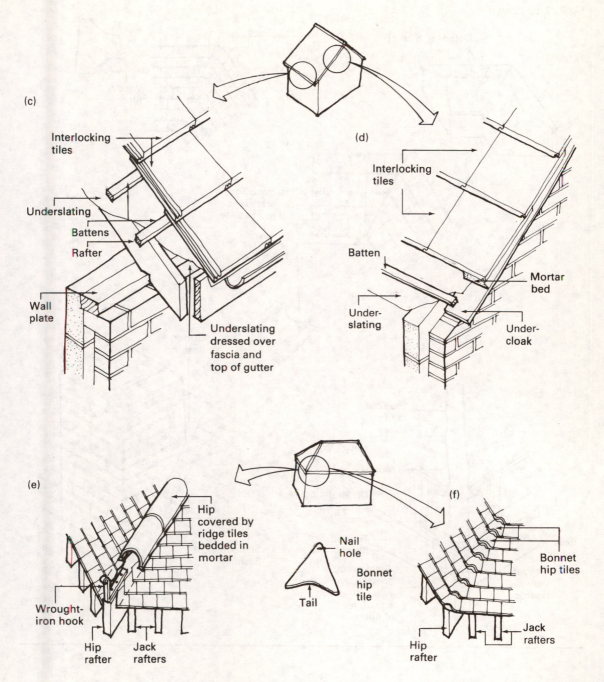

(c)

Interlocking tiles

Underslating

Battens

Rafter

Wall plate

Underslating dressed over fascia and top of gutter

(d)

Interlocking tiles

Batten

Mortar bed

Under-slating

Under-cloak

(e)

Hip covered by ridge tiles bedded in mortar

Wrought-iron hook

Hip rafter

Jack rafters

Nail hole

Bonnet hip tile

Tail

(f)

Bonnet hip tiles

Hip rafter

Jack rafters

Fig 3.23 continued overleaf

121

Fig 3.23 continued

(g)

Valley tiles

Valley tile

Details g, h and j

(h)

Interlocking concrete tiles cut to fit valley

Trough valley tiles

(j)

Tiles cut to fit valley

Mortar

Underslating turned up over fillet

Timber fillet

Sheet metal dressed over valley boards and up face of fillets

Jack rafter

Timber valley boards

Valley rafter

(k)

Gutter flashing

Stepped flashing

Apron flashing

Slates   Slate is a hard rock which is quarried and split into thin layers out of which the roofing slates are cut. Roofing slates are available in a wide variety of sizes varying from 300 × 200 mm to 600 × 350 mm. Like plain tiling, slates have no side laps, therefore alternate courses of slates must be laid with staggered side joints.

The size of lap between courses varies in accordance with the pitch of the roof: since water will run less rapidly off a shallow-pitched roof, a greater lap is required than that of a steeply pitched roof. A roof of 35° pitch might require a 75 mm lap, whilst a pitch of 25° might require a lap of 90 mm.

Each slate is fixed to the batten with two nails. The nail holes are located either in the centre of the slate ('centre nailed' – Fig. 3.24(a)(i)) or at the top of the slate ('head nailed' – Fig. 3.24(a)(ii)). The advantage of centre-nailed slating is that since the bottom edge of the slates are near the nails, they are less likely to be lifted by the wind. Less expensive slate substitutes are composed of materials such as crushed slate bonded with resin or cement with natural and synthetic fibres.

Slated ridges   Slate ridges may be formed in a similar manner to tiled roofs (see Fig. 3.23(b)) – although traditional techniques sometimes formed the ridge with stone ridges.

Slated eaves   Like plain tiled roofs, a double course of slates is necessary at the eaves to prevent rain entering between the side joints.

Slated verges   A slate undercloak is used to support the verge slates which are bedded in mortar, in a similar manner to the tiled verge illustrated in Fig. 3.23(d).

Slated hips   Hips may be formed by using clay ridge tiles using a similar detail to the tiled hip illustrated in Fig. 3.23(e). Traditional methods included the use of stone tiles or the use of lead rolls or lead dressed beneath mitred slates (Fig. 3.24(b)).

Slated valleys   Valleys are usually formed from sheet metal in similar manner to the tiled valley shown in Fig. 3.23(j).

Junctions with walls   Details of junctions with walls or chimneys are similar to those of the tiled roof illustrated in

Fig. 3.24 Slate roofs
(a) Nailing of slates
 (i) Centre nailed
(ii) Head nailed
Fig. 3.24(b) Hip detail

Fig. 3.23(k), using stepped, apron and gutter flashings. Sometimes the stepped flashing is dressed under the slates in separate pieces of lead known as 'soakers'. Each soaker is dressed up the wall and the stepped cover flashing is dressed down over the top of the soakers.

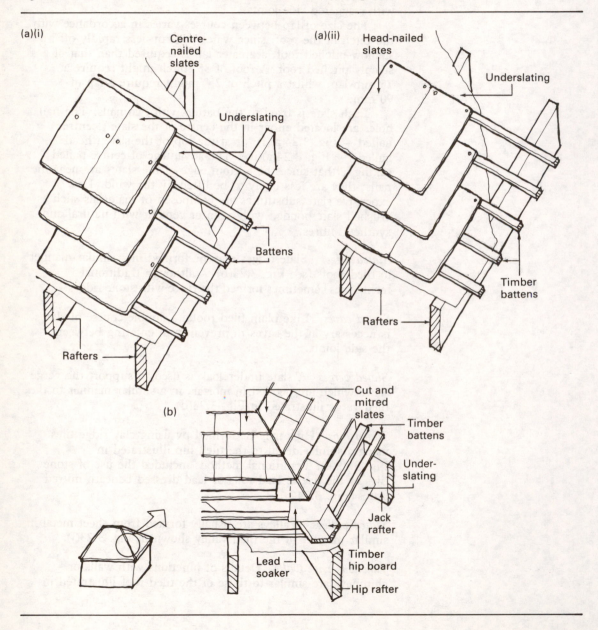

(a)(i)

Centre-nailed slates

Underslating

Battens

Rafters

(a)(ii)

Head-nailed slates

Underslating

Timber battens

Rafters

(b)

Cut and mitred slates

Timber battens

Under-slating

Jack rafter

Timber hip board

Lead soaker

Hip rafter

Profiled (corrugated) sheets

Corrugated sheets are frequently used for large-span pitched roof coverings, since:

- tiles and slates are usually required to be laid at a pitch of 20° and more which, in the case of a large-span roof, would increase the volume of the building, and consequently increase the building and heating costs;
- tiles and slates are relatively heavy and require a supporting structure with close spacing, in the form of rafters and battens. This would result in an expensive and heavy roof structure.

Corrugated sheets are frequently used for covering large-scale pitched roofs of factory and agricultural buildings employing 'portal frame' structures. Here the use of battens and rafters is eliminated by fixing the corrugated sheets direct to purlins (see key diagram to Fig. 3.25(a)). Table 3.5 compares typical profiled sheets of aluminium, fibre cement, galvanised steel and plastic. The corrugations:

- provide rigidity, which enables comparatively thin material to be used in large sheet form;
- form 'gutters' at the bottom of each corrugation, which carry rain down the surface of the roof.

Since corrugated sheets provide inadequate thermal insulation for buildings in which people work, they are often laid in 'sandwich' form – a layer of fibreglass or other insulation material is sandwiched between a profiled outer sheet and an inner sheet, which may be flat, or of flatter profile (Fig. 3.25(a)).

The sheets are laid with side and end laps, and are fixed to the purlins by means of either:

- hook bolts (if fixed to steel or concrete purlins, Fig. 3.25(c)(i));
- screws or nails (if used in conjunction with wooden purlins, Figs. 3.25(c)(ii) and (iii)).

To avoid the ingress of rain, the fixings are made at the tops of the corrugations. A plastic washer is installed between the head of the nail, bolt or screw and the corrugated sheeting, and a plastic cap is fitted over the head to prevent the penetration of rain.

Details Figures 3.25(d), (e) and (f) show typical details at the eaves, verge and ridge of a corrugated sheet roof.

Figure 3.25(d) illustrates the use of an eaves filler piece which closes the corrugations at the eave.

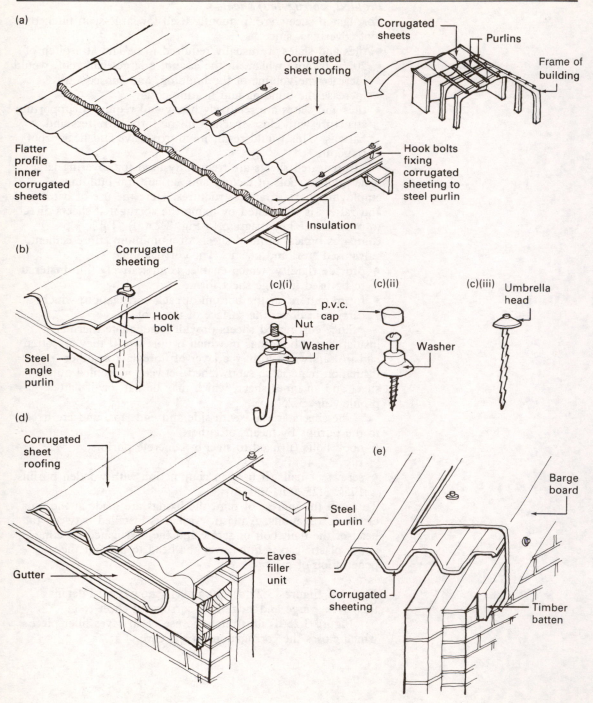

(a)

Corrugated sheets

Purlins

Frame of building

Corrugated sheet roofing

Flatter profile inner corrugated sheets

Hook bolts fixing corrugated sheeting to steel purlin

Insulation

(b)

Corrugated sheeting

Hook bolt

Steel angle purlin

(c)(i)

p.v.c. cap

Nut

Washer

(c)(ii)

Washer

(c)(iii)

Umbrella head

(d)

Corrugated sheet roofing

Gutter

Steel purlin

Eaves filler unit

(e)

Barge board

Corrugated sheeting

Timber batten

126

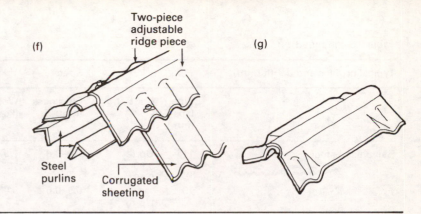

Fig. 3.25 *Corrugated roofing*
(a) *Sandwich construction*
(b) *Hook-bolt fixing*
(c) (i) *Hook bolt*
    (ii) *Screw*
    (iii) *Nail*
(d) *Eaves detail*
(e) *Verge detail*
(f) *Adjustable ridge piece*
(g) *Two-piece adjustable ridge and ventilation piece*

Table 3.5 Characteristics of profiled sheet roofing

Typical profiles and depth (mm)	Sizes				Comments
	Side laps (mm)	L × B (m)	Thickness (mm)	Weight (kg/m²)	
Aluminium	25	Max. 10.00 × 0.8–1.0	0.7–1.2	2.35–4.00	Aluminium is a metal obtained from bauxite clay with a small percentage of manganese added
	19	Max. 10.50 × 0.67–1.02	0.9–1.2	3.30–4.50	Pitch normally down to 15°, but if single lengths are used, down to 5°
	29	Max. 10.50 × 0.97–1.02	1.0–1.2	4.00–4.50	Obtainable in self-finish, enamelled or anodised finish. Curved sheets available. Good resistance to corrosion. Non-combustible
Fibre cement	134	1.225–3.050 × 0.782	5	15	Mixture of Portland cement, water and organic fibre. Pitch normally down to 10°, or with certain precautions, 5°. Side laps should be sealed with mastic.
	70	1.525–3.050 × 1.086	6	17	Non-combustible and largely maintenance free. Cranked ridge, curved eaves and coloured sheets available

127

Table 3.5 continued

Typical profiles and depth (mm)			Sizes		Comments
	Side laps (mm)	L × B (m)	Thickness (mm)	Weight (kg/m²)	
Galvanised steel	114	Up to 3.60 × 0.88	0.4–0.8	7.6–11.70	Steel sheeting dipped in molten zinc which forms a protective layer
	19	Up to 3.80 × 0.88	0.5–0.6	9.00–10.00	Non-combustible Available finished with paint or a variety of factory applied finishes Curved sheets available
Plastics	51	Up to 5.50 × 0.76	1.37	Approx. 2.00	Plastic sheets available in p.v.c. and g.r.p.
	45	Up to 5.50 × 0.93	1.37	Approx. 2.00	Available as transparent sheets (often used for daylighting in plastic or other corrugated roofs), or in a range of colours Can be easily sawn or bent Behaviour in fire varies with different types of plastic

The verge detail (Fig. 3.25(e)) shows the use of a 'barge-board' which covers and protects the junction between the gable wall and verge of the roof. Ridge details may be made with two-piece adjustable ridges as illustrated in Fig. 3.25(f). Alternatives include:

- single sheet ridge pieces;
- ridge pieces formed with a raised portion running along the corrugations to provide ventilation (Fig. 3.25(g)).

FLOORS

Functional requirements

Floors are the horizontal structure within a building that are designed to support their own weight (dead load) and most of the live loads inside the building imposed by the occupants, their possessions and equipment. Floors are either 'ground floors', which bear directly on the subsoil, or 'upper floors', which bear on the external, and sometimes also the internal walls of a building.

The functional requirements of a floor include: (*a*) structural stability; (*b*) resistance to penetration of moisture; (*c*) provision of fire resistance; (*d*) provision of thermal insulation; (*e*) provision of sound insulation.

Structural stability

Ground floors   Ground floors are usually of either suspended construction or solid construction.

In the case of a suspended ground floor, a concrete slab is laid on a bed of 'hardcore' (a well-compacted layer of material such as broken bricks or stones). A series of low walls ('sleeper walls') are built on the slab, and timber joists (typically of 100×50 mm cross-section) span between wall plates placed on the top surface of the sleeper walls. The spacing of the sleeper walls is arranged to ensure that deflection of joists is kept within acceptable limits (see Fig. 3.27(a)).

Solid ground floors consist of a slab of concrete laid on a bed of hardcore (Figs 3.26(a) and (b)). The thickness of a ground-floor slab is usually a minimum of 100 mm. Concrete slabs may be reinforced with mesh reinforcement if the bearing capacity of the subsoil is inadequate, or if the slab forms a raft foundation (see Fig. 2.4(b), Ch. 2).

Upper floors   Upper floors are suspended between the walls of a building, and consequently they must be sufficiently stiff and strong to carry their dead and live loads without causing an unacceptable degree of deflection in the floor. Upper floors of domestic-scale buildings usually consist of either:
- timber joists built into the external walls (see Fig. 3.28(a));
- reinforced *in situ* concrete slabs (Fig. 3.29(c));
- precast concrete planks or beams (Figs 3.30(a) and (b)).

The loads imposed on the floor, and the span of the floor must be taken into account in deciding either:

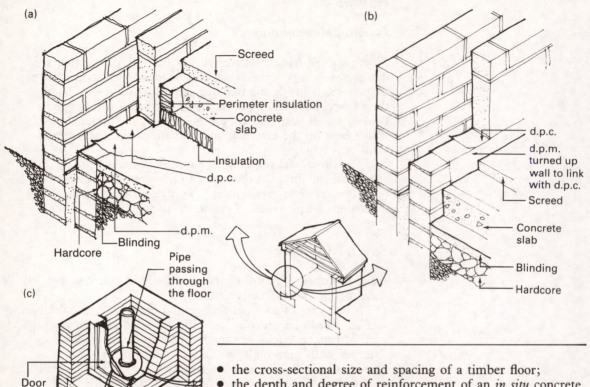

(a)
Screed
Perimeter insulation
Concrete slab
Insulation
d.p.c.
Hardcore
Blinding
d.p.m.

(b)
d.p.c.
d.p.m. turned up wall to link with d.p.c.
Screed
Concrete slab
Blinding
Hardcore

(c)
Pipe passing through the floor
Door to duct
Floor slab
Pipe
Pipe sleeve
Non-combustible fire-stopping material

Fig. 3.26 Solid floors
(a) Damp-proof membrane below slab
(b) Damp-proof membrane below screed
(c) Detail of pipe passing through solid floor

- the cross-sectional size and spacing of a timber floor;
- the depth and degree of reinforcement of an *in situ* concrete floor;
- the profile and depth of a precast concrete floor.

Resistance to moisture penetration

The problem of moisture penetration arises in ground floors because moisture from the soil is capable of passing through the pores in the concrete slab. In suspended ground floors, a d.p.c. is laid on the top of each sleeper wall which prevents moisture from rising up through the concrete and brick into the floor timbers. In solid floors, a 'damp-proof membrane' ('d.p.m.': a layer of impervious material) is installed in the floor by either:

- forming the slab in two separate layers and sandwiching the d.p.m. between them;
- laying the d.p.m. on top of the slab and applying a screed over it (Fig. 3.26(b)), which must be of sufficient depth to ensure that the pressure of moisture rising through the slab does not cause the d.p.m. or screed to blister;
- laying the d.p.m. on compacted sand beneath the concrete slab (Fig. 3.26(a)).

Fire resistance

The upper floors of a building can be an important factor in the behaviour of fire within a building.

The Building Regulations 1991 (approved document B) specify minimum periods of fire resistance for 'elements of structure'. A floor (other than the lowest floor of a building) is such an element of structure. Table 3.6 extracts some of the figures from the regulations.

The periods of fire resistance specified are determined by:
- the 'purpose group' of the building (broadly the function of the building);
- the height of the building or 'compartment' above or below ground.

A compartment, in essence, is a part of the building enclosed by 'compartment walls' and 'compartment floors'. Such walls and floors must fulfil certain conditions in the regulations which aim to reduce the possibility of fire spreading from the compartment to other parts of the building. The regulations specify where compartments are necessary, their maximum dimensions and the degree of fire resistance required.

Table 3.7 illustrates examples of floor constructions taken from a Building Research Establishment report: Guidelines for the construction of fire resisting structural elements (HMSO 1982). These floors might satisfy the various periods of fire resistance required for floor constructions. Pipes and 'ducts' (enclosures or casings which contain pipes or cables) could form a weak link in the ability of a floor to restrict the spread of fire. Measures to restrict the passage of fire through pipes and ducts include:
- restricting the diameter of the pipe;
- making the pipe of non-combustible material;
- enclosing the pipe in a non-combustible enclosure;
- packing the space between the pipe and floor with non-combustible material (Fig. 3.26(c)).

Thermal insulation

The problem of thermal insulation in floors is mainly confined to ground floors, although upper floors which:
- incorporate ceiling heating systems;
- are partially exposed to the open air (such as where part of the floor forms a balcony for the storey above),

should be provided with suitable thermal insulation.

Thermal insulation of ground floors The thermal insulation of upper floors within a building is usually adequate, since the

Table 3.6　Minimum periods of fire resistance (minutes)

Purpose group	Basement storey		Ground or upper storey			
	Depth of lowest basement (m)		Height of top floor above ground			
	more than 10	less than 10	less than 5	less than 20	less than 30	more than 30
1. Residential (domestic):						
a) flats	90	60	30*	60**	90**	120**
b) dwelling houses	—	30*	30*	60	—	—
2. Residential:						
a) Institutional	90	60	30*	60	90	120†
b) other	90	60	30*	60	90	120†
3. Office:						
not sprinklered	90	60	30*	60	90	not permitted
sprinklered	60	60	30*	30*	60	120†
4. Shop and commercial:						
not sprinklered	90	60	60	60	90	not permitted
sprinklered	60	60	30*	60	90	120†
5. Assembly and recreation:						
not sprinklered	90	60	60	60	90	not permitted
sprinklered	60	60	30*	30	60	120†
6. Industrial:						
not sprinklered	120	90	60	90	120	not permitted
sprinklered	90	60	30*	60	90	120†

 * 60 minutes for compartment walls separating buildings
** 30 minutes for floor within maisonette unless floor contributes to the support of the building
 † 90 minutes for elements not forming part of the structural frame

Source: From Building Regulations 1991, approved document B, Table A2

rooms below the floor are usually of sufficiently high temperature to cause minimal heat loss through the floor. However, additional thermal insulation would be required for the upper floors described below:

- floors which cantilever over the walls of a building, such that the underside of the floor is exposed to the external air;
- floors built over garages or other cold or unheated rooms;
- floors which incorporate floor or ceiling heating systems.

Unless the ground floor is sufficiently large in area insulation will be necessary to achieve the Building Regulation requirement of 0.45 W/m^{2}K.

Solid floors might be typically insulated by laying the screed on material such as expanded polystyrene. This should be turned up at the edges to prevent 'cold bridging' (Fig.

Table 3.7 Fire resistance of various floor constructions

Period of fire resistance (hr)

½-hour fire resistance

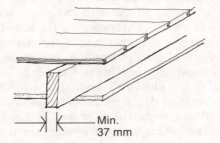

Min. 37 mm

(1) Floor finish:
minimum 15 mm tongued and grooved boarding, plywood or chipboard

Structure:
timber floor joists, minimum 37 mm thick

Ceiling:
12.5 mm plasterboard, minimum 5 mm neat gypsum plaster finish

(2) Floor finish:
minimum 21 mm tongued and grooved flooring, plywood or chipboard

Structure:
timber floor joists, minimum 37 mm thick

Ceiling:
12.5 mm plasterboard with taped and filled joists

1-hour fire resistance

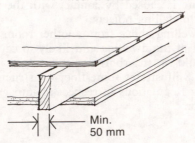

Min. 50 mm

(1) Floor finish:
minimum 15 mm tongued and grooved boarding, plywood or chipboard

Structure:
timber floor joists, minimum 50 mm thick

Ceiling:
30 mm plasterboard, joints staggered. Exposed joints taped and filled

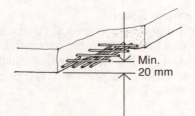

Min. 20 mm

(2) Reinforced concrete, minimum 95 mm thick with minimum 20 mm cover to lowest reinforcement

Source: From BRE: Guidelines for the construction of fire resisting structural elements

133

3.26(a)). Suspended ground floor insulation is dealt with on page 140 and Fig. 3.27(d).

Sound insulation

The problem of sound transmission within a building occurs mainly as a result of elements in the construction such as floor or walls being set in vibration by the sound, and transferring the vibration to the other side of the element. The effectiveness of such elements to prevent sound transmission depend largely on their 'mass' (or weight), since a heavy material is less likely to vibrate than a lightweight one. As well as incorporating dense materials in the floor construction, sound insulation may be improved by means of 'discontinuous construction'. This involves forming the floor (or other element) in two separate skins, in such a way that sound is prevented from travelling from one side of the floor to the other (see the example of a 'floating floor' in Table 3.8).

The Building Regulations 1991 (approved document E) require that floors separating

(a) a dwelling below from another dwelling above, or from other rooms that could cause noise nuisance should resist both 'airborne' and 'impact' sound. Airborne sound is caused by sources such as speech or radio whilst impact sound is caused by contact with the floor by, for example, feet or falling objects.

(b) a dwelling above from another room below, which is not part of the dwelling and likely to cause noise nuisance, should resist airborne sound only.

Table 3.8 illustrates some floor constructions that satisfy the Building Regulations in respect of impact and airborne sound insulation.

Ground floors

Solid ground floors

A solid ground floor is basically a concrete slab which is laid on the ground, between the external walls of the building. The elements of the floor construction usually consist of:
(a) hardcore; (b) concrete slab; (c) d.p.m.; (d) screed (Figs 3.26(a) and (b)).

Hardcore Hardcore consists of hard pieces of material such as broken brick, stone or concrete. Its function is to provide a flat stable base on which the floor may be laid.

Table 3.8   Sound resistance of floor constructions

Floor type	Floor structure	Sound resistance
Concrete base with soft covering Soft covering Hollow beams 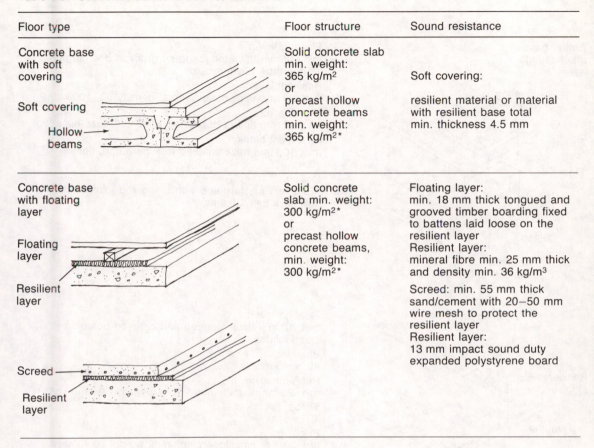	Solid concrete slab min. weight: 365 kg/m^2 or precast hollow concrete beams min. weight: 365 kg/m^2*	Soft covering: resilient material or material with resilient base total min. thickness 4.5 mm
Concrete base with floating layer Floating layer Resilient layer Screed Resilient layer	Solid concrete slab min. weight: 300 kg/m^2* or precast hollow concrete beams, min. weight: 300 kg/m^2*	Floating layer: min. 18 mm thick tongued and grooved timber boarding fixed to battens laid loose on the resilient layer Resilient layer: mineral fibre min. 25 mm thick and density min. 36 kg/m^3 Screed: min. 55 mm thick sand/cement with 20–50 mm wire mesh to protect the resilient layer Resilient layer: 13 mm impact sound duty expanded polystyrene board

Prior to laying the hardcore, the floor area is excavated to a depth of 150–230 mm to remove the topsoil. The hardcore is laid, usually to a thickness of 100 mm, and compacted. In most circumstances, a layer of 'blinding' (a layer of sand or ash) is laid on the hardcore, and brought to a level surface.

Concrete slab   Concrete (which is discussed in Ch. 7) consists of a mixture of cement, 'aggregate' and water. Aggregates include sand ('fine aggregate') and stone or gravel ('coarse aggregate').

A typical mix for a concrete ground-floor slab might be 1 : 3 : 6 (one part cement, three parts fine aggregate and six parts coarse aggregate). The concrete is laid on the blinding to

Table 3.8 (continued)

Timber base
with floating
layer

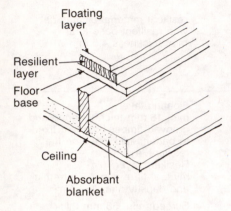

Floating layer:
min. 18 mm thick tongued and grooved boarding on
19 mm thick plasterboard

Resilient layer:
min. 25 mm thick mineral fibre, density 80−100 kg/m³

Floor base:
min. 12 mm timber boarding nailed to timber joists

Absorbent blanket:
min. 100 mm thick unfaced rock fibre min. density
10 kg/m³

Ceiling:
two layers plasterboard with staggered joints, total
thickness min. 30 mm

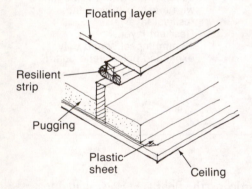

Floating layer:
min. 18 mm thick tongued and grooved boarding with
glued joints

Battens:
45 × 45 mm

Resilient strip:
min. 25 mm thick mineral fibre, density 80−140 kg/m³

Timber joists:
45 mm thick

Pugging:
dry sand or limestone chips min. weight 80 kg/m²

Ceiling:
19 mm dense plaster on expanded metal

* Including screed and ceiling finish if bonded to concrete.
Source: From Building Regulations 1991, approved document E, Section 2.

a minimum thickness of 100 mm, and brought to a level surface.

Unless the floor slab forms a raft foundation (see Fig. 2.4(b), Ch. 2), the concrete is laid with the edge of the slab butting against the internal surfaces of the external walls. In conditions where the subsoil is inadequate (such as weak or made-up ground), the floor may be reinforced with square mesh reinforcement.

Damp-proof membrane (d.p.m.) The d.p.m. is an impervious layer of material sandwiched in the floor construction. It is usually positioned either:

- on top of the blinding (Fig. 3.26(a));
- on top of the concrete slab (Fig. 3.26(b)).
    Materials frequently used for d.p.m.s include:
- polythene sheeting, which is available in 4 m wide rolls. The joints are overlapped by 150 mm, and may be joined with adhesive tape;
- solutions based on bitumen and latex, which are usually applied by brush.

Whilst either polythene or latex/rubber d.p.m.s may be laid under a screed, only polythene or other sheet materials are suitable for laying on top of the blinding.

If the level of the d.p.m. is the same as the d.p.c. in the walls, the d.p.m. is brought over the wall and below the d.p.c. (Fig. 3.26(a)). If the d.p.m. is either higher or lower than the d.p.c., the d.p.m. will be dressed vertically up (or down) the wall surface and linked with the d.p.c. Figure 3.26(b) illustrates a d.p.m. dressed vertically up the wall surface in order to link with the d.p.c.

Insulation See page 130 and Fig. 3.26(a).

Screed A sand/cement screed is laid over the floor slab in order to provide a smooth, level surface on which the floor finish may be laid.

When the d.p.m. is laid under the screed, a minimum of 50 mm thickness of screed is necessary to ensure that any moisture rising up through the concrete slab does not push up the d.p.m. and screed.

If floor finishes of different thickness are to be laid in different rooms, the thickness of screed might be varied from room to room so that the final floor finishes are all at the same level.

Suspended timber ground floors
Suspended timber ground floors are used less commonly today since they are more expensive in terms of material costs and the time taken to build them. The basic components of a suspended ground floor are:

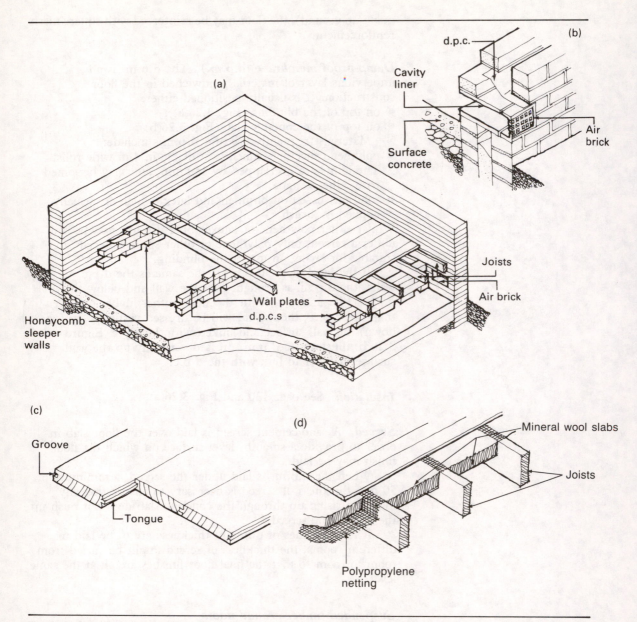

Fig. 3.27 Suspended ground floor
(a) General view of suspended
    floor
(b) Detail of air brick
(c) Tongued and grooved
    boarding
(d) Insulation

- surface concrete;
- sleeper walls and ventilation system;
- timber floor construction and insulation.

An example of a suspended timber ground floor is illustrated in Fig. 3.27(a).

Surface concrete   The surface concrete is the base on which the suspended floor is built, and is formed in a similar manner to a concrete ground-floor slab. Typical surface concrete construction might comprise of 100 mm thickness of well-compacted hardcore, finished with sand blinding and a 100 mm thick slab of concrete.

The surface concrete should be laid with its top surface no lower than the ground level, otherwise moisture from the ground could rise through the slab under pressure, and accumulate on top of the surface concrete.

Sleeper walls and ventilation system   A series of low walls are built on the surface concrete to provide support for the floor joists. These are known as 'sleeper walls'. Since the joists and wall plates are of timber, it is essential that the space between the top of the surface concrete and the joists is ventilated, otherwise dry rot could occur in the timber. Dry rot is discussed in Table 4.2, Chapter 4 (p. 157).
Ventilation is achieved by:
- the provision of air bricks;
- 'honeycombing' the sleeper walls.

Air bricks.   These are ventilation grilles of cast iron or 'terra cotta' (a very hard clay substance) made in sizes which enable them to be bonded into a brick wall. The air bricks are installed in the external wall and allow air to pass from the outside of the building into the underfloor space. Typical spacing of air bricks might be 215 × 65 mm air bricks at intervals of 2 m.

When air bricks are installed in a cavity wall, a 'cavity liner' is required to allow the air to pass through both leaves of the wall. A d.p.c. must be installed above the air brick and liner to prevent moisture travelling across the liner to the inner leaf (Fig. 3.27(b)). If the air bricks are obstructed by, for example, the building of a new extension with a solid floor, then air vents should be provided in the extension wall connected to pipes under the new solid floor, which convey air to the existing floor space.

Honeycombing.   As shown in Fig. 3.27(a), the sleeper walls are 'honeycombed' (built with voids between the bricks) which allows the passage of air between them. A d.p.c. is bedded in mortar on top of the wall to prevent moisture rising up the sleeper walls and into the timbers of the floor. The spacing of the sleeper walls depends on the cross-sectional size of the joists and may be between 1.2 and 1.8 m.

Timber floor construction and insulation   Timber wall plates of 100 × 75 mm or 75 × 50 mm cross-section are bedded in mortar on top of the d.p.c.s of the sleeper walls.

The timber joists, which are typically of 100 × 50 mm cross-section, are fixed with nails to the wall plates and are typically spaced at intervals of 400 mm. The joists are not built into the external walls, but rest on sleeper walls built parallel to the external walls with a gap of 38–50 mm separating the external and sleeper walls.

A suspended ground floor might be typically insulated by:

- laying mineral wool slabs below the boarding supported by polypropylene netting (Fig. 3.27(d));
- fitting expanded polystyrene boards between the joists supported by battens nailed to the joists.

The floor surface might typically consist of either:

- timber boards, which are usually 'tongued and grooved' (Fig. 3.27(c)) of 16, 19 or 21 mm thickness;
- sheets of chipboard of 18 or 22 mm thickness.

Upper floors

One of the major factors involved in the choice of an upper floor construction, is that of the span of the floor. Like flat roof construction, upper floors with small spans of up to 6 m or so may be constructed of either timber, cast _in situ_ concrete or precast concrete. Timber is the most commonly used material for domestic upper floors, since it is light in weight, involves relatively simple site plant and uses 'dry' construction techniques. Timber is limited, though, in its spanning ability, and floor spans of more than 4.5–5.0 m involve timbers that might be uneconomically large in cross-section. With spans greater than this, a cross beam might be necessary, similar to those used in timber flat roof construction (see Figs 3.7(d), (e), (f) and (g)). For floors in buildings such as flats or maisonettes, timber is at a disadvantage compared to concrete, in being both combustible and light in weight with subsequent loss of fire resistance and airborne sound insulation (see Tables 3.7 and 3.8).

Cast *in situ* r.f.c slabs, due to their heavy weight, have the advantages of good fire resistance and airborne sound insulation, and they may be formed into complex and irregular floor plan shapes.

The weight of a cast *in situ* concrete upper floor, however, could also be a disadvantage, since to support the slab, the walls might necessitate more substantial construction than walls supporting a timber floor. Compared with precast concrete floors, *in situ* slabs involve more site construction time in forming the shuttering, laying the reinforcement, placing the concrete and waiting for the concrete to cure sufficiently before it can be walked on.

Precast concrete units which are available in a wide variety of solid or hollow, plank, beam or channel form, often require no formwork and form an instant working platform once they are placed in position on the walls. Unlike *in situ* concrete or timber floors, precast units should be used in standard lengths in order to achieve maximum economy from them, consequently they are less suitable for irregular floor plans.

Timber upper floors

Timber upper floors consist of a series of timber joists which may span from either:
- external wall to external wall;
- external wall to internal loadbearing wall (as is the case in Fig. 3.28(a)).

Table 3.9 is extracted from approved document A of the Building Regulations 1991. The table indicates permissible spans for floor joists in relation to:
- their cross-sectional size
- their loading.

The table is based on 'SC3' timbers (basically general structural grade). If 'SC4' timbers (basically special structural grade) are used, slightly greater spans are permitted. Joists may be either:

(a) *'Built in' to the external walls.* When joists are built in (as shown in Fig. 3.28(a)) a small gap should be left between the sides of the joist and the walling in order to allow ventilation. The ends of the joists should be treated with a suitable preservative to prevent rot occurring. Floor joists are usually built in to buildings with cavity walls.

(b) *Attached to the walls by joist hangers.* The use of joist hangers is more common for buildings with solid walls (see Fig. 3.7(c)). If the joists are also supported by an

Table 3.9 Permissible spans for floor joists*

Size of joist (mm)	Spacing of joists (mm)		
	400	450	600
38 × 97	1.72	1.56	1.21
38 × 122	2.37	2.22	1.76
38 × 170	3.28	3.10	2.69
50 × 97	1.98	1.87	1.54
50 × 147	3.13	3.01	2.69
50 × 220	4.64	4.47	3.91
75 × 195	4.68	4.52	4.13

* These spans are for floors with a loading of 0.25–0.50 kN/m^2, excluding the mass of the joist.
Source: From Building Regulations 1991, approved document A, Table A1.

Fig. 3.28 *Suspended timber upper floors*
(a) *General view*
(b) *Support for wall parallel to joists*
(c) *Support for wall perpendicular to joists*
(d) *Opening in timber upper floor*
(e) *Tusk tenon joint*

internal wall (as is the case in Fig. 3.28(a)) the joists will bear on a wall plate. The joists are often lapped over the wall plate to avoid the use of excessively long timbers.
If a lightweight, non-loadbearing wall is to be built off a timber floor, the wall will be supported by either:
• two joists nailed together (in instances where the wall to be supported is parallel with the joists – Fig. 3.28(b));
• a 'sole plate' (in instances where the wall to be supported is perpendicular to the direction of the joists – Fig. 3.28(c)).

(a)

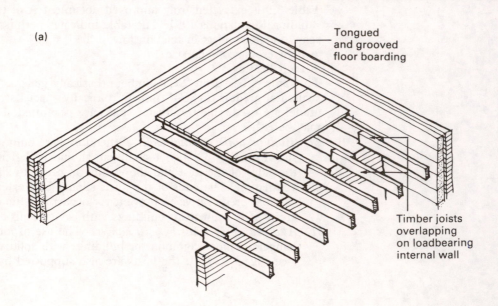

Tongued and grooved floor boarding

Timber joists overlapping on loadbearing internal wall

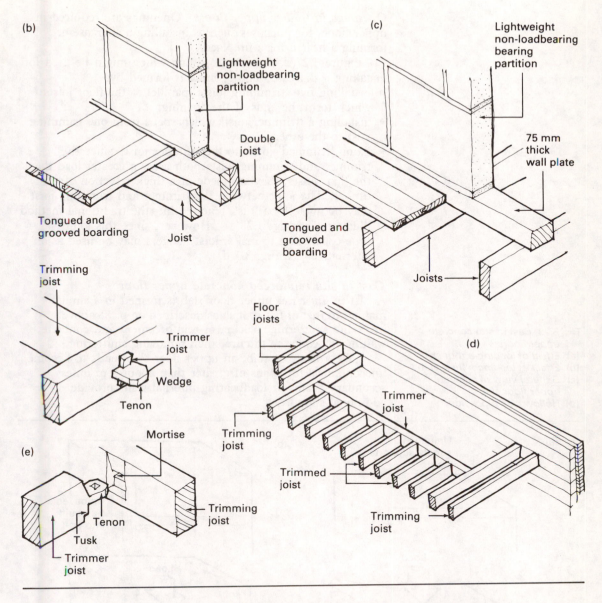

(b)

Lightweight non-loadbearing partition

Double joist

Tongued and grooved boarding

Joist

Trimming joist

Trimmer joist

Wedge

Tenon

(e)

Mortise

Tenon

Tusk

Trimmer joist

Trimming joist

(c)

Lightweight non-loadbearing bearing partition

75 mm thick wall plate

Tongued and grooved boarding

Joists

Floor joists

(d)

Trimmer joist

Trimming joist

Trimmed joist

Trimming joist

A sole plate is a length of timber, typically 75 mm in depth, which is nailed across the joists. The floor surface might typically be either of 16, 19 or 21 mm thick tongued and grooved floor boards (Fig. 3.27(c)) or 18 or 22 mm thick chipboard.

Openings in timber upper floors Openings are required in upper floors for purposes such as installing a staircase or forming a hearth for a fireplace.

Figure 3.27(d) shows an opening which might be used for installing a staircase. The opening is formed by:

- installing two 'trimming joists' parallel to the floor joists which form the ends of the opening;
- installing a 'trimmer joist' which spans from one trimming joist to the other;
- fixing 'trimmed' joists to the trimmer joist. Since the trimmer and trimming joists both carry a greater load than the floor joists, they are made of a thicker section of timber. The joint between the trimmer and trimming joist may be made by a 'tusk tenon' joint (the traditional method illustrated in Fig. 3.28(e)). However, since this joint is time-consuming to make, joist hangers may be used as an alternative (see Fig. 3.7(d)).

Cast in situ reinforced concrete upper floor

A cast *in situ* r.f.c. upper floor slab is formed in a similar manner to that of flat roof slabs described on p. 88 above, involving shuttering, placing of reinforcement and concrete, waiting for concrete to cure, then removing shuttering.

Like a flat roof slab, an upper floor slab tends to become uneconomical for spans of greater than about 5 m unless secondary beams or loadbearing internal walls provide additional support.

Fig. 3.29 *Cast* in situ *concrete upper floors*
(a) *Effect of loading a roof slab*
(b) *Effect of loading a floor slab*
(c) *General view of concrete upper floor slab*
(d) *Hollow clay block floor*

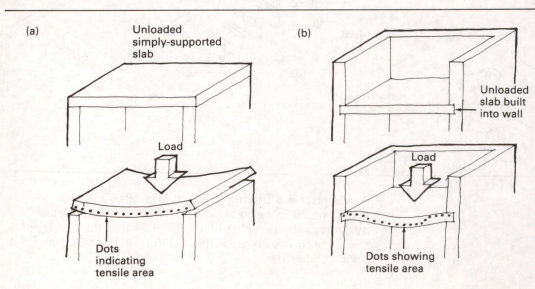

(a) Unloaded simply-supported slab

Load

Dots indicating tensile area

(b) Unloaded slab built into wall

Load

Dots showing tensile area

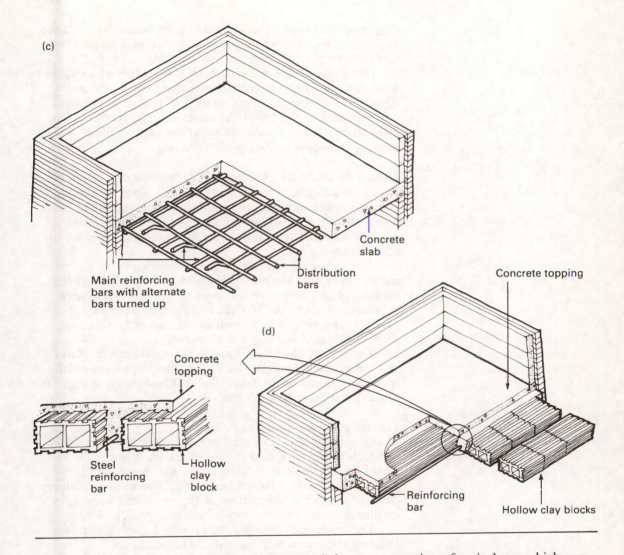

(c)

Concrete slab

Main reinforcing bars with alternate bars turned up

Distribution bars

Concrete topping

(d)

Concrete topping

Steel reinforcing bar

Hollow clay block

Reinforcing bar

Hollow clay blocks

The steel reinforcement consists of main bars, which run parallel to the shortest walls of the building, and distribution bars which run perpendicular to the direction of the main bars. For a small floor slab, main bars might be in the order of 12 mm diameter, at 250 mm intervals, and distribution bars 8 mm diameter at 350 mm intervals.

One of the differences between a flat roof slab and an upper floor slab, is that whilst a roof slab might be 'simply supported' (resting on top of the walls – Fig. 3.29(a)), a floor slab is restrained at its edges by the weight of the walling above it (Fig. 3.29(a)). Whilst the loading of a simply-supported slab causes tensile stress in the bottom of the slab, an upper floor slab, when loaded, causes tensile stress to occur also in the top at the edges of the slab (Fig. 3.29(b)).

Consequently the reinforcement at the edges of an upper floor slab must be designed to resist these tensile stresses. Figure 3.29(c) illustrates one method of providing tensile reinforcement at the top of the edges of an upper floor slab. Here alternate bars are turned up at the edge of the slab towards the upper surface of the concrete.

Hollow block floors   Figure 3.29(d) shows a cast *in situ* floor system which, like precast concrete floors, relies on placing the concrete in the compressive area near the upper surface of the floor, whilst the steel reinforcement is held by a minimal quantity of concrete in the tensile area near the lower surface of the floor.

The hollow blocks, which are made of clay, are placed in parallel rows on a temporary platform. The steel reinforcing bars are placed between these rows and concrete is poured between and above the blocks. The profiled edges of the blocks causes ribs to be formed in the concrete, which are sufficient to hold the clay blocks in place.

The hollow blocks are usually 300 × 300 mm in plan size, and are available in depths of 75–250 mm depending on the span and loading of the floor. The depth of concrete above the blocks must be at least 25 mm.

Precast concrete floors

A wide variety of precast concrete flooring units are available, of which Figs 3.30(a) and (b) are examples. Figure 3.30(a) shows an inverted 'U' or 'channel section' beam, whilst Fig. 3.30(b) shows a hollow beam unit. The main reinforcing bars are situated at the bottom of the units, and sometimes smaller diameter bars are located in the top of the units.

The beam ends are solid, to provide a good bearing for building the beams into the walls. The beams are manufactured in a factory, transported to site, hoisted into position on the walls, and the space between the beams is filled with fine concrete. A sand/cement screed is then laid on top of the units to form a smooth flat surface on which the floor finish is laid.

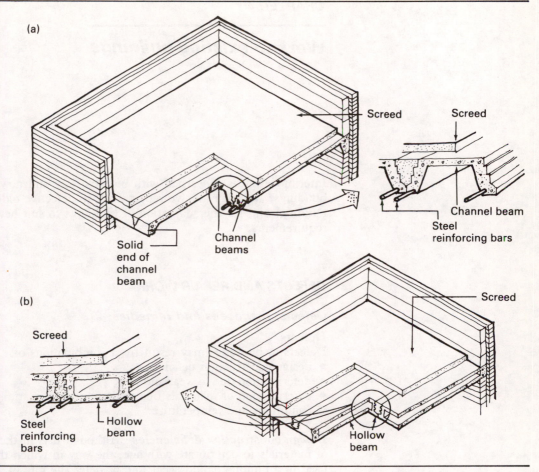

Fig. 3.30 *Precast concrete upper floors*
(a) *Precast concrete channel beams*
(b) *Precast concrete hollow beams*

Work to existing buildings

Interior design is concerned with both existing and new buildings, and may involve repair work to defective older buildings and for the adaptation of buildings to suit new requirements.

DEFECTS AND REPAIR WORK

Defects, diagnoses and remedies

Defects
Defects in buildings may result from combinations of:
- design/construction deficiencies;
- deterioration due to wear and tear;
- deterioration of materials due to dampness, climatic conditions and temperature.

Design/construction deficiencies Although it is in the nature of materials to deteriorate with age, the way in which they are used in a building could accelerate or deter the rate of deterioration.

The use of materials in a well-designed and constructed building involves:

(a) The designer specifying materials and designing details that are:
- able to withstand such effects as temperature, climatic conditions, fire, chemical attack, etc.;
- compatible with other building materials and details;
- suitable for their structural purpose;
- able (where necessary) to be inspected, cleaned and maintained.

(b) The contractor and those supervising the work in providing materials, quality of work and construction details that can conform with the design drawings and relevant British Standards, Codes of Practice, etc.

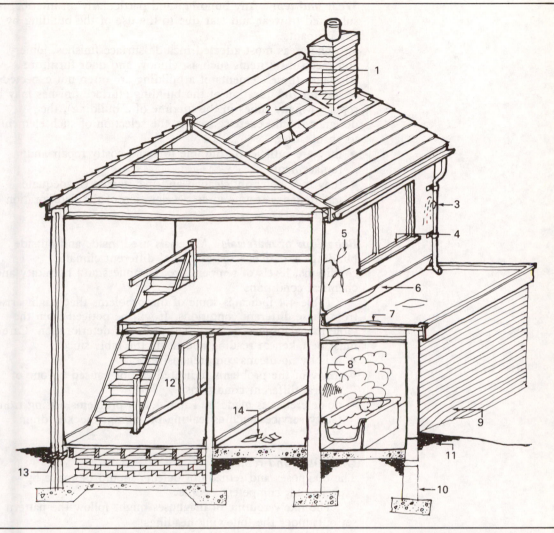

Fig. 4.1 *Dampness in buildings*

1 *Rain penetration to roof through defective chimney flashing.*

2 *Rain penetration to roof through slipped roof tiles.*

3 *Staining on wall due to water from cracked downpipe.*

4 *Poor paintwork allows ingress of moisture causing wet rot in window frame.*

5 *Cracked rendering in wall allows rainwater to penetrate through solid wall.*

6 *Rain penetration through wall due to defective flashing*

7 *Lack of vapour barrier in roof allows condensation to push up roof finish*

8 *Condensation encourages mould growth on wall.*

9 *Slipping brick courses caused by failed foundation.*

10 *Failed foundation caused by swelling/shrinkage of subsoil.*

11 *Soil above d.p.c. level allows moisture from subsoil to penetrate through wall.*

12 *Poor ventilation and high humidity below stairs encourages insect infestation*

13 *Soil blocking up air brick causing lack of ventilation and high humidity in floor space giving rise to dry rot*

14 *Lack of d.p.m. allows moisture from subsoil to push up floor finish.*

Wear and tear Any building, and particularly an interior, is subjected to wear and tear due to the use of the building by its occupants.

The areas most affected include surface finishes, joinery and mechanical items such as window and door furniture. Although these elements of a building are often not expected to last for the duration of the building (surface finishes may be replaced many times in the lifetime of a building), the designer should exercise care in the selection of such elements to ensure that:
- they are sufficiently durable to avoid costly repair and replacement;
- they may be easily cleaned and maintained (inadequate maintenance results in an acceleration of the deterioration of a material).

Behaviour of materials Materials used inside and outside buildings experience the effects of different climatic conditions, levels of temperature, dampness and humidity and chemical conditions.

Table 4.1 indicates some of the problems that might arise from these different conditions. It will be noticed from the table that dampness is a primary cause of deterioration. Care must be taken in reading and using the table since:
- it is by no means comprehensive;
- many of the problems indicated can be caused by one of several different conditions;
- no reference is made in Table 4.1 to problems arising from faulty services such as leaking plumbing, blocked drains, etc.

Diagnoses and remedies
The diagnoses and remedy of defects should be made by experienced, competent people.

The likely routine of diagnoses might follow the pattern set out under the following headings:

Inspection A careful and exhaustive inspection is made of the building. Observations are recorded in the form of annotated sketches. The inspector will try to ascertain the root causes of problems, drawing on his/her own experience. Where clues to the cause are not apparent, or are confusing, the inspector might remove parts of the structure (such as floor boards or areas of plaster) to facilitate further inspection. Other diagnostic indications might result from:
- measurements taken by moisture meters to check if building elements have an above-average moisture content;

Table 4.1 Some of the conditions causing building defects

Condition	Defects arising
Climatic	
(a) Heavy rainfall, followed by dry spell (or vice versa)	1. Increase and decrease in subsoil volume due to changes of moisture content could cause failure of foundations and cracks/slippage of brickwork (see 'Foundations and underpinning', p. 171)
(b) Frost action	1. Crumbling of brickwork 2. Deterioration of roof tiles
Humidity, temperature and dampness	
(a) Condensation	1. Condensation in a room rising up through a flat roof structure (if no vapour barrier was present) could cause blistering of the roof finish 2. Condensation in rooms such as kitchens and bathrooms could cause damp patches on walls and ceilings or mould growth
(b) Changes in moisture content of materials due to temperature differences	1. Timber of joinery items becomes deformed and distorted 2. Fine cracks could appear in plasterwork of internal partitions due to shrinkage of timber or concrete blocks on which the plaster is applied 3. Floor boards (particularly if not tongued and grooved) could curl up because of shrinkage/expansion due to changes in moisture content. Wood-block floor finishes may lift for similar reasons
(c) Dampness	1. Wet rot could occur in external and internal joinery and structural timbers causing the timber to become soft and deformed 2. Dry rot (which requires a high moisture content) could destroy joinery and structural timbers, and this may spread over masonry construction 3. Discoloration of internal finishes due to mould growth 4. Breakdown of surface finishes of joinery (particularly when finished with veneers or plastic laminates due to breakdown of adhesive) 5. Large areas of paint peeling off walls due to presence of moisture in the walls 6. Plastic floor finishes lifting off or blistering due to moisture from subsoil rising through ground-floor slab, where slab has missing or defective d.p.m. 7. Damp occurring low down around perimeter of ground-floor walls due to damp rising up wall from the subsoil. This is usually due to the d.p.c. in the wall being defective or missing 8. Patches of damp appearing on the inside of cavity walls due to moisture penetrating across the cavity, where wall ties have been incorrectly installed (e.g. where mortar has fallen and dropped across the drip of the wall tie)
Chemical	
(d) Efflorescence	1. Chemicals within the structure of brickwork may be brought to the surface in the form of a white powder (known as 'efflorescence'). Dampness and subsequent drying out of the brickwork encourages efflorescence
(e) Electrolysis	2. Deterioration of metals due to 'electrolytic' action (see Ch. 7, p. 353).

- devices to check whether cracks in a building are 'live' or dormant;
- taking samples of material for laboratory analysis;
- making enquiries with local organisations (such as the local authority district surveyor) to check on the geological and historical background of the area.

Report   A report will be made in order to present the client with an idea of the problems associated with the building, together with recommendations for remedying faults. Again, it must be stressed that diagnosis and remedies should be carried out by experienced, competent people, otherwise inappropriate remedies could deepen the problem and cause unnecessary expense to the client. The report might well be accompanied by explanatory drawings.

Specification   A detailed specification will be prepared, listing precisely what work is required to be carried out.

Copies of the specification and drawings will then be sent to usually between three and six contractors, who will submit prices ('tenders') for carrying out the work.

Dampness in buildings

Many of the defects arising in buildings may be attributed to dampness, which might occur due to: (*a*) condensation; (*b*) rising damp; (*c*) rain penetration; (*d*) leakage in plumbing and drainage systems. Figure 4.1 illustrates some of the common occurrences of dampness in buildings.

Condensation

Air is able to hold moisture in the form of water vapour, and as the temperature of the air is increased, so its capacity to hold moisture is increased. However, there is a limit to the amount of moisture that air, of any particular temperature, may hold; if warm air containing moisture is cooled, a temperature will be reached at which the air will be 'saturated' (holding its maximum quantity of moisture). If the air is cooled further, moisture in the form of drops of water will be expelled from the air.

This dampness is known as 'condensation'. Condensation will occur in buildings due to the temperature of air in a room being raised, which increases its capacity to hold moisture. In such rooms as bathrooms and kitchens, where heat and moisture are present, the air will absorb moisture.

This air cools when it touches a cool surface such as a window. The air is cooled, saturation point is reached, then condensation occurs. Condensation may be reduced by introducing ventilation or installing insulation.

Ventilation   Ventilation will encourage moisture-laden air to be extracted, and dry air to be introduced into the building.

In rooms such as bathrooms or kitchens, where ventilation by open windows may be unacceptable or inadequate, it might be necessary to install a system of mechanical ventilation.

Insulation   Ventilation may cause draughts, or a drop in temperature, and in such conditions it may be preferable to install additional thermal insulation on cool surfaces. This might be achieved by:

(*a*) Double glazing windows (see Ch. 2 p. 68: Double windows and double glazing).

(*b*) Injecting cavity wall insulation. The common means of insulating an existing cavity wall is to inject urea formaldehyde foam through holes drilled in the outer skin. To reduce the health risk to occupants due to the escape of toxic formaldehyde fumes, approved document D of the Building Regulations 1991 requires that the inner skin of the wall is built of brick or blockwork, that the wall is assessed for its suitability for foam injection and that the installer, the material used and the installation system complies with relevant British Standards.

(*c*) Lining internal walls with insulative material, such as expanded polystyrene lining sheets. (Care should be taken with this material, since it is highly inflammable.)

(*d*) Lagging metal pipework.

Rising damp

Rising damp is moisture which becomes absorbed into the material of foundations, ground-floor slabs and walls below ground level where moisture rises up to the inner surface of the wall or floor. This is caused by moisture from the subsoil becoming absorbed into the pores of the concrete or brickwork and being carried up vertically by capillary attraction. This may result in staining of the internal finishes, or the finishes becoming dislodged from the wall or floor surface.

Rising damp may be caused by:

- the lack of a d.p.c./d.p.m. in a wall/floor;
- a defective d.p.c. or d.p.m.;
- the d.p.c. being bridged by soil or rubble outside the wall, which allows moisture to be absorbed into the wall above d.p.c. level.

Installing a d.p.m. in an existing floor   Where no d.p.m. exists in the floor, it may be considered worth while to install one (see 'Damp-proof membrane, Ch. 3, p. 137). Since d.p.m.s usually require at least 50 mm thickness of screed above them, it is likely that doors and other joinery items may require adjustment to suit the new floor level. It is important that the d.p.m. should link to the existing d.p.c.s in walls in the manner shown in Figs 3.26(a) and (b) (Ch. 3).

Installing a d.p.c. in an existing wall   There are several methods for installing d.p.c.s in existing walls, which are based on traditional methods or the newer, chemical injection processes.

The traditional method involves cutting through a horizontal mortar joint around the entire outside wall. The mortar joint chosen should be at least 150 mm above the ground level. The cut is made in lengths of 1 m to avoid upsetting the stability of the wall. The d.p.c., which is of a rigid material (such as slate), is loaded with mortar and pushed into the cut joint. Temporary wedges are also inserted, then withdrawn later once the mortar has cured sufficiently. Finally, the joint is pointed with mortar. The main drawbacks of the traditional system are:

- it is expensive;
- it tends to be messy;
- it is time-consuming;
- it could be hazardous to use the technique on structurally unsound walls.

The most commonly used system involves the injection of water-repellent solutions into the wall. A series of closely spaced holes are drilled into the wall, and a solution (usually based on silicone) is either injected into the holes under high pressure, or transfused under low pressure into the wall, forming a relatively impermeable area of wall at d.p.c. level. This system is cheaper, cleaner and quicker than the traditional system, and may be useful for dealing with smaller areas of rising damp.

Rain penetration

There are many possible defects in wall and roof surfaces which give rise to rain penetration, and the quantity of water passing through to the internal surfaces of the building may be increased substantially by the effect of wind driving the rain. Some of the more common occurrences of rain penetration, together with possible remedies for rectifying the faults, are summarised below.

Roofs

Flat roofs   Widespread cracks and splits in asphalt and built-up felt roof finishes could indicate that the finish is past its useful life, in which case the roof finish should be stripped and replaced. However, occasional splits might be indicative of movement in the roof finish, in which case it might be possible to form a flexible joint in the split.

Proprietary brush-on solutions based on silicone and bitumen are available. Although some manufacturers of such solutions claim that the solutions may extend the life of the roof by many years, it might be prudent to consider such treatments for only a means of temporary repair. Occasional splits in metal roof finishes may be repaired by such means as welding the split, or replacing the bay of the roof finish in which the split occurs.

Pitched roofs   Where large numbers of roof tiles or slates are slipping, cracked or missing, it could indicate that the roof finish is beyond its useful life. The nibs may be breaking off tiles, nail holes in slates may be cracked or the nails holding the slates corroded. In such instances, the entire roof finish may require replacement.

Occasional broken tiles or slates may be replaced.

Flashings   Where flashings have pulled away from the mortar joint, it may be possible to rake out the joint, dress the flashing back in and repoint the joint. Cracked or split flashings may require complete replacement.

Walls

Solid walls   A common cause for rain penetration in walls is due to the development of cracks between the junction of bricks and mortar which form capillaries for the passage of water to the inside of the wall face. Sometimes the problem is increased by deterioration of the mortar.

Remedies for rain penetration in walls include:
(a) Repointing the mortar joints, which may need to be preceded by raking out any soft and crumbling mortar.
(b) Water-repellent solutions, usually colourless, and based on silicone, may be applied to the surface of the wall. The solution is brushed on, and becomes absorbed into the pores of the wall.

(c) Colourless sealants, based on plastic solutions, are brushed on the wall face to seal the pores of the wall. Sealants tend to be more visible than repellents.

(d) Masonry paints similarly seal the wall surface and are available in a range of colours. The character of a building could be visually changed since the textural and colour effect of mortar and brickwork would be lost.

(e) Rendering (a mix of cement, lime and sand with water) may be applied to the wall to form either a smooth or textured surface, or the wall might be clad with materials such as timber or tiling (see Fig. 2.7(a), Ch. 2). Either of these treatments could radically effect the visual character of the building.

Cavity walls   Rain penetration through cavity walls may be caused by a combination of:

- the outer leaf failing in a similar manner to a solid wall;
- the cavity being bridged by such means as accumulation of mortar on cavity wall ties or d.p.c.s around openings.

The problem may be remedied by locating the cause of penetration, opening up the wall and rectifying the fault, or by application of the remedies listed under 'Solid walls' (items (b) – (e) above).

Timber decay

Wood is a natural material which is able to provide nourishment for certain species of insect and fungi. In nature, these insects and fungi perform the task of breaking down dead wood and ultimately destroying it. Timber used in buildings is, in effect, 'dead wood' and is therefore subject to insect and fungal attack.

Wood-destroying fungi

Fungi reproduce by means of tiny spores which are formed in vast numbers in the fruiting body of the fungus plant. The spores are released from the fruiting body and, being extremely light in weight, are carried in the air to settle and germinate on damp timber to form new fungus plants. The spores develop strand-like tubes ('hyphae') which push into the wood and branch out to form a mass of hyphae (known as 'mycelium' which is, in fact, the fungus).

Cellulose forms a substantial part of the cellular structure of timber and provides the source of nourishment for the fungus, which eventually produces fruiting bodies for the development of more spores. The extraction of cellulose

Table 4.2 Characteristics and treatments of wood – destroying fungi

Fungus	Characteristics			Summary of treatment
	General	Appearance of fungus	Appearance of attacked wood	
True dry rot	• Not found outside buildings • The most devastating of wood-destroying fungi • Can attack wood with moisture content as low as 20% but prefers 25% • Attacks softwoods and occasionally hardwoods • Strands able to penetrate through brickwork and across steel in search of more timber	• Early growth: white cotton-wool-like masses • Later growth: forms grey skin on wood, sometimes with yellow or lilac tinges • Hyphae: grey strands, may be as thick as 5–6 mm • Fruiting body: pancake shaped with white edges and brick-red centre. Fruiting bodies usually seen on brick or plaster work near the fungus, where spores are easily released	• Light brown • Dry and brittle • Deep cracks along and across grain divide wood up into rectangular shapes	• Open up plasterwork and other materials to reveal the full extent of the attack • Locate and remedy the cause of dampness that provide ideal germination conditions. Dry out building • Cut away and burn all infected timber, including sound timber up to 600 mm beyond the extent of fungal growth • Strip away plaster and other materials that may contain strands, including sound plaster and materials up to 600 mm beyond the extent of fungal growth • Sterilise non-combustible materials in the vicinity. Brickwork containing strands should be sterilised by feeding fungicide into holes drilled into the walls • Treat all sound timbers in the vicinity with wood preservative • Replace the removed timbers with sound timber which has been treated with preservative
Cellar fungus	• Found in very damp wood inside or outside buildings • One of the most common of wood-destroying fungi	• Often no evidence of external growth • Sometimes the hyphae may be seen on the surface of the wood as thin black/brown strands • Fruiting body rarely found in buildings	• Dark brown • Cracks occurring along the grain • Sometimes a veneer of sounder wood is left on the surface	• Locate and remedy the cause of dampness which has allowed the fungus to germinate • Dry out building • Cut out all affected timbers and replace with sound timber which has been treated with preservative • If fungus has been discovered in its early stage of growth, it may be possible to arrest its growth by application of fungicide after eliminating the cause of dampness and drying out the infected area
Mine fungus	• Occasionally found in very damp buildings • Attacks softwoods	• Hyphae: pure white strands • Fruiting body: flat white plates with visible pores	• Light brown • Cracks along and across the grain, but less pronounced than true dry rot	

gradually reduces the timber to a brittle, weak condition.

Table 4.2 outlines the characteristics of three species of fungi that might be encountered in a building and a summary of the treatment necessary to eradicate the fungus. It will be noticed from Table 4.2 that the moisture content necessary for growth of fungus is at least 20 per cent for true dry rot, and at least 30 per cent for cellar and mine fungus. The moisture content of timber found in sound buildings is likely to be in the region of 14 per cent.

Insect attack

Table 4.3 outlines the distinguishing features of some of the wood-destroying insects that may be found in buildings, commonly known as woodworms (which are winged beetles).

Although the life cycle of the insects varies in its duration, the stages of life cycle are similar:

Table 4.3 Distinguishing features of wood-destroying insects

Insect		Colour	Length of beetle (mm)	Size of exit hole	Length of life cycle	Woods attacked	Emergence of adults	Appearance of bore dust
Common furniture beetle		Dark brown	3–5	1.5 mm	2 or more years	Hardwood and softwood	May–September	Fine egg-shaped pellets
Death-watch beetle		Dark brown	8–9	3 mm	4–5 years but can be up to 10	Hardwood	March–June	Flattish round pellets
Lyctus (powder post) beetle		Reddish brown	Approx. 5	1.5 mm	1–2 years	Hardwood	May–September	Fine flour-like dust
House longhorn beetle — Black spots		Dark brown	Approx. 20	Oblique slits 6–9 mm wide	3–11 years	Softwoods	July–September	Fine dust and short cylindrical pellets
Weevils — Projecting snout		Brown	3–5	Oblique slits and holes 1 mm wide	7–9 months	Hardwood and softwood	Any month	Fine egg-shaped pellets

(a) A female beetle flies to a piece of timber, and lays eggs in a crack or crevice of the wood.

(b) An egg hatches, and a grub emerges which bores its way into the wood by eating the wood and excreting the wood in the form of pellets or dust. The pellets or dust may often be seen lying in small heaps near the infested timber.

(c) The grub may live in the wood for several years, after which it will bore its way towards the surface of the timber. Just below the surface of the timber, the grub hollows out a chamber and forms itself into a chrysalis, where it gradually develops into a beetle.

(d) Once fully developed, the beetle eats its way out of the timber, mates and lays its eggs. The beetle may fly off to lay its eggs and spread the infestation.

Treatment of wood attacked by insects

The steps taken to eradicate an insect attack are described below.

(a) A thorough inspection of the affected areas is made to:
- determine which beetle is causing the damage (treatments might vary for different beetles);
- the extent of the infestation is ascertained (it may be necessary to remove items such as floor boards, panelling, etc.);
- an assessment of structural timbers is made to determine whether replacement is necessary.

(b) Badly infested timbers are removed and burnt.

(c) Surface finishes, which may hinder the application and penetration of insecticide, are removed.

(d) The timbers are treated with suitable insecticides. This may take the form of brush or spray application, or injection of fluid into the flight holes.

In the case of furniture, where it may be unsatisfactory to remove surface finishes, insecticide may be brushed on untreated areas such as under drawers and the backs of furniture, and fluid injected into flight holes.

Modern insecticides are available which:
- will kill insects at any stage of their life cycle;
- will last for the duration of several life cycles for preventing further attack;
- may also act as a fungicide.

(e) Replacement timbers are fixed which have been treated with insecticide.

Timber preservatives

Preservative fluids are toxic chemicals in solution which protect timber from insect and fungal attack. Some of the types and application methods of preservatives are given under the headings below.

Types of preservative

1. *Tar-oil preservatives (creosote):* a low-cost preservative, used externally, particularly for fences and posts in contact with the ground. It is difficult to apply paint to timber treated with tar-oil preservative. Most are non-corrosive to metals.
2. *Organic solvent preservatives:* these consist of solutions of toxic chemicals such as pentachlorophenol or copper naphthenate. They are suitable for internal and external use and are non-corrosive to metals. Pigments may be added which enhance the colour and grain of the timber or they sometimes may be painted over.
3. *Water-borne preservatives:* chemicals such as copper chrome arsenic are dissolved in water to produce preservatives, most of which are non-corrosive. They are suitable for internal or external use and may be overpainted.

Application methods

Pressure impregnation   This factory-based process involves placing the timber in a tank, extracting the air and forcing in the preservative. The process causes the preservative to penetrate deep into the timber and is suitable for conditions where the timber is at greatest risk to attack.

Hot and cold tank method   The timber is placed in a tank of preservative which is heated up, then allowed to cool. This method is simpler than pressure impregnation and is usually used for timbers in external locations.

Dipping and steeping   The timber is either dipped in to the preservative and withdrawn (dipping) or left in the solution for some time (steeping). The method is simple to operate, but is not recommended for applications where a high risk of attack is expected.

Brushing and spraying   The fluid is flooded on to the surface of the timber by brush or spray, but does not provide a deep penetration of preservative.

It is important that where the timber is to be cut after application of the preservative, that extra preservative is applied to the vulnerable exposed face, since even using the pressure-impregnation method, it is unlikely that the preservative will penetrate through the entire thickness of the timber.

ADAPTATION OF BUILDINGS

Interior designers are often called upon to design adaptations to existing buildings. This may involve:
- conversion of the functional use of buildings (e.g. from a warehouse to a local museum or from a redundant church to a community centre);
- to modify an interior (e.g. to improve the display facilities and circulation area of a shop or to provide new accommodation for staff in an office building).

Conversion and modification work might involve:
- making structural alterations within a building; and/or
- extending a building.

Structural alterations inside buildings

Design procedure

Structural alterations within a building usually involve forming 'openings' in a part of the existing structure such as removing a whole, or part of a wall, floor or ceiling.

In deciding to remove any part of an existing building structure to form a new opening, the following procedure should be taken:

(a) Carefully measure the building in sufficient detail to enable plans of the existing building to be drawn.

(b) Take thorough measurements in the vicinity of the new opening noting such details as sizes of timbers, thickness and materials of the structure.

(c) Take vertical measurements of the building to enable elevations and sections to be drawn of the building. Establish such details as thicknesses of floors, relationship of floor heights to ground level.

(d) A photographic survey of the building can prove useful in checking materials and details of the building.

(e) From this information and the drawings produced, it should be possible to gauge the structural significance of the different parts of the building, such as which walls are loadbearing and how the load of the roof is carried.

(f) Where it is decided to remove loadbearing parts of the structure, it will usually be advisable to consult a structural engineer, who, with the use of the information and drawings given him, will be able to calculate the sizes and materials used to support the new opening.

(g) The designer can now produce detailed drawings of the new opening to enable:

- the local authority to issue notification of approval under the Building Regulations;
- the builder to provide a price for the work and carry out the alterations.

Building procedure

The building procedure in, for example, forming an opening involves the following operations (which are based on situations where the floor joists run perpendicular to the wall being removed):

- providing temporary support for the structure;
- removing the existing structural materials;
- installing the means of permanent support; then
- removing the means of temporary support.

This procedure is illustrated in Fig. 4.2 where a loadbearing wall is to be removed. In this particular case, the wall supports the floor joists above (in the manner shown in Fig. 3.28(a), Ch. 3) and also a wall (in the manner shown in Fig. 3.28(c), Ch. 3).

- the temporary support is installed which consists of a length of timber laid over the ground-floor finish ('sole plate'), a series of adjustable steel props which rest on the sole plate and a length of timber ('head plate') which rests on the tops of the steel props. The props are adjusted in height to ensure that the head plate is firmly wedged under the ceiling finish (Fig. 4.2(a));
- the wall is demolished (Fig. 4.2(b));
- the permanent means of support is installed – in this case a steel beam (Fig. 4.2(c)).

Provision must be made at each side of the opening for a 'pier' (column built in to the wall) to support the beam. In the case of a beam carrying a light load the piers may be formed by leaving a vertical strip of wall ('nib') either side of the opening. In the case of a heavy load a more substantial pier may be built each side of the opening with a foundation formed beneath each pier. If the material of which the wall is constructed is of insufficient compressive strength to withstand the load of the beam, it may be necessary to install a concrete block ('padstone') beneath the bearing area of each side of the beam. This will spread the load over the area of the pier.

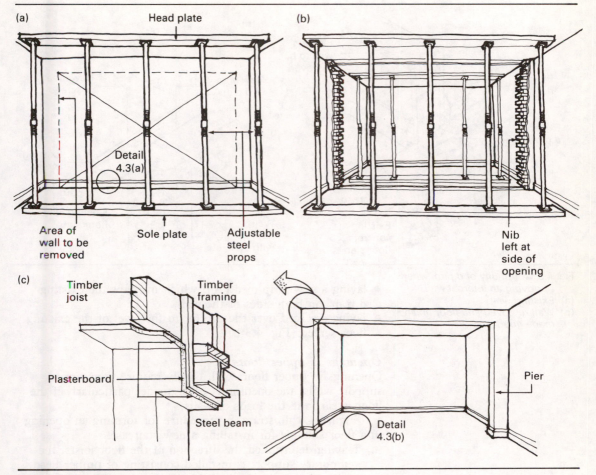

Fig. 4.2 *Forming an opening in an internal wall*
(a) *Temporary support installed*
(b) *Wall removed*
(c) *Completed opening*

Removing a ground-floor wall

Problems could occur in removing a ground-floor wall due to the disruption of the d.p.c. in the wall. Figure 4.3(a) shows a loadbearing wall in a building with a concrete ground floor and a d.p.m. installed below the floor screed: when the wall is removed, water could penetrate from the soil, through the portion of wall left below the ground, and up into the floor. This problem may be overcome by:

- removing the wall to below the level of the floor, and also removing the screed adjacent to each side of the wall, being careful to leave the d.p.m. intact;
- building up the wall with concrete to the level of the concrete floor slab;

163

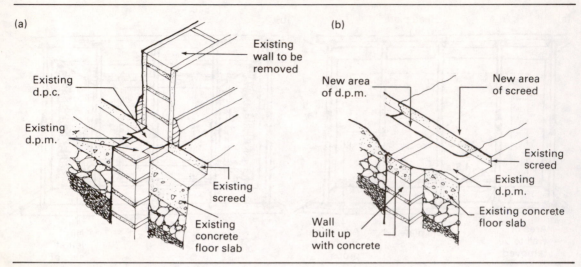

(a)

Existing
wall to be
removed

Existing
d.p.c.

Existing
d.p.m.

Existing
screed

Existing
concrete
floor slab

(b)

New area
of d.p.m.

New area
of screed

Existing
screed

Existing
d.p.m.

Existing concrete
floor slab

Wall
built up
with concrete

Fig. 4.3 *Continuity of d.p.m. when
removing an internal wall*
(a) *Existing wall*
(b) *Wall removed and floor finish
made good*

- laying a new strip of d.p.m. which overlaps the existing
  d.p.m. on both sides of the wall;
- laying a screed over the d.p.m. to the level of the existing
  floor screed (Fig. 4.3(b)).

Openings in upper floors

Openings in upper floors will usually require temporary
support whilst the opening is being made, particularly if the
floor is carrying the loads of walls.

Figure 4.4 illustrates a procedure for forming an opening
in a floor suitable for installing a new staircase.

(*a*) Having determined the direction of the floor joists, the
temporary support is installed consisting of timber head
and sole plates and adjustable steel props (Fig. 4.4(a)). If
the joists were running in the opposite direction, it may
have been possible to dispense with temporary supports.

(*b*) The ceiling finish and appropriate lengths of joist are
removed, and joist hangers are attached to the ends of the
trimmed joists.

(*c*) A trimmer joist is installed which is built into the two end
walls and fits against the joist hangers (Fig. 4.4(b)).

(*d*) The ceiling finish is 'made good', that is, the rough edge
of the ceiling finish is brought to a smooth, clean edge.

Removing chimneys

Figure 4.5 shows a simplified diagram of a chimney structure
consisting of:

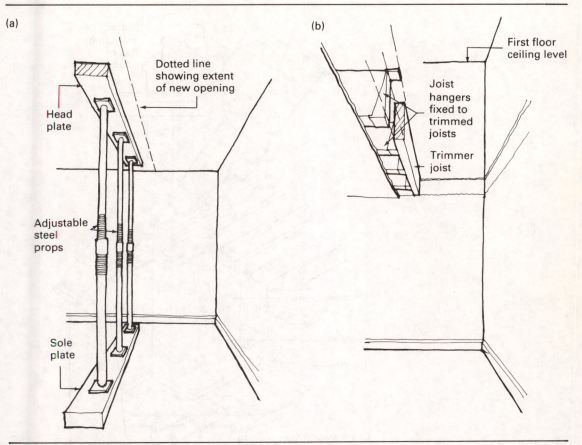

(a)

Head
plate

Dotted line
showing extent
of new opening

Adjustable
steel
props

Sole
plate

(b)

First floor
ceiling level

Joist
hangers
fixed to
trimmed
joists

Trimmer
joist

Fig. 4.4 *Forming an opening in a floor/ceiling*
(a) *Temporary support installed*
(b) *Portion of floor removed*

- a concrete hearth which provides a base on which the fire is burnt and extends into the room to provide protection of the floor against fire;
- a fireplace which contains the fire;
- a chimney breast, which is the portion of the chimney structure projecting from the wall and houses;
- the flue, which is a passageway through which the smoke travels;
- a chimney stack, which is the portion of chimney which projects above the roof finish to discharge the smoke into the air.

Where the chimney structure is built into an external wall, the chimney may be removed, starting at the top of the stack and working downwards. The opening in the roof caused by removal of the stack must be made good by such means as

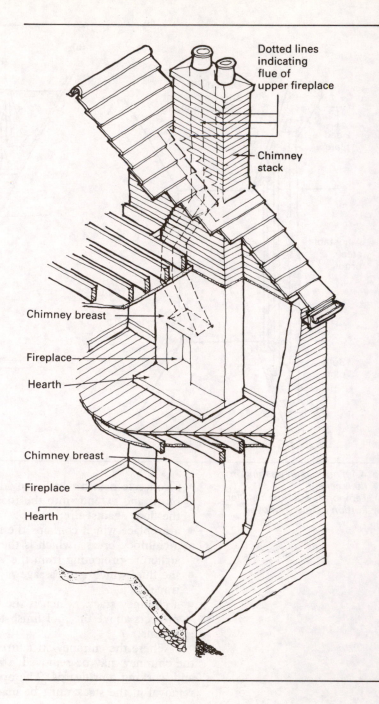

Dotted lines indicating flue of upper fireplace

Chimney stack

Chimney breast

Fireplace

Hearth

Chimney breast

Fireplace

Hearth

Fig. 4.5 *Parts of a chimney*

166

installing new lengths of rafter to bridge the opening, laying a piece of roofing felt over the opening which laps the existing roofing felt, then fixing new battens and roofing tiles/slates. The openings in the floors caused by removal of the chimney breasts must be made good by such means as fixing new lengths of floor joist to span the opening, then fixing floor boards above and plasterboard/plaster below the joists. Where the chimney breasts have been removed, the brickwork will usually be made good, and plaster applied to the surface then new lengths of skirting board fixed at floor level.

Where the chimney to be removed is built into a 'party wall' (a wall which separates one building from another, such as between terraced houses), the owner may be required to enter into an agreement with the owner of the adjoining building. Under the terms of the agreement, the owner of the building in which the chimney is to be removed will be liable for damage occurring in the adjoining owner's house which may result from the removal of the chimney.

Where it is necessary to remove a chimney breast from one floor of a building without removing the chimney breasts and stack above it, some form of lintel or supporting structure will be required. In such circumstances, it would be wise to consult a structural engineer in order to determine the structural support necessary for supporting the chimney above the removed portion of chimney.

Extending a building

Figure 4.6 (a, b, c) illustrates three of the methods commonly adopted for extending a building:
Fig. 4.6(a) – extending from the ground floor of the building;
Fig. 4.6(b) – extending above the building;
Fig. 4.6(c) – extending into the roof of the building.
Constructional problems peculiar to extensions concern:
- the *structural stability* of both the extension and the existing building;
- ensuring a *weather-tight junction* between the extension and existing building;
- *forming the opening(s)* between the existing building and the extension.

Extending from the ground floor
Extending a building from the ground floor involves the entire range of building operations, including the forming of new foundations and ground floor.

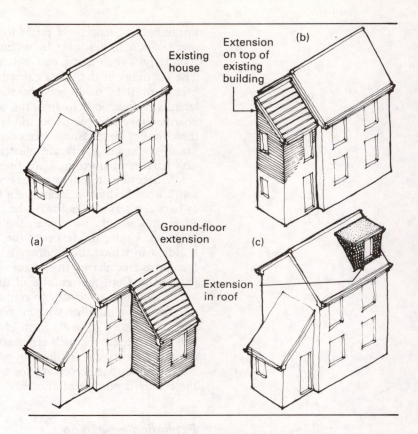

Existing house

Extension on top of existing building

(b)

(a)

Ground-floor extension

Extension in roof

(c)

Fig. 4.6 *Types of extension*
(a) *Extension from ground floor*
(b) *Extension above a building*
(c) *Extension in the roof*

The extension may be only single storey (such as Fig. 4.6(b)) or may also extend vertically to the upper storeys of the building.

Structural stability One of the major factors in ensuring structural stability of the extension is the depth of its foundations. If the foundations are less deep than the foundations of the existing building, settlement could occur due to shrinkage and slipping of the subsoil below the extension foundations (Fig. 4.7(a)). For this reason, it is usual for the extension foundations to be laid at a similar depth to those of the existing building. Exceptions may be made for such cases as a lightweight extension (e.g. a glazed porch) to an existing traditional building structure.

Weather-tight junction

Walls Extension walls may be 'toothed in' to the existing walls (Fig. 4.7(b)(i)). Alternate bricks removed from the existing wall allow the extension wall to be bonded in. Galvanised or stainless-steel 'wall extension profiles' provide a less time-consuming alternative. These are bolted to the existing wall, and new wall mortar joints are bedded in standard wall ties which clip into the profile (Fig. 4.7(b)(iv)).

Roof Figure 4.7(b)(ii) shows possible routes for rain penetration through the existing wall into both the roof timbers of a flat-roofed extension and the internal face of the existing wall. This problem will usually be overcome by inserting a d.p.c. in the existing wall at a level of at least 150 mm above the new roof level. The d.p.c. will extend out of the wall and be dressed down over the upturned felt at the edge of the new roof (Fig. 4.7(b)(iii)). To avoid undermining the structural stability of the wall, the d.p.c. is installed in short lengths of, say, three bricks.

Forming the opening An advantage of a single-storey ground floor extension is that the building of the extension can progress to an advanced stage before the opening to the existing building is made. This minimises the disruption caused to the occupants of the building.

The design of the opening should follow the same process as previously described for forming openings in internal walls.

The means of temporary support for the opening is similar to that of an internal opening using a system of head and sole plates and props.

However, whereas the support system illustrated in Fig. 4.2 relied on supporting the ceiling and floor joists, an external wall requires a series of timber supports (needles) to be inserted through holes made in the wall. The needles are supported each side of the opening by plates and props (Fig. 4.7(c)).

Extending above a building
Extending above a building will necessitate removal of the existing roof finish, and in the case of timber roofs the structure will also probably be removed. In these instances, a temporary cover will be required which might comprise of plastic sheeting fixed to scaffold tube or timber framing. This will provide protection of the existing structure until the new roof, walls and glazing is installed.

Structural stability
Before deciding to extend above a building, it is essential to check that the existing foundation and walls are capable of withstanding the loads imposed by the new extension, since increasing the loadbearing capacity of the

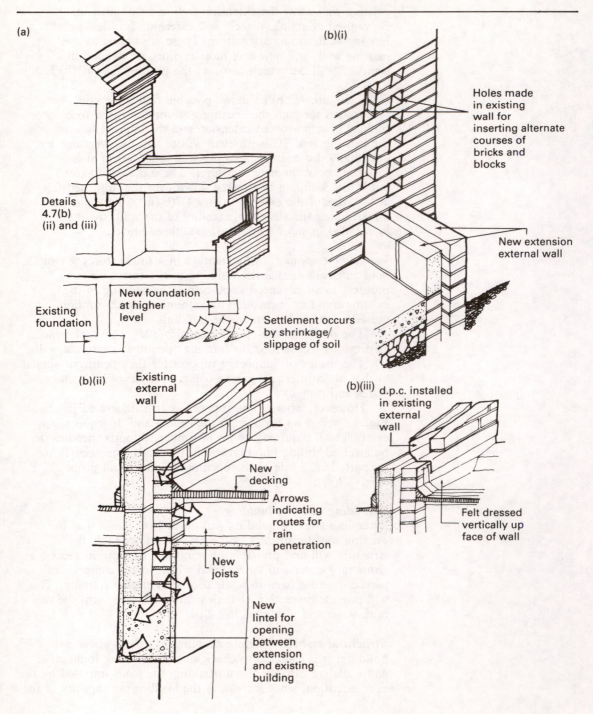

(a)

Details 4.7(b) (ii) and (iii)

Existing foundation

New foundation at higher level

Settlement occurs by shrinkage/ slippage of soil

(b)(i)

Holes made in existing wall for inserting alternate courses of bricks and blocks

New extension external wall

(b)(ii)

Existing external wall

New decking

Arrows indicating routes for rain penetration

New joists

New lintel for opening between extension and existing building

(b)(iii)

d.p.c. installed in existing external wall

Felt dressed vertically up face of wall

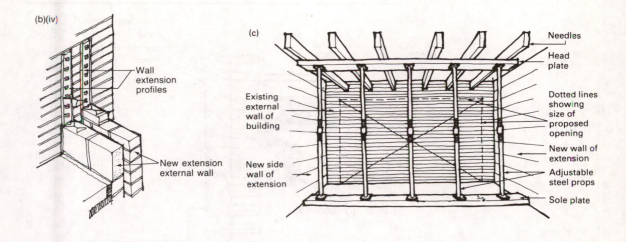

(b)(iv)

Wall extension profiles

New extension external wall

(c)

Needles

Head plate

Existing external wall of building

Dotted lines showing size of proposed opening

New side wall of extension

New wall of extension

Adjustable steel props

Sole plate

Fig. 4.7   Details of extension from ground floor
(a) *Foundation of ground-floor extension*
(b) *Prevention of rain penetration*
 (i) *Toothing in extension wall*
(ii) *Potential causes of rain penetration at junction with roof*
(iii) *Detail of d.p.c. and flashing*
(iv) *Wall extension profiles*
(c) *Temporary support for forming the opening viewed from inside the new extension*

foundations and walls could increase the cost of the extension very substantially. It is also necessary to check whether the existing roof joists (in the case of a timber flat roof) or ceiling joists (in the case of a pitched roof) are of suitable size and spacing to act as floor joists.

Foundations and underpinning   A process known as 'underpinning' may be used to increase the loadbearing ability of the foundations (the same process is also used where foundations have failed due to settlement) and involves the installation of a new foundation below the level of the existing one.

In the traditional method of underpinning, a trench is dug in short sections (typically of 1200 mm width) around the perimeter of the wall; the depth of the trench is the same as the level of the new foundation bottom. The sequence of digging the short sections of trench is such that no two adjacent sections of trench are dug consecutively. This reduces the possibility of settlement or fracture of the foundation.

Figure 4.8 illustrates the underpinning which has been carried out in one of the short sections. The section of trench has been dug, the new foundation installed, a new length of wall built between the new and existing foundations with concrete placed between the back of the wall and back face of the excavation, and the projecting edge of the existing foundation has been removed.

Walls   Inadequate walls may be strengthened by building an extra leaf of brickwork which is tied into the existing wall

171

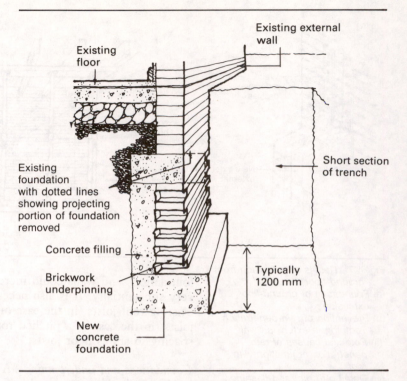

Existing floor

Existing external wall

Short section of trench

Existing foundation with dotted lines showing projecting portion of foundation removed

Concrete filling

Brickwork underpinning

Typically 1200 mm

New concrete foundation

Fig. 4.8 *Traditional underpinning*

of the building. In the case of a solid wall, a cavity could be left between the new and existing walls in order to increase weather resistance and thermal insulation.

Floors Where existing ceiling or roof joists are inadequate to act as floor joists, they might be removed and replaced by more substantial joists. If the original roof timbers and roof tiles or slates are in sound condition, they could be carefully removed and set aside for reusing in the new roof.

Weather-tight junction

Walls The extension walls will be connected to those existing as illustrated in Fig. 4.7(b)(ii) or (iv).

Roof Where the extension roof abuts the existing roof, the problem of rain penetration must be overcome. This may be achieved in a similar manner to that described for ground-floor extensions and in Figs 4.7(b)(ii) and (iii). Where the extension roof is pitched, and is of similar pitch and height to the existing roof, the roof finishes may be made continuous. This involves:

- removing the existing verge tiles;
- laying the felt and battens over the new roof and over the top of the existing gable wall – the felt should overlap the existing felt, and the battens should be laid to the same gauge as the existing battens;
- laying the new roof with tiles or slates of similar size and profile to continue the existing roof finish over the extension.

Extending in a roof

In deciding whether to extend into a roof, consideration should be given to the Building Regulations 1991 regarding means of escape: approved document B. Small loft extensions (less than 50 m^2 in area) in the roof of an existing two storey house which comprises no more than two habitable rooms and involves no increase in existing roofline require:

(i) fire-resisting enclosure of the ground and first floor stairs giving access to external exit;

(ii) self-closing doors to doorways between habitable rooms and the existing stair;

(iii) any glazing to existing stair enclosure to be fire-resisting;

(iv) a new stair to the loft complying with approved document K;

(v) fire separation of the loft by provision of fire-resisting door at the top or bottom of the new stair.

Structural stability Extensions in roofs of domestic buildings (sometimes known as 'loft conversions') are usually constructed in timber.

The building procedure for a simple extension in a roof such as that illustrated in Fig. 4.6(c) might be:

- increasing the effective depth of the ceiling joists (if necessary) to take the extra loads of the new floor;
- providing temporary support for the ceiling, forming the opening in the existing ceiling and installing a new staircase to provide access to the extension;
- removing the roof finish in the area of the new extension, and ensuring temporary covering is available to protect the existing structure from rain penetration;
- installing a trimmer beam to support the top portion of the existing roof and new roof joists and providing any other necessary permanent supports for the existing roof;
- installing the timbers to form the framework for the walls and extension roof;
- removing the existing rafters and any other parts of the existing roof structure necessary for forming the opening;
- cladding the walls and roof and glazing the windows;
- the basic extension shell is now complete and the

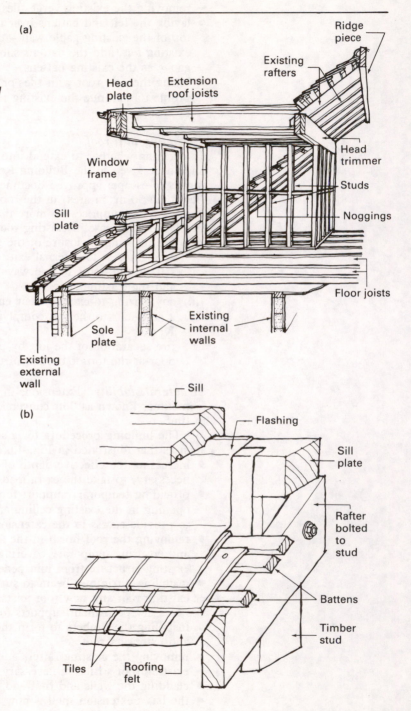

Fig. 4.9 *Extension in a roof*
(a) *General view*
(b) *Junction between front wall
   and existing roof*
(c) *Junction between front wall
   and extension roof*
(d) *Junction between existing and
   new roofs*

(a)

Ridge piece

Existing rafters

Head plate

Extension roof joists

Head trimmer

Window frame

Studs

Noggings

Sill plate

Floor joists

Existing internal walls

Sole plate

Existing external wall

(b)

Sill

Flashing

Sill plate

Rafter bolted to stud

Battens

Timber stud

Tiles

Roofing felt

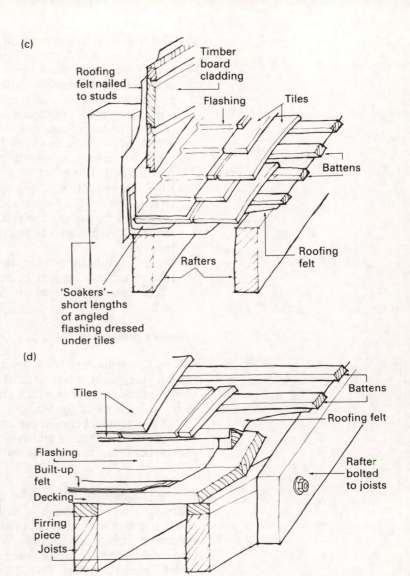

(c)

Roofing
felt nailed
to studs

Timber
board
cladding

Flashing

Tiles

Battens

Roofing
felt

Rafters

'Soakers' –
short lengths
of angled
flashing dressed
under tiles

(d)

Tiles

Battens

Roofing felt

Flashing

Built-up
felt

Decking

Firring
piece

Joists

Rafter
bolted
to joists

installation of thermal insulation and internal finishes and fittings may be installed.

Figure 4.9(a) illustrates a simple extension into a roof, the elements of which are:

(a) A head trimmer consisting of two 150 × 50 mm timbers bolted together which is supported at each end by the gable walls. The existing rafters have been nailed to the trimmer.

(b) The wall structure consists of timber sole plates fixed to the floor joists and vertical timber studs jointed into the plate. The size of these timbers might be 100 × 50 mm, or if an extra thickness is required in the walls, 150 × 50 mm. A head plate of similar dimensions is fixed on the top of the studs. Horizontal nogging pieces between the studs provide stiffness to the structure.

(c) The window frame is screwed to a sill plate and the lintel above it ('head plate') consists of two 150 × 50 mm timbers bolted together.

(d) Roof joists of 100 × 50 mm span between joist hangers attached to the head plate and head trimmer. The joists are bolted through to the sides of the trimmed rafters to reduce the tendency of the rafter feet to slide over the trimmer.

(e) Where purlins are cut to form the opening, they will require suitable support by means such as timber framing which transfers the load of the purlin to a loadbearing wall.

Weather protection

Walls The walls may be clad with timber boarding, vertical tile hanging, sheet metal or other materials. Boarding or tiles are fixed to battens which are nailed over roofing felt to the framework of the walls.

The junctions between the existing roof and the walls may be made with a flashing which is dressed over the roof finish and vertically up beneath the wall cladding (Figs 4.9(b) and (c)).

Roof The roof will often be of flat construction with sheet or board decking and a finish of built-up felt roofing. The detail at the eaves and verge might be similar to that illustrated in Fig. 3.12(b) (Ch. 3). At the junction of the extension and existing roof, the roof felt is dressed up under the existing tiles (Fig. 4.9(d)).

Forming the opening The opening into the existing ceiling for installation of the staircase will be made in a similar manner to that illustrated in Fig. 4.4. The opening in the existing roof should require no temporary support since the permanent supports are installed prior to the removal of the existing roof timbers.

CHAPTER 5

Internal structure

PARTITIONS

Functional requirements

A partition is a wall, within a building, which divides the interior space into a series of rooms or areas. Table 5.1 summarises broad categories into which partitions might fall. Usually loadbearing and permanent non-load-bearing partitions are built *in situ*, whilst demountable and movable partitions are prefabricated in factories and assembled or fitted on the site.

Partitions may be designed to satisfy different requirements, for example a load-bearing wall might be required to carry part of the floor and roof loads of a building, whereas a lightweight office partition might be required to provide simply a psychological barrier defining a person's workspace.

Structural stability is the only functional requirement common to all the categories of partitions. Partitions may also frequently be required to offer sound insulation, fire resistance, flexibility of planning and provision for electrical services.

Structural stability

Loadbearing partitions A loadbearing partition carries some of the structural load of the building, which might include:
- part of the roof load (see Fig. 3.2(b)(iii), Ch. 3); and/or
- part of the floor load (see Fig. 3.28(a), Ch. 3).

Loadbearing partitions may also act as stiffeners for a building (see Fig. 1.14(c), Ch. 1). Like an external wall, a loadbearing partition could buckle if the 'slenderness ratio' (the ratio between the height of a partition and its thickness) is too great. Corners and junctions with other partitions will reduce the tendency of the partition to buckle.

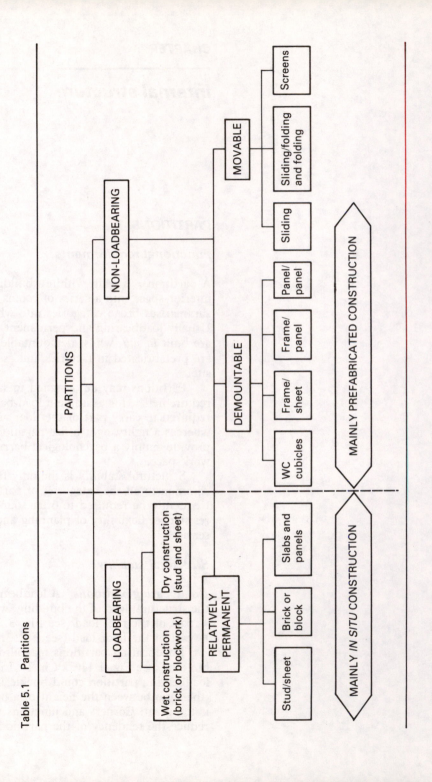

Table 5.1   Partitions

Non-loadbearing partitions Non-loadbearing partitions are
only required to carry their own weight ('self-weight') plus the
weight of any fitments such as shelves or hanging cupboards
which might be fixed to them.

In situ partitions can be easily designed with sufficient
strength to enable fitments to be fixed to them. However, if a
prefabricated partition system is to be used, it is essential to
know:

* whether the partition provides adequate support and
  strength for fitments;
* whether a range of accessories, fittings and furniture is
  available, which is specifically designed for attachment to
  the partition system.

Sound control

Sound is caused by rapid fluctuations of air pressure which
effect the mechanism of the ear and are translated in the brain
as audible sounds. Sound may be measured by instruments
which gauge the increase in air pressure caused by the sound.

The 'decibel' (dB) scale is usually used to express these
differences of sound pressures. The decibel scale is logarithmic
which enables the dB units to be approximately related to
perceptible differences in sound levels. Zero dB is the level at
which sound is just perceptible, whilst 130 dB causes painful
discomfort to the ear.

Table 5.2 indicates typical background noise levels that
might be expected in various rooms.

Table 5.2 Background noise levels for rooms

Room	Sound level (dB)
Concert hall	30
Classroom	30–35
Conference room	30–35
Bedroom	30–35
Office	40–45
Libraries	40–45
Living room	45
Restaurant	50
Shops	55

Sound waves hitting the surface of a partition will cause it
to vibrate. The vibration will travel through the thickness of
the partition causing the opposite surface to vibrate, and the
sound is thus transmitted to the other side of the partition.

Table 5.3 Sound reduction for various forms of partition construction

In situ partitions

1. Single sheet or slab partitions	Sound reduction (dB)
13 mm thick fibre insulation board	18
25 mm thick blockboard	27
22 mm thick tongued and grooved timber boarding	24
9.5 mm thick plasterboard	25
3 mm thick steel sheet	30
Single glazing with 3 mm thick glass	20
Double glazing: two sheets of 6 mm thick plate glass with 200 mm air gap. Cavity reveal lined with absorbent material	45

2. Timber or metal framework ('stud') partitions, with finish on both sides of:	Sound reduction (dB)		
	Single studding	Double studding	Double studding with absorbent quilt in the cavity
13 mm thick fibre insulation board	20–22	22–25	24–28
19 mm thick blockboard	30	32–33	34–36

The ability of a material to resist vibration depends on its density. Consequently a partition constructed of dense material would provide better resistance to sound transmission than a lightweight partition. Partitions, particularly of the non-loadbearing variety, might be constructed largely of lightweight materials. To improve reduction of sound transmission, 'discontinuous construction' might be used. This involves constructing the partition in two separate skins which makes it more difficult for vibrations caused by sound to be transmitted from one surface to the other. Sometimes a sound-absorbent material such as glass-fibre matting might be suspended between the skins (see Fig. 5.5(a)), further to reduce sound transmission.

Table 5.3 continued

9.5 mm thick plasterboard	30	32–33	34–36
Three coats plaster on expanded metal lathing	35–37	37–40	39–43

3. Single-leaf partitions	Sound reduction (dB)
60 mm thick slab consisting of two sheets of 9.5 mm plasterboard either side of a cardboard egg-crate core	26
76 mm thick concrete blocks (density 1000–1500 kg/m³ clinker aggregate): unplastered	23
As above, but with 12.5 mm plaster on both sides	41
As above, but using 100 mm thick blocks	43
102.5 mm thick brickwork: unplastered	35–40
As above, but with 12.5 mm thick plaster on both sides	45
As above, but using 215 mm thick brickwork	50

Prefabricated partitions

A wide variety of different construction systems available with widely varying sound-reduction characteristics. Some manufacturers claim figures in excess of 40 dB sound reduction. Check with manufacturer's technical literature

Table 5.3 shows sound-reduction figures that might be achieved by various forms of partition construction.

The intrusion of sound into a room becomes a nuisance when it approaches a similar level to the background noise of the room. To prevent the likelihood of this happening, the partitions between quiet rooms (such as rest rooms, studies or interview rooms) and other rooms should be designed to provide suitable sound reduction. Partitions offering a sound-reduction figure in the order of 45 dB would usually be suitable for such purposes.

For less quiet rooms, such as general offices or social rooms, partitions providing 30 dB sound reduction would usually be sufficient to reduce sound from adjacent rooms to

Table 5.4(i)   Fire resistance of internal loadbearing wall constructions

Construction	Finish	Fire resistance (hrs)
Timber framing minimum 44 mm thick spaced at maximum 600 mm intervals	12.5 mm plasterboard (both sides) with taped and filled joints, nailed with 40 mm galvanised nails at 150 mm centres	$\frac{1}{2}$
100 mm reinforced concrete wall with minimum 25 mm cover to reinforcement	None required	$\frac{1}{2}$
Timber framing minimum 44 mm thick spaced at maximum 600 mm intervals	Lining both sides of minimum 25 mm plasterboard with taped and filled joints in two layers, each layer nailed with galvanised nails at 150 mm centres	1
Solid masonry wall minimum 90 mm thick, or 75 mm for non-loadbearing wall	None required	1
120 mm reinforced concrete wall with minimum 25 mm cover to reinforcement	None required	1

Source: BRE report Guidelines for the construction of fire resisting structural elements. HMSO, 1982.

an acceptable level.

At the planning stage of designing an interior noisy rooms should, if possible, be kept well away from quiet rooms.

Glazed areas and door openings in a partition could severely impair the sound reduction achieved by a partition. This problem may be alleviated by means such as

- the provision of double glazing to areas of glazed partition and glazed doors (see 'Double glazing', Ch. 2, p. 68, see also Table 2.11);
- providing sound-resistant seals to door frames (see Fig. 5.15).

Resistance to fire

Partitions, together with ceilings and floors, are the basic components that can assist or deter the rate at which fire will spread within a building. Partitions may resist the spread of fire by:

- providing fire resistance which deters combustion and/or collapse;

Table 5.4(ii) Guide to fire resistance of non-loadbearing wall constructions

Framework	Covering material (both sides)	Fire resistance (hrs)
63 × 50 mm softwood studs at 610 mm centres with noggings at 1200 mm centres space between studs filled with mineral wool insulation	6 mm non-combustible building board	$\frac{1}{2}$
As above, but no insulation	9 mm non-combustible building board	$\frac{1}{2}$
As above, but space between studs filled with mineral wool insulation	9 mm non-combustible building board	1
48 × 35 × 0.5 mm steel channel-section studs at 610 mm centres, space between studs filled with mineral wool insulation	9 mm non-combustible building board; same material used for 75 mm wide cover strips fixed behind each horizontal joint	$\frac{1}{2}$
As above	As above, but in addition same material used for a 50 mm wide cover strip fixed to both outer faces of the channel	1

* Individual manufacturers' data should be checked. The regulations require fire resistance ratings to be determined by British Standard tests described in BS 476: part 8: 1972. This is gradually being superseded by BS 476: parts 20−23. Part 22 specifically concerns tests to determine the fire resistance of non-loadbearing elements, which includes partitions.

● resisting the spread of flames across the surface.
Both factors are referred to in the Building Regulations 1991, approved document B.

Fire resistance The regulations require large buildings to be divided into 'compartments' comprising walls and floors, of specified fire resistance. (See also 'Fire resistance p. 33.)

Table 3.6 provides examples of the fire-resistance ratings required for various 'elements of structure' (which includes compartment walls). The table also shows the dimensional limitations beyond which a building must be divided into compartments.

Table 5.4(i) provides examples of the fire-resistance ratings of loadbearing internal wall constructions taken from Building Research Establishment report Guidelines for the construction of fire resisting structural elements. Table 5.4(ii) provides a guide to ratings for non-loadbearing internal walls based on typical manufacturers' literature.

Resistance to the spread of flame across the surface of the partition A British Standards test (BS 476 : Part 7 : 1971) gauges the rate that flames spread along the surface of samples of materials. The materials are designated as Classes 1, 2, 3 or 4. Class 1 provides the lowest rate of flame spread and Class 4 the most rapid (British Standards Institution).

Approved document B of the Building Regulations 1991 refers to classifications 1 and 3 but also introduces a Class 0 which is more demanding than Class 1. Class 0 materials or products are either composed entirely of material of limited combustibility or are Class 1 materials with specified minimum fire propagation characteristics. Table 5.5 indicates the classification of some common internal partition surfaces based on Table A8 of approved document B.

Table 5.6 indicates the minimum surface spread of flame requirements from the regulations. The classifications relate to walls and ceilings. For the purposes of the regulation, *wall* includes

- the surface of any glazing (except glazed doors);
- parts of ceilings sloping at an angle of more than 70° to the horizontal.

Doors, door and window frames, architraves, skirtings and fireplace surrounds are not included in the definition.

Flexibility of planning

The plan layout of many interiors such as factories or offices may be required to be changed from time to time to accommodate different numbers of staff, or to make different functional use of the space. The use of (*a*) demountable partitions; (*b*) sliding/folding or retractable partitions; (*c*) screens; enable such changes to be made.

Demountable partitions Many prefabricated partition systems are simple to install, and may be dismantled at a later date and reassembled in a different layout.

It is important that the heating, lighting and decorative schemes are suitably designed/selected so that different plan layouts may be achieved without expensive alterational work to the heating, lighting or decoration. This is usually achieved by

Table 5.5 Surface spread of flame ratings for various materials

Class 0 (includes Class 1) materials)	Brickwork Blockwork Ceramic tiles Concrete Plasterboard — painted or unpainted Wood wool cement slabs Mineral fibre tiles or sheets with cement or resin binding
Class 3*	Timber or plywood with density of minimum 400 kg/m^3 painted or unpainted Chipboard or hardboard treated or untreated Standard glass reinforced polyesters
*	Timber products can be brought up to Class 1 by application of appropriate treatments

Source: From the Building Regulations 1992, approved document B Table A8.

use of a 'planning grid' and a modular planning treatment in order to co-ordinate the dimensions and location of heating, lighting and decorative treatments with the dimensions of the demountable partition units. Demountable partition systems include: (a) frame/sheet systems; (b) frame/panel systems; (c) panel/panel systems (see Figs. 5.18(a), 5.18(b) and 5.19 respectively).

Sliding/folding and retractable partitions (see Figs. 5.19–21) Sometimes it may be necessary to design an interior in which several small rooms may be opened up to form one large space. This may be achieved by installing either sliding/folding or retractable partitions, whereby the partitions may be rapidly withdrawn and reinstated.

Screens (see Fig. 5.22) Screens are often used for open-plan interiors or for temporary displays and exhibitions. The screens are usually less than full room height and are often supported on feet attached to the base of the screen, or by clipping screens together and arranging them in an angular plan shape to provide stability.

Provision of electrical services

Solid partitions In small buildings with brick or blockwork partitions, the electrical installation is usually carried out by:

Table 5.6   Classification of linings*

Location	Class
Small rooms less than 4 m² in area in residential buildings Small rooms less than 30 m² in area in non-residential buildings	3
Other rooms Circulation spaces within dwellings	1
Other circulation spaces including common areas of flats and maisonnettes	0

*Parts of walls may be of lower classification (but not lower than 3) if the total area of those parts does not exceed one half the floor area of the room and the total is less than 20 m² (residential buildings) and 60 m² (non-residential buildings).

Source: From the Building Regulations 1991, approved document B.

- fixing the cables (which consist of live, neutral and earth wires surrounded by plastic sheathing) to the wall surface with cable clips;
- fixing 'capping' (metal or plastic cover strip) over the cable;
- plastering over the capping to conceal the wiring installation (Fig. 5.1(a)).

   Larger installations may employ either:

(a) 'Conduit' (tubing often of 20 or 32 mm diameter steel – see Fig. 5.1(b) – which is fixed to the partition surface). The cables are installed by drawing them through the conduit.

(b) 'Trunking' (lengths of box section plastic or sheet metal fixed to the wall surface). The trunking is fixed to the wall surface often at either worktop height (Fig. 5.1(c)) or skirting level (Fig. 5.1(d)). The cables (and sometimes telephone and other cables) are housed in the trunking. The front surface of the trunking usually consists of a series of clip-on panels which provide easy access to the wiring installation.

Hollow partitions   In hollow partitions, such as those of stud and sheet construction, it is usual for the cables to be located in the void between the surfaces of the sheet material. Unless the studs are perforated, it will be necessary to drill holes in the studs to enable the cables to be passed along the length of the partition.

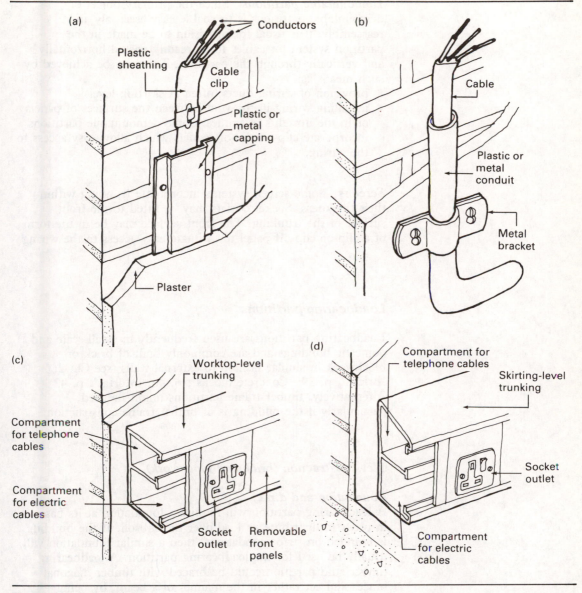

Fig. 5.1 *Installation of electric cables*
(a) *Capping*
(b) *Conduit*
(c) *Worktop-level trunking*
(d) *Skirting-level trunking*

Demountable partitions Since the primary object of demountable partitions is to enable easy assembly and reassembly, it is usual for provision to be made in the partition system for cables to be readily passed horizontally and vertically through the partition. This may be achieved by such means as:

- provision of wiring ducts (often at skirting level);
- installing wiring in the void between the surfaces of panels or in the uprights of frames. Some demountable partitions incorporate clip-on/clip-off panels which allow easy access to the wiring.

Screens Some screen systems incorporate trunking within their thickness. Socket outlets may be fitted to the front surface of the trunking. The front surface may be in the form of a clip-on/clip-off panel to facilitate easy access to the wiring.

Loadbearing partitions

Loadbearing partitions are used frequently in small-scale and domestic buildings and are commonly built of brick or blockwork in similar fashion to external walls (see Ch. 2: 'Bricks', p. 39; 'Concrete blocks', p. 40; 'Mortar', p. 42). Alternatively, timber frame partitions might be used, particularly if the building is of timber frame construction.

Wet construction (brick and blockwork)

Foundation and d.p.c.

A loadbearing partition will require a foundation at its base in order to distribute its load over the subsoil. If the building is founded on strip foundations, then a similar foundation will usually be used for the loadbearing partition. A loadbearing timber stud partition could be braced with timber diagonal braces and act rather in the manner of a beam, by being securely fixed to loadbearing side walls. Consequently, the loads bearing on the partition are largely transferred to the side walls and a foundation may be unnecessary.

However, since a loadbearing partition is likely to be less wide than an external wall, its foundation will also be less wide. Typical foundation widths might be: (*a*) for a 102.5 mm wide wall: 460 mm; (*b*) for a 215 mm wide wall: 600 mm.

If the building is founded on a raft foundation (see Fig. 2.4(b), Ch. 2) then the partition will be built off the raft. In the area of raft below the partition:

- additional reinforcement will be added near the bottom surface of the raft;
- the thickness of the raft may be increased.

Like an external wall, water from the subsoil could travel up an internal loadbearing partition, therefore a d.p.c. must be installed. The d.p.c. will be linked:

- with the d.p.c. of the external walls;
- with the d.p.m. of the floor.

Lintels Lintels for brick or block loadbearing partitions resemble those for external walls, however they are less wide and require no design features to resist the weather. Figure 5.2 illustrates some types of lintels available and Table 5.7 gives information concerning their dimensions. Alternatively, brick arches could be constructed (see Fig. 2.15, Ch. 2). For spans in excess of 2,000 mm a steel 'I' section beam RSJ might be used.

Dry construction (timber stud partitions – Fig. 5.3)
Stud partitions are frequently used in conjunction with timber framed buildings and can be designed to be loadbearing.

Stud partitions consist of a series of vertical timber members ('studs') and will be placed at intervals suitable to:

- withstand the loads imposed on them;
- suit the size of sheet material used for the partition surface (e.g., for 900 mm wide plasterboard, studs might be at 300 or 450 mm centres, or for 1200 mm wide plasterboard, studs might be at either 300 or 400 mm centres).

The studs are fixed to a horizontal timber member (sole piece) at the bottom on the partition, and to a horizontal (head piece) at the top. Short horizontal pieces of timber (nogging pieces) are fixed tightly between the studs in order to provide

- rigidity of the partition;
- suitable horizontal fixing pieces to which the sheet coverings are nailed.

Foundations and d.p.c. If a strip foundation is used for the building, the sole piece of the partition will usually be laid on

Table 5.7 Typical dimensions for internal lintels

Lintel	Typical dimensions (mm)				
	Depth	Span	End bearing	Width	
Cast *in situ* r.f.c. (Fig. 5.2(a))	140	Up to 1200	140	100	Typical reinforcement: one 16 mm diameter bar
	215	Up to 2000	215	100	Typical reinforcement: one 20 mm diameter bar
Prestressed concrete (Fig. 5.2(b))	65	Up to 2500	150	100	
Steel (Fig. 5.2(c))	25	Up to 900	100	108	Box-type loadbearing steel lintels available for heavier loadings and spans up to 4500 mm

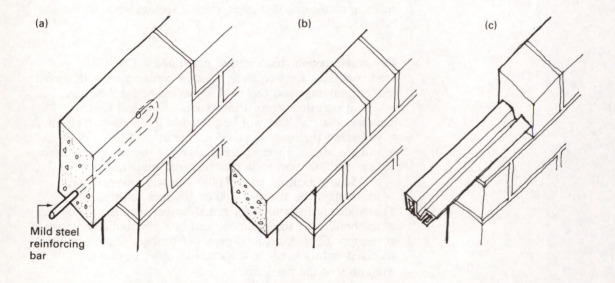

(a)   (b)   (c)

Mild steel
reinforcing
bar

Fig. 5.2 *Internal lintels*
(a) *Cast* in situ *concrete lintel*
(b) *Prestressed concrete lintel*
(c) *Pressed galvanised steel lintel*

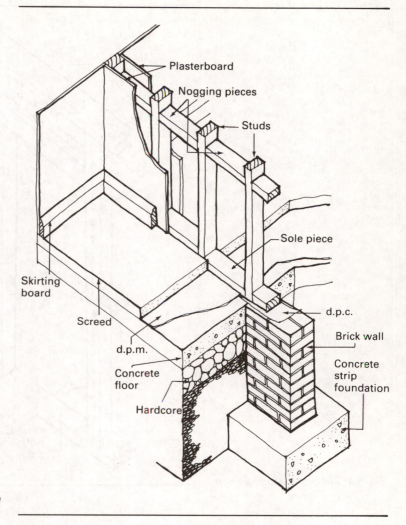

Fig. 5.3 *Loadbearing stud partition and strip foundation*

a low brick wall founded on a strip foundation (Fig. 5.3). The d.p.c. lies beneath the sole plate to prevent moisture rising through the slab and into the timbers.

If the building is founded on a raft foundation, extra reinforcement will be applied in the bottom area of the raft slab beneath the partition and the depth of the slab might be increased.

Lintels (Fig. 5.4)   Lintels will usually be formed by a double

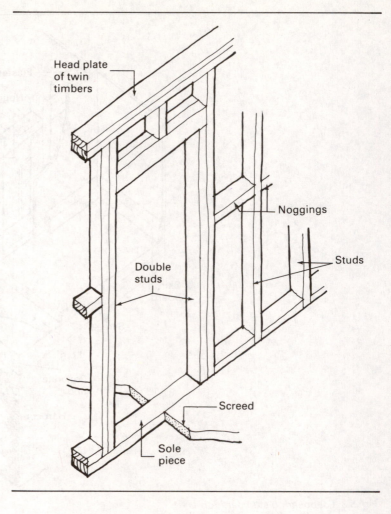

Head plate
of twin
timbers

Noggings

Studs

Double
studs

Screed

Sole
piece

Fig. 5.4 *Opening in stud partition*

horizontal timber member at the head of the opening fixed at
the jambs to either:

- double studs;
- to thick studs, where a single thick horizontal member will
  be jointed to the studs with a mortise and tenon joint (see
  Ch. 7, p. 268, for details of timber joints).

Relatively permanent non-loadbearing partitions

Construction methods employed for relatively permanent non-loadbearing partitions include: (*a*) brick or blockwork; (*b*) stud and sheet; (*c*) slabs and panels.

Although some of these methods (particularly certain forms of stud and sheet partitioning) are capable of being easily dismantled, they are not demountable in the strict sense of the term, since they are not specifically designed for easy dismantling and re-erection.

Brick or blockwork

Relatively permanent brick or blockwork non-loadbearing partitions are built in a similar fashion to loadbearing brick or block partitions, but differ in the following respects:

(*a*) No foundation is necessary, since the partition carries only its self-weight. The partition is built directly off the concrete ground-floor slab, and the d.p.m. in the floor prevents moisture from the subsoil rising up through the slab and into the partition.

(*b*) Lintels may be lighter in weight and/or smaller in cross-sectional size.

(*c*) Whilst loadbearing partitions might be constructed of half-brick thickness or 100 mm thick concrete blockwork, relatively permanent non-loadbearing partitions may be constructed in 75 or 50 mm thick blockwork.

Unless a high degree of sound reduction is required, lightweight concrete blocks will usually be used in preference to bricks, since being larger, the partition can be built faster.

Stud and sheet

Stud and sheet relatively permanent non-loadbearing partitions resemble loadbearing stud and sheet partitions, but differ in the following respects:

(*a*) Since the partition carries only its self-weight, no foundation is necessary. The sole plate of the partition may be fixed directly to the concrete ground floor slab, and like a brick or block partition, the floor d.p.m. will protect the partition timbers from the ingress of moisture from the ground.

(*b*) Timber lintels for small openings may be of similar cross-sectional size to the nogging pieces.

(*c*) The cross-sectional size of studs and nogging pieces may be smaller than those of loadbearing partitions, and the distance between studs and noggings may be increased.

Fig. 5.5 *Stud partitions*
(a) *Double studding*
(b) *Metal stud partitioning*

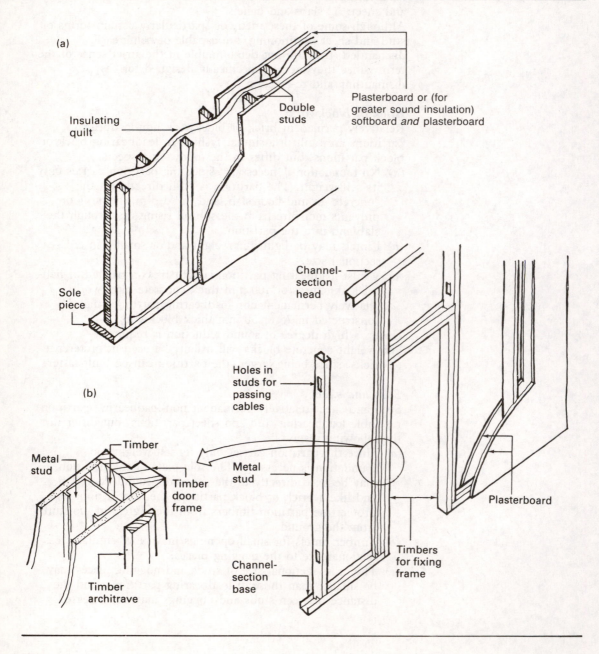

(a)

Insulating
quilt

Double
studs

Plasterboard or (for
greater sound insulation)
softboard *and* plasterboard

Sole
piece

Channel-
section
head

Holes in
studs for
passing
cables

(b)

Metal
stud

Timber

Timber
door
frame

Metal
stud

Timber
architrave

Channel-
section
base

Timbers
for fixing
frame

Plasterboard

Discontinuous construction As previously mentioned in this chapter (p. 180, above), 'discontinuous construction' may be used to increase the degree of sound reduction from one side of the partition to the other. This may be achieved by constructing two independent stud partitions with an insulating quilt draped between them (Fig. 5.5(a)). The studs of one side of the partition are placed between the intervals of the other side in order to reduce the overall thickness of the partition.

Metal stud partitions As an alternative to timber studs, metal stud partition systems are available (Fig. 5.5(b)). The system uses channel-section galvanised mild steel studs which are screwed to channel section head and base pieces. Plasterboard sheets are screwed to each side of the studs. Door openings are formed by:

- two studs which are placed either side of the opening to form the jambs and a channel section lintel fixed between them at the head of the opening;
- fixing a timber to each stud and beneath the lintel;
- fixing the door frame to the timbers.

Figure 5.5(b) shows the timbers fixed to the studs in preparation for fixing the door frame, whilst the detail shows the door frame fixed to the timber and an 'architrave' (a timber moulding designed to cover the joint between the plasterboard and the frame).

Panel and slab construction

Panel and slab materials offer the advantage over wet construction of being built of large units which can be speedily erected. Systems used include: (*a*) plasterboard in either laminated or timber form; (*b*) slab materials.

Plasterboard

Laminated partitions Laminated plasterboard partitions consist of three layers of plasterboard bonded together with adhesive or bonding compound, which are fixed to small-section timber battens screwed to the floor and ceiling (Fig. 5.6(a)). The dimensions of the panels are: thickness 50 or 65 mm; heights 2350–3200 mm.

The construction process involves the following:

(*a*) Timber battens are fixed at floor and ceiling level, door frames are installed and electrical wiring is installed in position so as to line up with the centre layer of plasterboard.

(*b*) One of the outer faces of plasterboard is nailed to the battens.

(*c*) The centre layer of plasterboard is cut to fit between the battens and around the electrical services. Adhesive or bonding compound is applied to the inner face of the already fixed outer layer of plasterboard and the centre layer is pressed against the adhesive.

(*d*) Adhesive or bonding compound is applied to the fixed centre layer of plasterboard and the final layer is pressed against the adhesive and nailed to the battens.

Cellular core partitions Plasterboard cellular core partitions are prefabricated and are available in the following range of sizes: thickness 50, 57 and 63 mm; heights from 1800 to 3600 mm.

The panels consist of a lightweight cellular core with a layer of plasterboard bonded to each side (Fig. 5.6(b)). Cellular core partitions are installed by small-section timber battens being screwed to the floor and ceiling to which the panels are nailed. Battens are also inserted vertically between adjacent panels.

Fig. 5.6 *Panel and slab partitions*
(a) *Laminated plasterboard partition*
(b) *Cellular core partition*
(c) *Slab partition*
(d) *WC partitions*

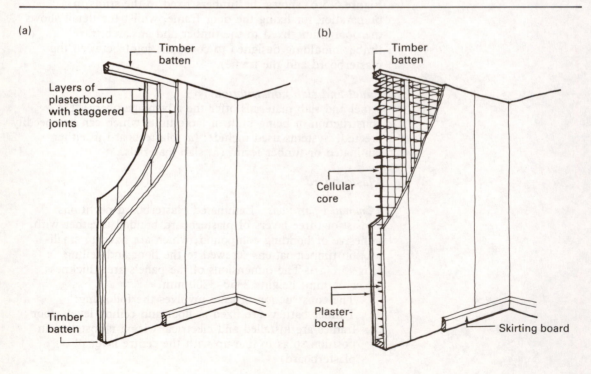

(a)

Timber batten

Layers of plasterboard with staggered joints

Timber batten

(b)

Timber batten

Cellular core

Plasterboard

Skirting board

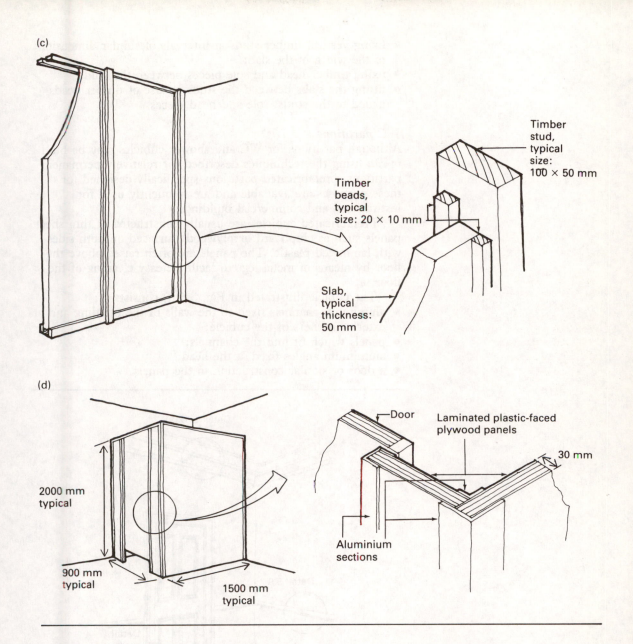

(c)

Timber stud, typical size: 100 × 50 mm

Timber beads, typical size: 20 × 10 mm

Slab, typical thickness: 50 mm

(d)

2000 mm typical

900 mm typical

1500 mm typical

Door

Laminated plastic-faced plywood panels

30 mm

Aluminium sections

Slab materials   Strawboard slabs, and woodwool slabs, which are commonly used for roof deckings (see Table 3.2, Ch. 3) may also be used for constructing partitions.

Figure 5.6(c) shows a method of installing a slab partition by:

- fixing vertical timber studs at intervals of similar dimension to the width of the slab;
- fixing timber head and sole pieces between the studs;
- fitting the slabs between the studs by use of timber beads nailed to the studs, sole and head pieces.

WC partitions

Although partitions for WC and shower cubicles may be built *in situ* using the techniques described for relatively permanent partitions, prefabricated partitions specifically designed for these purposes are available and are frequently used for institutional and commercial buildings.

Prefabricated cubicles are usually constructed of thin sheet panels such as chipboard or plywood, surfaced on both sides with laminated plastic. The panels are often raised above the floor by means of metal legs to facilitate easy cleaning of the floor.

The cubicle illustrated in Fig. 5.6(d) consists of:
- aluminium channels fixed to the walls of the building and at external corners of the cubicle;
- panels which fit into the channels;
- aluminium angles fixed at the head;
- a door of similar construction to the panels.

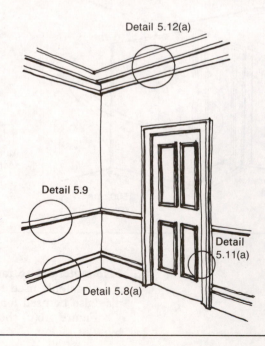

Fig. 5.7 *Skirtings, dados, cornices and architraves*

Skirtings, dados, cornices and architraves

These are strips of material which might be fixed to the surface of a partition in locations which are:
- liable to result in unsightly junctions between the partition surface and other elements of the internal structure of the building; and/or
- liable to damage of the surface finish of the partition due to, for example, scuffing by shoes or furniture.

Figure 5.7 indicates the positions of skirtings, dados, cornices and architraves.

Skirtings

Skirtings are designed to:
- prevent damage to the surface of the partition due to scuffing by, for example, shoes or floor-cleaning equipment;
- conceal any unsightly gaps that might occur between the floor finish and the bottom of the partition.

Skirtings may be installed either flush with the partition surface (Fig. 5.8(c)) or projecting out from the partition surface (Figs 5.8(a), (b), (d) and (e)).

Timber is a commonly used material for skirtings. Figure 5.8(a) shows a traditional timber skirting with complex mouldings fixed to timber 'grounds' (battens nailed to the wall, prior to plastering). Timber skirtings today tend to be less deep, of simple profile, and are commonly nailed directly to the partition (Fig. 5.8(b)).

Skirtings may be made of other materials and are often of the same material as the floor finish (Figs 5.8(c), (d) and (e)).

Dado rails

Sometimes the lower portion of a partition is visually distinguished from the upper portion by the use of a 'dado rail' or 'chair rail'. In traditional interiors, the 'dado' (or lower portion of wall) was given a decorative surface treatment such as embossed wallpaper, or timber boarding, whilst the upper portion was given a different treatment. Traditional dados were of complex profile which would relate to the skirting and other moulded rails (Fig. 5.9).

Dado rails are sometimes used today in interiors which make use of 'built-in' furniture (furniture which uses the structure of the building for its support – see Ch. 6). In Fig. 5.10 the dado rail, together with the cornice and skirting,

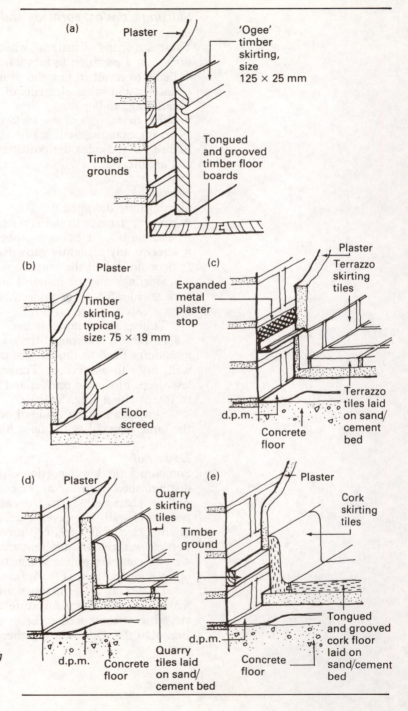

Fig. 5.8 *Skirting details*
(a) *Traditional timber skirting*
(b) *Typical modern timber skirting*
(c) *Terrazzo tile skirting*
(d) *Quarry tile skirting*
(e) *Cork skirting*

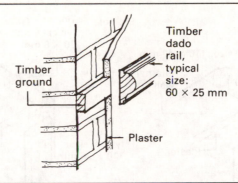

Fig. 5.9 *Traditional dado rail*

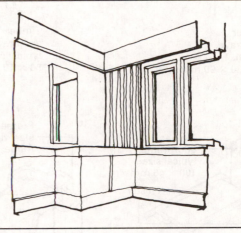

Fig. 5.10 *Modern use of cornice, dado and skirting*

has the effect of unifying the elements of furniture in a room. The dado rail may be of hollow construction and act as a trunking for electric and other cables (see Fig. 5.1(c)).

Cornices

Traditionally, the junction between the top of the partition and the ceiling was made with a 'cornice' of moulded timber or plasterwork (Fig. 5.12(a)).

Cornices are still sometimes used either:

- as a decorative element such as the plasterboard 'coving' illustrated in Fig. 5.12(b), which is obtainable in standard sizes of 100 or 127 mm depth and is held to the ceiling and partition with adhesive;
- to conceal services such as lighting or heating installations or to form a pelmet for curtain track, as in Fig. 5.12(c). In these cases, the cornice might be constructed of a timber framework covered in some form of board material such as plasterboard or timber.

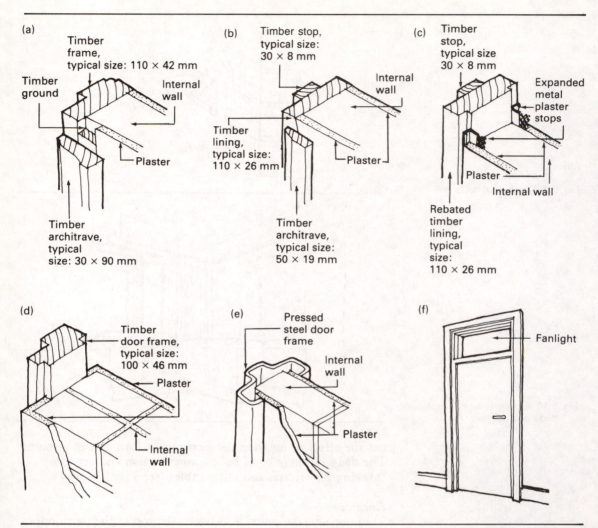

Fig. 5.11 *Interior timber door frame details*
(a) *Traditional architrave*
(b) *Typical modern architrave*
(c) *Negative detail*
(d) *Internal timber door frame*
(e) *Internal pressed steel door frame*
(f) *Internal door with fanlight*

Architraves

Architraves are designed to cover the unsightly junction that can occur between a timber door frame and the plaster of a partition.

Architraves are usually made of timber, the profile of which is suited to the profile of the skirting architrave (Figs 5.11(a) and (b) show a traditional and modern architrave). The architrave is usually thicker in section than the skirting board so as to provide a visually acceptable junction between the two. In the past, an 'architrave block'

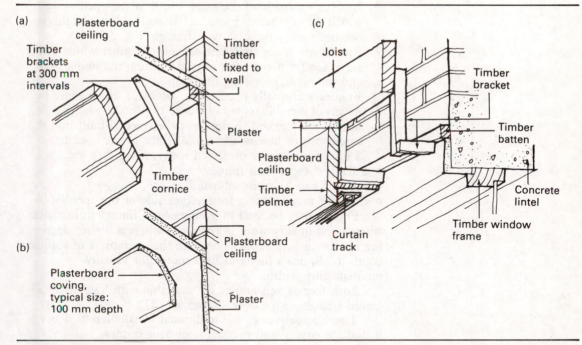

(a)
Plasterboard ceiling

Timber brackets at 300 mm intervals

Timber batten fixed to wall

Plaster

Timber cornice

(b)
Plasterboard coving, typical size: 100 mm depth

Plasterboard ceiling

Plaster

(c)
Joist

Timber bracket

Timber batten

Plasterboard ceiling

Timber pelmet

Concrete lintel

Curtain track

Timber window frame

Fig. 5.12 *Junction between wall and ceiling*
(a) *Timber cornice*
(b) *Plasterboard coving*
(c) *Timber pelmet concealing curtain track*

was fixed at the bottom of the architrave, against which the architrave and skirting board were terminated.

An alternative door frame to partition treatment is to use a 'negative detail', such as that illustrated in Fig. 5.11(c). Here, the plaster is brought to an edge adjacent to the door frame by use of an expanded metal 'plaster stop', which enables the plaster to be brought to a firm edge. Since timber is liable to swell or shrink in accordance with the temperature and humidity of a room, a visually annoying gap could occur between the timber frame and the plaster. If a rebate is cut in the frame adjacent to the plaster, an artificial gap is created which is of sufficient size to enable swelling or shrinkage of the gap to occur without being visually perceptible.

Internal door frames, linings, doors and ironmongery

Internal door frames and linings

Internal doors may be 'hung' (or hinged) to either 'frames' or 'linings'. The basic differences between the two are as follows:
(a) Timber frames are of comparatively substantial cross-section size, whilst linings are usually of smaller size, typically 26 mm in thickness.

(*b*) Linings are made of the same width as the opening in which they are fixed (Figs. 5.11(b) and (c)), whilst this is not necessarily the case with frames (Fig. 5.11(d)).

(*c*) Frames are made from one section of timber whilst linings are made from two (one section forming the lining itself and the other providing the 'stop').

(*d*) Frames are usually fixed to the jambs of the opening by means of metal lugs (see Fig. 2.25(c), Ch. 2), whereas linings are screwed to timber 'pallet pieces' built into the brickwork or blockwork of the jambs of the opening.

Linings tend to be preferred today, since they are:

● usually cheaper than frames;
● the stop can be easily adjusted on site;
● the door may be hung from either side of the opening.

Frames may be used in preference to linings in instances where good-quality work is required, where a higher degree of fire resistance is necessary or where the opening is of sufficient depth to require a lining of uneconomic or visually unsatisfactory width.

Both frames and linings are available with 'fanlights' (a glazed area above the door — Fig. 5.11(f)).

Two alternatives to the traditional installation of doors include 'doorsets' and pressed steel door frames.

Doorsets   Doorsets are designed to reduce the amount of work carried out on site. Doorsets consist of a complete door frame, door and ironmongery which is all assembled in a factory. The doors and frames may also be partially or completely finished in order to reduce site work still further. A disadvantage of doorsets is that they are vulnerable to damage due to site work which is still in progress after the doorset has been installed.

Pressed metal door frames (Fig. 5.11(e))   Pressed metal door frames are available in several widths to suit a variety of common depths of opening. The frames are usually zinc plated and supplied with one coat of primer.

Steel frames offer the advantage of:

● not shrinking or swelling to the degree that poorer-quality timber frames or linings might;
● requiring no lintel if installed in a non-loadbearing partition.

Internal doors
Internal doors may be of similar construction and of similar appearance to the external doors illustrated in Figs. 2.26(a–c)

(Ch. 2). However, since internal doors are not required to provide weather resistance, and normally require a lesser degree of security, they are usually of lighter weight construction than external doors. Typical internal door sizes are shown in Table 5.8.

Table 5.8 Typical internal door sizes

Sizes (mm)		
Height	Width	Thickness
	610	
1981	762	
	838	35
2032	813	

Flush doors of the type illustrated in Fig. 2.26(b)(iii) are used occasionally for cheap-quality work, although panelled doors of the type illustrated in Figs 2.26(a)(ii) and (iii) are also commonly used. Traditional-style panel doors similar to those in Figs 2.26(a)(iii) and (iv) are still manufactured for internal use.

Other types of door which are intended primarily for internal use include:
(a) sliding doors; (b) sliding/folding doors; (c) sound-resistant doors; (d) fire check doors.

Sliding doors   Sliding doors offer the advantage over swing doors of not using up floor space whilst being opened; however, they do take up workspace (Fig. 5.13(a)(i)). Figures 5.13(a)(i), (ii) and (iii) show some of the different ways by which a sliding door could operate.

There are many different types of sliding-door mechanism available, although those designed for domestic or lightweight doors mostly involve an overhead track from which the door is hung. The mechanism illustrated in Fig. 5.13(b) involves:
(a) Two sets of nylon wheels housed in brackets which screw to the top edge of the door. A slotted bolthole in the bracket enables the height of the wheels to be finely adjusted.
(b) The door is hung by locating the wheels in a track which is fixed above the door opening.
(c) To ensure that the door remains vertical during its sliding action, a channel, which is inserted in the bottom edge of the door, passes over two guides. The guides are screwed to the floor, near to each jamb of the opening.

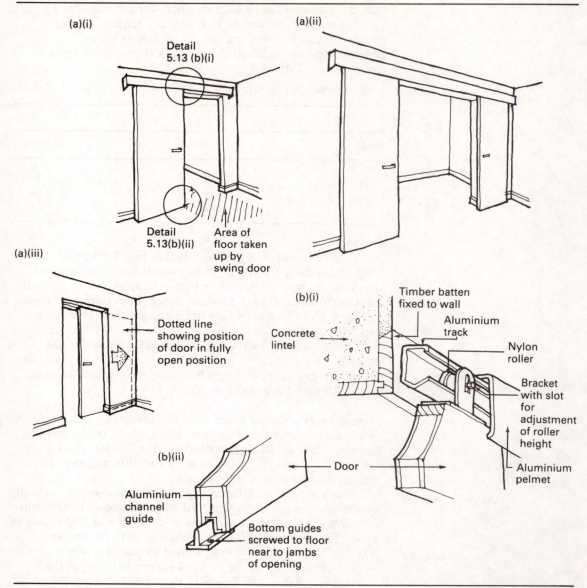

Fig. 5.13 *Sliding doors*
(a)
  (i) *Single sliding door*
 (ii) *Double sliding doors*
(iii) *Door sliding into cavity in wall*
(b) *Mechanics of sliding door*
 (i) *Top of door*
(ii) *Bottom of door*

(d) Usually a pelmet of wood or metal is fixed above the opening to conceal the track and runners.

For large and heavy doors, the wheels and track are sometimes situated at the bottom of the door, whilst the guides and channel are situated at the top.

Sliding/folding doors Sliding/folding doors consist of two or more 'leaves' (individual doors) which open and close in a concertina fashion. An example of such a door is illustrated in Fig. 5.14(a)

A major advantage of these doors is that, unlike a straight sliding door, no wall space is necessary at each side of the door opening to accommodate the door when open. However, it should be borne in mind when designing the opening that space is necessary at each jamb for the doors to stack when they are fully open.

Figure 5.14(b) shows some of the different possible configurations of sliding/folding doors. Notice that:
● centre folding doors, if hinged at one end to the door frame, require a half-width leaf adjacent to the jamb;
● end folding doors project into the room when opened.

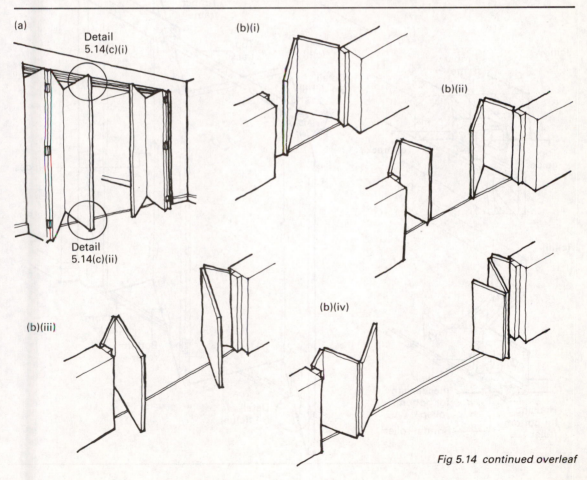

(a)

Detail 5.14(c)(i)

Detail 5.14(c)(ii)

(b)(i)

(b)(ii)

(b)(iii)

(b)(iv)

Fig 5.14 continued overleaf

Fig 5.14 continued

(b)(v)

(b)(vi)

(c)(ii)

Timber frame

Top guide

Top track

Door

(c)(iii)

Top track

Housing for bottom roller

Swivel mechanism for bottom roller

Bottom roller

(c)(i)

Concrete lintel

Timber frame

Top track

Detail 5.14(c)(ii)

Hinges

Bottom track

Bottom roller

Detail 5.14(c)(iii)

Figure 5.14(c) shows the mechanism of a centre folding door which consists of the following features:

(a) Hinges: the first set of hinges is fixed between the jamb of the door frame and the stile of the half-width door. The next set of hinges link the stiles of the half-width door and the next full-width leaf and so on (see Fig. 5.14(c)(i)).

(b) A top track is fixed to the soffit of the door opening, and in line with this a bottom track is fixed in the floor.

(c) A top guide which locates in the top track is screwed to the centre top edge of alternate leaves.

(d) To the same leaves, a bottom roller is fixed which locates in the bottom track. To enable the leaves to slide and fold, the bottom roller is housed in a swivel mechanism. Thus the bottom roller may freely slide in the track whilst the leaf slides at an angle to the track.

Sound-resistant doors   Some internal doors based on a core of high-density material (such as laminated timber or chipboard) are able to achieve sound reduction figures in excess of 30 dB; however, the gaps between the door and door frame can seriously impair the sound reduction. To reduce this problem, the gaps should be sealed by a compressible, resilient material such as neoprene (a rubber material). The closing of the door compresses the neoprene, which forms an air seal between the frame and door (Fig. 5.15).

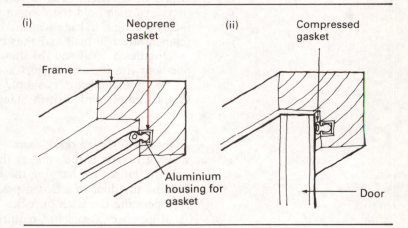

Fig. 5.15 *Door seal for sound resistance*
(i) *Door open*
(ii) *Door closed*

Fire-resisting doors   The Building Regulations 1991, approved document B, require certain doors to have a specified period of fire resistance. Examples of such doors include:

(*a*) Doors to openings in 'compartment walls' separating adjoining buildings. Such doors require fire resistance equal to that of the wall (which is considered as an 'element of structure' see Table 3.6) but not less than one hour.

(*b*) Doors to openings in 'compartment walls' (see 'fire resistance' p. 182). Compartment walls separating a flat or maisonette from a space of common use require a door with minimum fire resistance of 30 minutes. Other doors may require the same fire resistance as the compartment wall in which they are installed and need to be capable of restricting smoke leakage.

(*c*) Doors to openings in 'protected shafts' (stairways or lifts surrounded by fire-resisting walls), 'protected stairways' (stairways discharging through an exit to a place of safety) and 'protected corridors/lobbies' (corridors/lobbies protected by fire-resisting construction). Such doors require 20 or 30 minute fire-resistance (dependent on location) and must be capable of resisting smoke leakage.

(*d*) Any door in a wall separating a house from a garage requires half-hour fire resistance and should be installed at least 100 mm above the level of the garage door.

Fire-resisting doors are required to be self-closing by means of either 'door closers' (Fig. 5.16(c) and (d)) or by 'fusible links' (devices which hold the door in the open position by a cable incorporating a linkage which when exposed to a specified temperature will melt and thus close the door).

Figures 5.16(a) and (b) show examples of half-hour and one-hour fire-resistance doors and door frames.

Many manufacturers supply fire-resisting doors which should comply with British Standard 476 Part 22 (1987).

Ironmongery for fire check doors The ironmongery for internal doors generally is the same as that for external doors, namely, hinges for hanging the door, a lock or latch for holding the door in a closed position and handles or knobs for operating the latch or lock.

Fire check doors will require some form of automatic device to close the door. Such devices include: (*a*) floor springs; (*b*) door closers; (*c*) spring hinges. Factors to be considered in selecting a closing device include:

(*a*) *Cost*: floor springs are much more expensive than door closers or spring hinges.

(*b*) *Size and weight of door*: most closing devices are designed to operate on doors of a specified maximum size and weight.

(*c*) *'Check action'*: Many closing devices incorporate an adjustable mechanism which regulates or 'checks' the speed at which the door closes. This is necessary to ensure that the door will close with sufficient force to activate the door latch (if one is fitted) without causing the door to slam.

Floor springs (Fig. 5.16(c)) Floor springs are devices which both hang the door and provide the closing action. The components of floor springs consist of:

(*a*) A box which contains the spring and check mechanism and is covered by a metal plate. The box is recessed into the floor so that the plate is level with the floor finish. The spring activates a metal spindle which projects out of the plate.

(*b*) A metal shoe which is fitted to the bottom of the door stile locates in the spindle of the box.

(*c*) A 'top centre': two metal plates, one of which is screwed to the top of the door stile and the other is secured to the underside of the door frame. The upper plate has a pin which locates in a recess in the lower plate.

Floor springs are available for closing either single or double-action swing doors.

Door closers There are many different types of door closer which are installed either at the head of the door (overhead closers) or at the stile of the door (spring closers).

Overhead closers today are usually activated hydraulically, Fig. 5.16(d) shows a typical example. The hydraulic mechanism, which is housed in a metal casing screwed to the top of the door, activates a metal spindle. A cranked metal arm is attached to the metal spindle at one end, and to the head of the door frame at the other.

Concealed overhead closers are also manufactured, where the casing of the closer is housed within the thickness of the door top rail.

Spring closers consist of a barrel-shaped casing which contains a spring mechanism. In the spring closer illustrated in Fig. 5.16(e), the casing is recessed into the door stile, and the spring activates a chain which is attached to the door frame.

Fig. 5.16 *Fire-check doors*
(a) *Half-hour fire-resistance door*
(b) *One-hour fire-resistance door*
(c) *Floor springs*
 (i) *Door in closed position*
(ii) *Door in open position*
(d) *Overhead door closer*
 (i) *Door in open position*
(ii) *Door in closed position*
(e) *Spring closer*
(f) *Spring hinge*
(g) *Door selector*
 (i) *Doors closed in
     wrong order*
(ii) *Door selector
     arresting doors*
(iii) *Door selector*

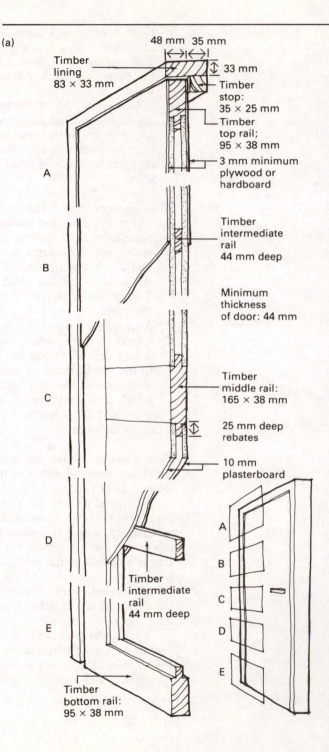

(a)

48 mm   35 mm

Timber
lining
83 × 33 mm

33 mm

Timber
stop:
35 × 25 mm

Timber
top rail;
95 × 38 mm

3 mm minimum
plywood or
hardboard

A

Timber
intermediate
rail
44 mm deep

B

Minimum
thickness
of door: 44 mm

Timber
middle rail:
165 × 38 mm

25 mm deep
rebates

10 mm
plasterboard

A

B

D

C

Timber
intermediate
rail
44 mm deep

D

E

E

Timber
bottom rail:
95 × 38 mm

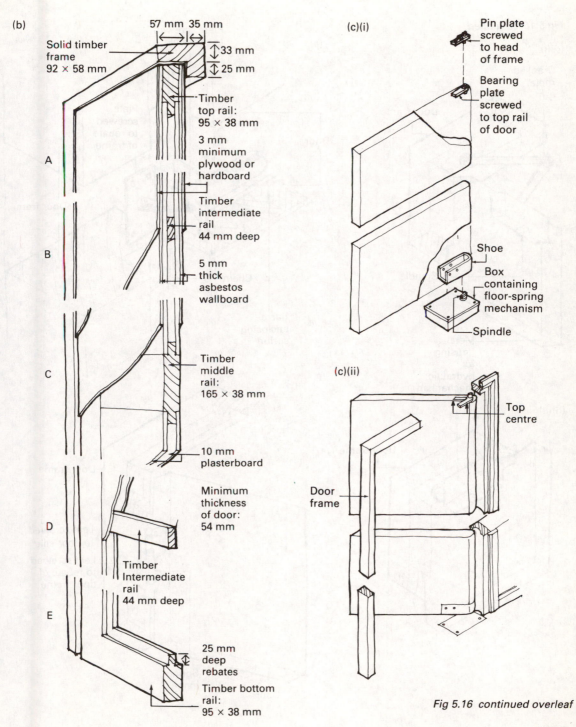

(b)

57 mm   35 mm

Solid timber
frame
92 × 58 mm

33 mm
25 mm

Timber
top rail:
95 × 38 mm

A

3 mm
minimum
plywood or
hardboard

Timber
intermediate
rail
44 mm deep

B

5 mm
thick
asbestos
wallboard

C

Timber
middle
rail:
165 × 38 mm

10 mm
plasterboard

Minimum
thickness
of door:
54 mm

D

Timber
Intermediate
rail
44 mm deep

E

25 mm
deep
rebates

Timber bottom
rail:
95 × 38 mm

(c)(i)

Pin plate
screwed
to head
of frame

Bearing
plate
screwed
to top rail
of door

Shoe

Box
containing
floor-spring
mechanism

Spindle

(c)(ii)

Top
centre

Door
frame

Fig 5.16 continued overleaf

213

Fig 5.16 continued

(d)(i)

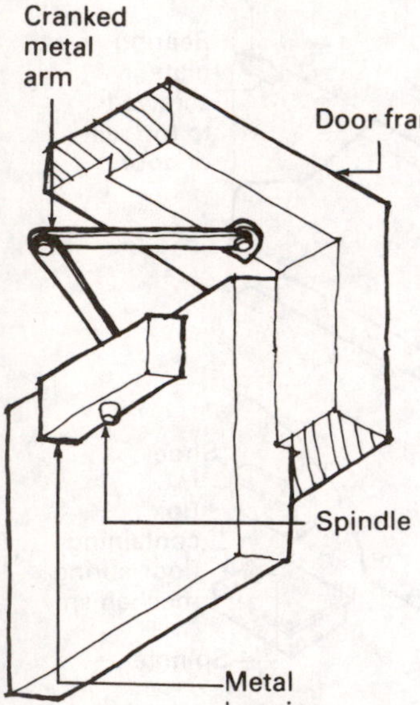

Cranked
metal
arm

Door frame

Spindle

Metal
housing
for
hydraulic
mechanism

(d)(ii)

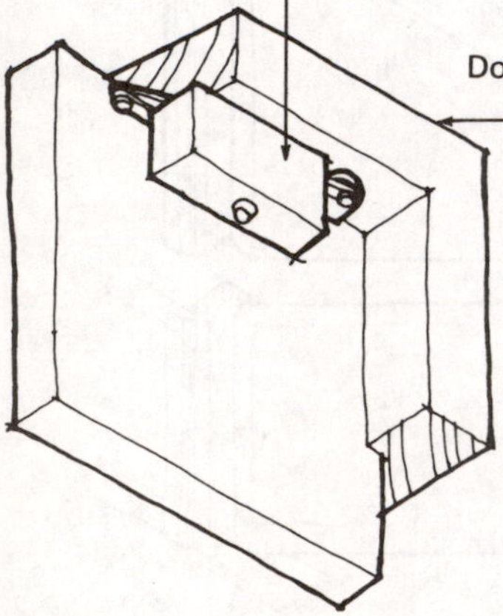

Door frame

(e)

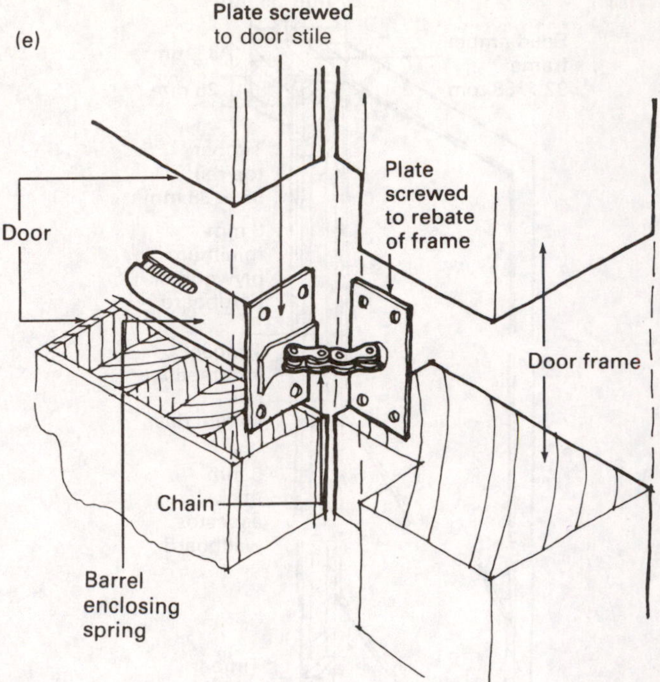

Plate screwed
to door stile

Door

Plate
screwed
to rebate
of frame

Door frame

Chain

Barrel
enclosing
spring

(f)

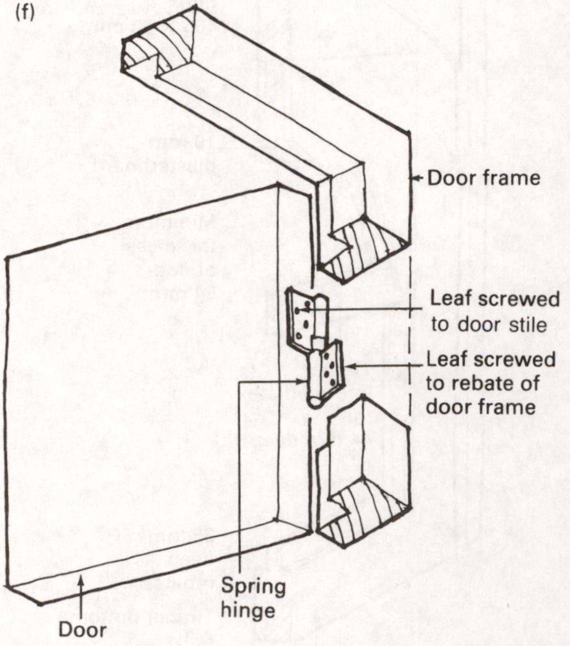

Door frame

Leaf screwed
to door stile

Leaf screwed
to rebate of
door frame

Spring
hinge

Door

Fig 5.16 *continued*

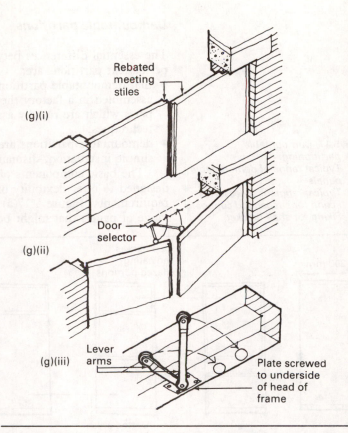

(g)(i)

Rebated meeting stiles

Door selector

(g)(ii)

(g)(iii)

Lever arms

Plate screwed to underside of head of frame

Spring hinges   Like floor springs, spring hinges are designed to both hang the door and provide the closing operation. Figure 5.16(f) illustrates a spring hinge consisting of a metal barrel located between two leaves of the hinge. The barrel houses a spring which activates one of the hinge leaves. This type of spring hinge is usually installed in pairs, one hinge providing the closing action and the other the check action.

Door selectors   Where door closers are fitted to pairs of doors with rebated meeting stiles, it is essential that the doors close in the right order otherwise one door will be caught against the rebate of the other (Fig. 5.16(g)(i)). In such instances, a 'door selector' (Fig. 5.16(g)(iii)) is screwed to the underside of the head of the frame. The selector consists of two spring arms of unequal length. Figure 5.16(g)(ii) shows the arms arresting the doors just before they close. The unequal length of the arms ensure that the doors close in the correct order.

Demountable partitions

The essential differences between demountable and relatively permanent partitions are:
- that demountable partitions are made and partially assembled in a factory then delivered to the site as a kit of parts which are rapidly assembled using dry construction techniques;
- demountable partitions are specifically designed for rapid, simple installation, dismantling and reassembly.

The basic components of frames and/or panels are designed to offer flexibility of functional and height requirements. Figure 5.17(a) shows elevations of a typical range of panels that might be offered by a manufacturer.

Fig. 5.17 *Demountable partitioning*
(a) *Typical range of demountable panels*
(b) *Typical range of accessories*
(i) *Fixing for suspended cupboard*
(ii) *Fixing for shelf bracket*

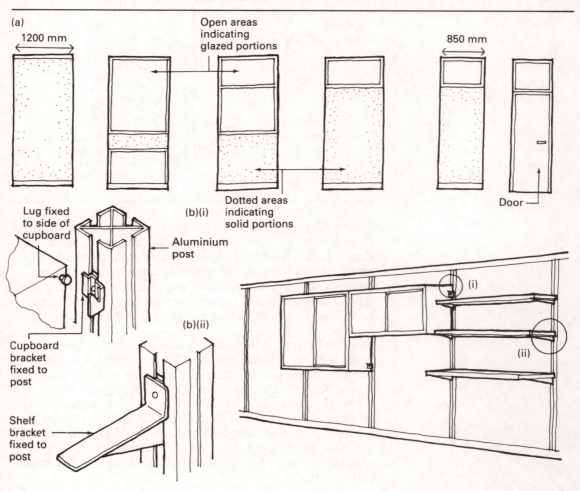

(a)

1200 mm

Open areas indicating glazed portions

850 mm

Dotted areas indicating solid portions

Door

Lug fixed to side of cupboard

Aluminium post

(b)(i)

Cupboard bracket fixed to post

(b)(ii)

Shelf bracket fixed to post

(i)

(ii)

Although most manufacturers offer a range of partition heights, some form of fine height adjustment is necessary to enable the partitioning system to be adapted to the height of a specific location. This adjustment is usually provided either at the head of skirting level of the partition, or by simply cutting the panels and vertical frame members to the required height. The head and skirting frame details also allow a degree of adjustment to the take account of uneven floors or ceilings.

Demountable partitions are often used in conjunction with suspended ceiling systems. In order to obtain a neat appearance at the junction of the suspended ceiling and partition the module size of the ceiling panels and partition widths should be compatible.

Since the partition panels are usually of lightweight construction, the panels are often unsuitable for the fixing of items such as shelves or hooks which might carry heavy articles. Many partition systems offer a range of accessories which attach to the vertical frame members (Fig. 5.17(b) shows some of the typical accessories that might be offered with a partitioning system).

Demountable partitions may be classified as either: (a) frame/sheet systems; (b) frame/panel systems; (c) panel/panel systems.

Frame/sheet and frame/panel systems
Both of these systems are based on frameworks consisting of head and skirting members with vertical members fitted between them.

Frame/sheet systems are infilled with sheet material (typically plasterboard). The system illustrated in Fig. 5.18(a) and details, Figs 5.18(a)(i), (ii) and (iii), consists of:
- a lightweight galvanised steel inner framework consisting of top and bottom channels and channel section posts fitted between them;
- one layer of plasterboard fitted each side of the framework;
- an aluminium exposed outer framework consisting of aluminium top channel, skirting and push-fit cover strips to conceal the junction between adjacent sheets of plasterboard.

Frame/panel systems consist of panels of material such as cellular core plasterboard held by vertical posts. The system illustrated in Fig. 5.18(b) and details, Figs. 5.18(b)(i), (ii) and (iii), consists of:
- a timber base section;
- an aluminium channel-section head;
- aluminium cruciform section posts with push-fit p.v.c. channels;

- cellular core plaster panels;
- aluminium skirtings which cover the junction between the panel and the base section. The voids beneath the base section may act as ducts for electric wiring.

Panel/panel systems

Panel/panel systems are, in effect, series of panels wedged between the ceiling and floor surfaces. Consequently, the panels require no supporting framework and very rapid demountability and re-erection is possible. Each panel incorporates a telescopic section in its head which may be retracted or extended in depth by means such as air pressure or spring-loaded clamp mechanisms.

The panels are of lightweight construction and, like framed demountable systems, manufacturers offer solid, glazed, semi-glazed and door panels. The system illustrated in

Fig. 5.18 *Frame/sheet and frame/panel demountable partitioning*
(a) *Frame/sheet system*
 (i) *Detail at head*
 (ii) *Detail at mid height*
 (iii) *Detail at floor level*
(b) *Frame/panel system*
 (i) *Detail at head*
 (ii) *Detail at mid height*
 (iii) *Detail at floor level*

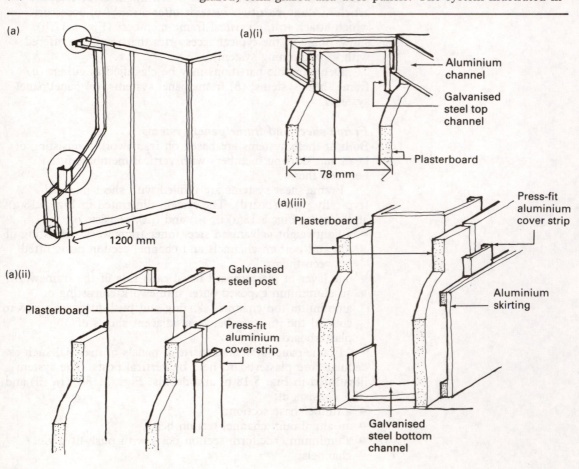

218

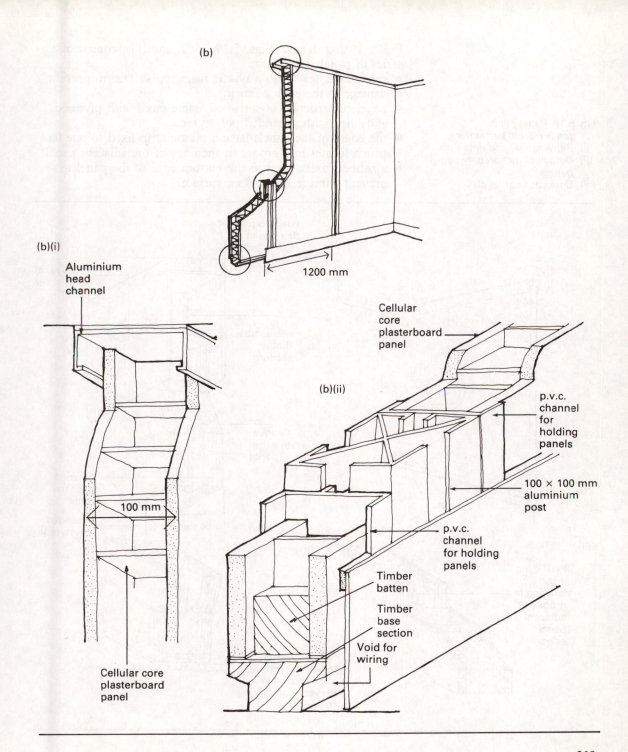

(b)

1200 mm

(b)(i)

Aluminium head channel

100 mm

Cellular core plasterboard panel

Cellular core plasterboard panel

(b)(ii)

p.v.c. channel for holding panels

100 × 100 mm aluminium post

p.v.c. channel for holding panels

Timber batten

Timber base section

Void for wiring

Fig. 5.19 and details, Figs. 5.19(i), (ii) and (iii), consist of a series of panels each with:
- retractable head with a plastic foam top surface to prevent damage to the ceiling finish;
- panel construction of softwood frame faced with plywood and filled with expanded polystyrene;
- the edges of the panels have a plastic strip fixed to one face which locates in a groove in the edge of the adjacent panel;
- a rubber insert fitted to the bottom edge of the panel to prevent damage to the floor surface.

Fig. 5.19 *Panel/panel demountable partitioning*
(i) *Retractable head detail*
(ii) *Detail of junction between panels*
(iii) *Detail at floor level*

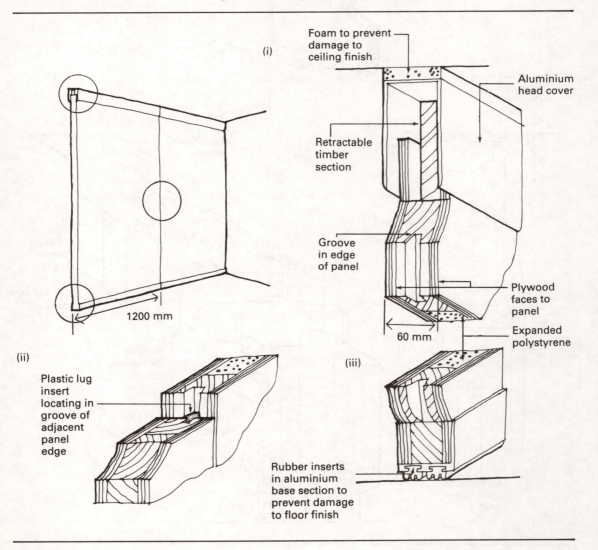

(i)

Foam to prevent damage to ceiling finish

Aluminium head cover

Retractable timber section

Groove in edge of panel

Plywood faces to panel

Expanded polystyrene

1200 mm

60 mm

(ii)

Plastic lug insert locating in groove of adjacent panel edge

(iii)

Rubber inserts in aluminium base section to prevent damage to floor finish

Movable partitions and screens

Movable partitions are used in instances where the walls of a room are frequently required to be opened up to form one large floor area.

Although panel/panel demountable partitions could perform this function, movable partitions may be opened and closed with minimum effort, since no lifting is necessary. Some acoustically designed movable partitions use:

- dense materials in panel construction;
- carefully designed details of moving parts and junctions;
- insulation quilts within the panels which are to achieve sound reduction figures in excess of 30 dB.

Screens are partitions of less than full floor to ceiling height and are frequently used for subdividing large areas such as open-plan offices, or for temporary exhibitions.

Sliding partitions

Sliding partitions consist of a series of panels which slide in tracks fixed to the floor (bottom tracks), and/or tracks fitted to the ceiling (top tracks). The folding/sliding mechanics of the partitions are similar to those of sliding/folding doors illustrated in Fig. 5.14(c)(i). At its most simple level, a sliding partition could operate in a similar fashion to the sliding doors illustrated in Fig. 5.13(a)(ii), although such systems are limited since a large space would be required to house the panels when in an open position. This problem is overcome by enabling the panels to be 'stacked' against the side walls, or in cupboards built into the side walls. Figures 5.20(a) and (b) show two of the many possible permutations for stacking the panels.

For stacking, it is often necessary for the panels to swing either at an angle or at a right angle to the track. The pivot swinging action of the partition illustrated in Fig. 5.20(a) is achieved by:

- fixing a top guide which locates in the top track centrally on the top of the panel;
- bottom rollers which are capable of being withdrawn from the bottom track;
- two castors which are fitted to the bottom of the panel which enable ease of movement whilst the panels are being swung or moved at an angle to the track.

By fixing the top guides towards one edge of the panel, the panel could swing in a hinged fashion as in Fig. 5.20(b).

Without a bottom track, the bottom edges of the panels would not be held in position vertically beneath the top track; however, the need for a bottom track may be eliminated by

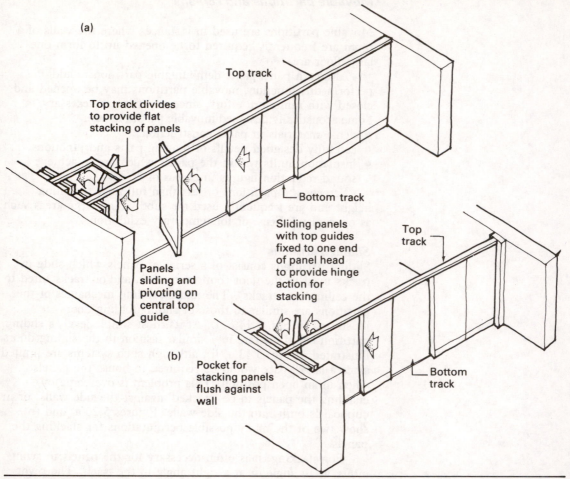

Fig. 5.20 *Sliding partitions*
(a) *Panels slide and pivot*
(b) *Panels slide and swing*

Labels in figure:

(a)

Top track

Top track divides to provide flat stacking of panels

Bottom track

Panels sliding and pivoting on central top guide

Sliding panels with top guides fixed to one end of panel head to provide hinge action for stacking

(b)

Top track

Bottom track

Pocket for stacking panels flush against wall

use of a telescopic head section to the panel. This device is similar to the panel/panel system illustrated in Fig. 5.19(i) where a top guide locates in the top track, and two castors are fitted to the bottom of the panel. Once the panel is in its desired position, the telescopic head may be raised which wedges the panel between the floor and ceiling.

Sliding/folding and folding partitions

Sliding/folding partitions Sliding/folding partitions operate in a similar manner to sliding/folding doors using similar mechanical components (hinges, top and bottom track, rollers and guides; see Figs 5.14(a), (b) and (c). However, whilst sliding/folding doors are designed to operate within a limited size of opening, sliding/folding partitions may be installed to open and close the entire height and width of a wall.

Since the weight of the panels of sliding/folding partitions is liable to be greater than the leaves of sliding/folding doors:
- the mechanical components will be of more robust construction;
- if the sliding/folding partition is particularly heavy, the weight of the panels might be carried by rollers fixed to the bottom edge of the panels and locating in a bottom track. Heavy panels hung by a top track could otherwise necessitate excessively substantial rollers and track.

Folding partitions Folding partitions are different in concept to sliding/folding or panel/ panel partitions since they are made of p.v.c. fabric attached to a framework of metal hinges which enables the partition to extend or collapse in an action similar to that of bellows (Fig. 5.21(a)). A major advantage of folding partitions is that they tend to be of lighter weight than sliding/folding or panel/panel systems, and consequently even large partitions may be operated by only top rollers running in a top track. No floor track or guides are necessary.

Figure 5.21(b) shows the components of a typical folding partition which consist of:
(*a*) A top track fixed at ceiling level.
(*b*) Sets of rollers which run in the top track and are attached to steel pins which also act as a pivot for the hinges.
(*c*) Hinges, which are in an 'X' configuration, are able to extend or collapse. Rows of hinges are arranged at the top, bottom and at intermediate levels of the partition.
(*d*) Rods, which connect the limbs of adjacent hinges, pass down the entire length of the partition, connecting the intermediate and bottom rows of hinges.
(*e*) The framework formed by the hinges and rods is covered on both sides by p.v.c. fabric.
(*f*) Aluminium channels which are fixed to the side walls. The outer edges of the p.v.c. fabric are attached to the channel.
(*g*) Aluminium strips which are attached to the inside (or 'lead') edges of the partition enable the partition to be closed effectively by either a latch or lock mechanism.

Like sliding/folding partitions, folding partitions may be folded back into recesses or cupboards in the side walls.

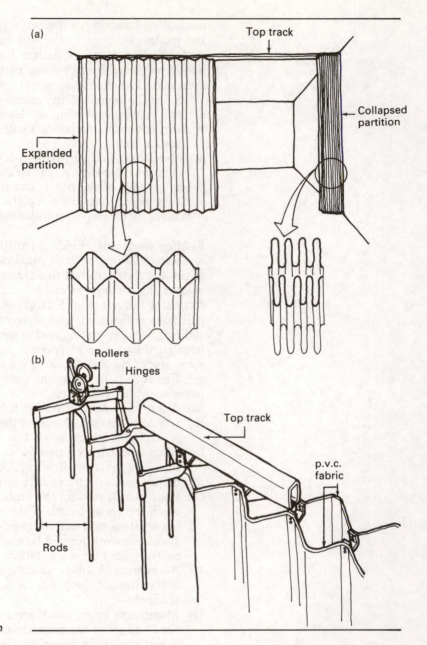

(a)

Top track

Collapsed partition

Expanded partition

Rollers

Hinges

(b)

Top track

p.v.c. fabric

Rods

Fig. 5.21 *Folding partition*
(a) *Folding partition*
(b) *Folding partition mechanism*

Screens

Screens are modular panels of less than full floor to ceiling height which may be linked together to form different partition configurations. Screens are either free-standing, in which case two feet are attached to the bottom edge (Fig. 5.22(a)), or they might link together to provide an angular formation which provides stability (Fig. 5.22(b)).

The screen is usually constructed of a metal or timber frame infilled with sheet material such as plywood or chipboard which is covered with fabric or thin sheet material.

Many screen systems are designed to link with furniture modules, where the screen width will relate to the furniture dimensions. A typical range of sizes for such screens might be as follows: width 500, 600 and 800 mm; height 1500 mm. The linkage between screens and furniture might be by:

- furniture resting against the side of the screen;
- furniture clipping to the vertical framework of the screen;
- furniture such as cupboards or shelves being suspended from the screen framework;
- furniture such as worktops being suspended between the frameworks of two screens (Fig. 5.22(b)).

Some screen systems are designed to provide such facilities as:

- a degree of acoustic control, by incorporating padding in the screen construction;
- providing trunking for electric and telephone wiring at skirting or dado level;
- providing fully or half-glazed screens.

Fig. 5.22 *Screens*
(a) *Freestanding screen*
(b) *Screen and furniture system*

(a)

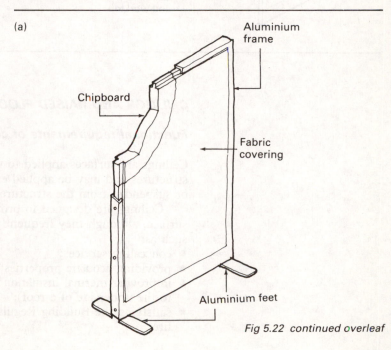

Aluminium frame

Chipboard

Fabric covering

Aluminium feet

Fig 5.22 continued overleaf

225

Fig 5.22 continued

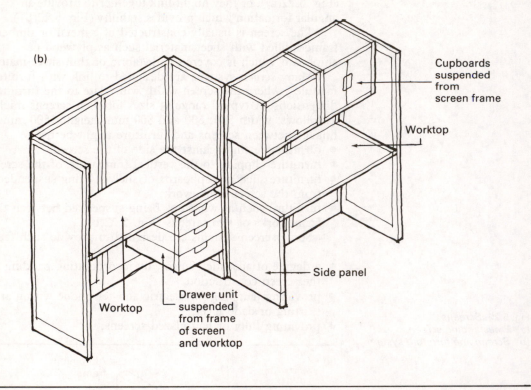

(b)

Cupboards suspended from screen frame

Worktop

Side panel

Worktop

Drawer unit suspended from frame of screen and worktop

CEILINGS AND RAISED FLOORS

Functional requirements of ceilings

Ceilings are surfaces applied to the underside of floor or roof structures and may be applied either directly to the structure, or suspended from the structure (Fig. 5.23 and Table 5.9).

Ceilings are designed to provide a visually acceptable surface, although they frequently serve additional functions such as:

- concealing services;
- providing acoustic properties;
- improving thermal insulation (in the case of ceilings applied to the underside of a roof);
- satisfying the Building Regulations in respect of safety in fire.

Table 5.9

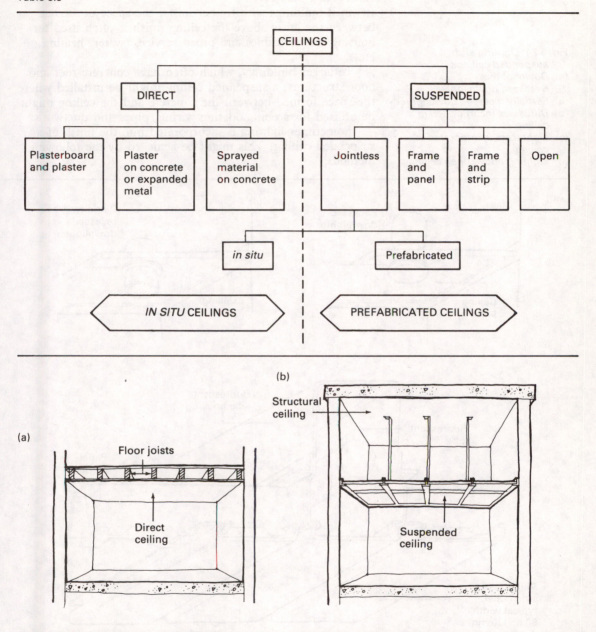

Fig. 5.23 *Ceilings*
(a) *Direct ceiling*
(b) *Suspended ceiling*

Fig. 5.24 *Lighting and
suspended ceilings*
(a) *Downlighters*
 (i) *Recessed downlighter*
(ii) *Partially recessed downlighter*
(b) *Diffussers incorporated in
ceiling*
(c) *Light track incorporated in
ceiling*
(d) *Lighting above open ceiling*

Concealing services

In small buildings with timber joists, the space occurring
between the joists above the ceiling finish is often used for
housing electric cables and piped services (water, heating,
etc.).

In larger buildings, which often have concrete roof and
floor structures, a suspended ceiling might be installed where
the space formed between the concrete and the ceiling might
be utilised for accommodating wiring, pipes and ductwork.

Sometimes lighting is incorporated into the finish of a
suspended ceiling. This might be achieved by the following
methods:

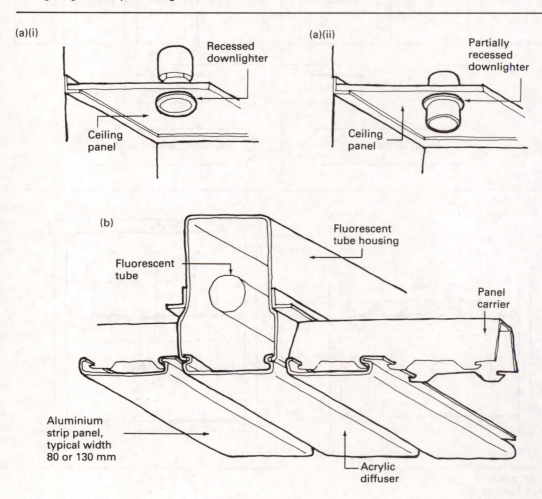

(a)(i)

Recessed
downlighter

Ceiling
panel

(a)(ii)

Partially
recessed
downlighter

Ceiling
panel

(b)

Fluorescent
tube

Fluorescent
tube housing

Panel
carrier

Aluminium
strip panel,
typical width
80 or 130 mm

Acrylic
diffuser

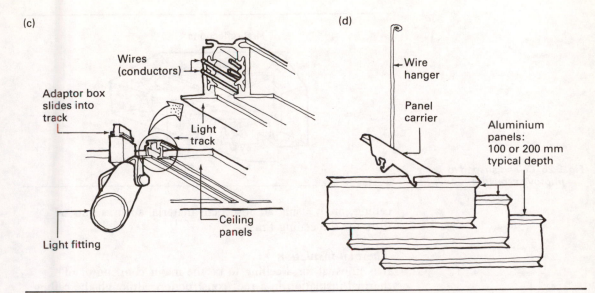

(c)

Wires (conductors)

Adaptor box slides into track

Light track

Light fitting

Ceiling panels

(d)

Wire hanger

Panel carrier

Aluminium panels: 100 or 200 mm typical depth

(a) Installing downlighters, where the bottom edge of light fitting is either flush with the ceiling finish or projecting below the ceiling finish (Fig. 5.24(a)).
(b) Incorporating 'diffusers' (semi-transparent panels for diffusing artificial light) within the ceiling finish (Fig. 5.24(b)).
(c) Incorporating light track between ceiling panels for use with suitable lamps or Figs. 5.24(c) or (d) forming a completely illuminated ceiling by installing lighting above a ceiling finish of either:
  • semi-transparent material;
  • a series of vertical hanging strips or plates (Fig. 5.24(d)).

Providing acoustic properties

In large rooms such as auditoria, lecture rooms, etc., ceilings are an important element in the acoustic characteristics of the room. In such rooms it is desirable that the audience should be able to hear by 'direct' sound (sound which reaches the ear directly rather than by reflection from surfaces). However, the first reflection of sounds can augment direct sound; the ceiling might be shaped to cause suitable first reflections (Fig. 5.25).

If the distance between a reflective surface and a listener is too great, the reflected sound might be heard as an echo, which could result in unintelligible sound. A high ceiling could cause such an effect. The problem could be reduced by either reducing the ceiling height by use of a suspended

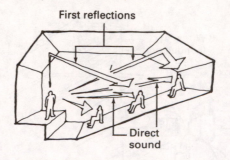

First reflections

Direct sound

Fig. 5.25 *Ceiling shaped to improve acoustics*

ceiling, or by using an absorbent material such as mineral fibre for the ceiling finish.

Thermal insulation

It is unusual for a ceiling to be the major contributor of thermal insulation in a roof construction, although the ceiling might improve thermal insulation by either use of insulative materials such as:

- foil-backed plasterboard which reflects heat back into the room;
- plaster using expanded mineral aggregates (exfoliated vermiculite or expanded perlite);
- the inclusion of materials such as expanded polystyrene in the construction of the ceiling finish by forming a cavity between the underside of the roof and the ceiling. To be effective, the cavity should be between 20 and 230 mm in depth.

Table 5.10 Maximum dimensions of cavities in non-residential buildings

Location of cavity	Class of surface exposed in cavity	Maximum dimension in any direction (m)
Between a roof and a ceiling	Any	20
Any other cavity	Class 0 or 1	20
	Other than Class 0 or 1	10

Source: From Building Regulations 1991, approved document B Table 14.

Safety in fire

Section 9 of approved document B of the Building Regulations 1991 concerns the spread of fire in 'concealed spaces'. Suspended ceilings are regarded by the regulations as such spaces and are dangerous in providing a route for the spread of smoke and flame which is concealed, unlike the more obvious causes of fire spread.

To hinder fire spread concealed spaces (or 'cavities') are required to be sealed at their edges by 'cavity barriers'. A cavity barrier consists of material bridging and sealing the cavity and is required to have a minimum half-hour fire resistance.

Large concealed spaces are required to be sub-divided by cavity barriers. Table 5.10 is adapted from Table 14 of approved doument B and indicates the maximum permissible dimensions for concealed spaces in non-residential buildings.

Cavity barriers are also required in a suspended ceiling above a wall to hinder fire spreading across the void above the wall although a compartment wall must be carried up to the underside of the floor.

Approved document B also prescribes surface spread of flame characteristics for ceilings, (see also 'Resistance to the spread of flame' p. 184 and Tables 5.5 and 5.6, both taken from approved document B). Table 5.5 lists examples of materials offering Class 0, 1 and 3 surface spread of flame, whilst Table 5.6 lists the classifications required by walls and ceilings for different types of building.

Direct ceilings

In most domestic buildings, or buildings with a low ceiling height, the ceiling is applied directly to the underside of the floor or roof structure (Fig. 5.23(a)).

Direct ceilings applied to timber joists will usually consist of sheets of plasterboard (typically of 1220 or 1370 mm × 406 mm wide) nailed to the underside of the joists. If the joists are spaced at 450 mm intervals, then plasterboard of 9.5 mm thickness will be used. For joist intervals of 600 mm, 12.7 mm thick plasterboard will be used.

Plaster may then be applied to the surface of the plasterboard in either one coat of 5 mm thickness (if the ceiling surface is flat and true) or in two coats of 10 mm total thickness for less true ceiling surfaces or for good-quality work.

Direct ceilings applied to concrete slabs will usually consist of plaster either of one coat 2–3 mm thick (if the slab surface is flat and true) or in two coats of 10 mm total thickness for less true slab surfaces or good-quality work. The nature of plaster is considered in 'Ceiling finishes' (Ch. 8, p. 420).

Suspended ceilings

Suspended ceilings consist of some form of framework suspended from the underside of the floor or roof structure by metal or timber hangers (see Fig. 5.23(b)). A ceiling finish is fixed or applied to the underside of the framework.

Suspended ceilings may be categorised as either: (a) jointless; (b) frame and panel; (c) strip; (d) open (see Table 5.9).

Jointless systems tend to be of *in situ* construction, whereas framed and open systems tend to be prefabricated in a factory, then delivered and rapidly assembled on the site.

Jointless ceilings will be in appearance similar to direct ceilings, whilst framed and open ceilings will expose either the ceiling framework, or the joints between panels or strips.

Where the ceiling is used to conceal services, it is important that suitable access is provided for inspection and maintenance of cables, pipes, ducts, etc. This represents little

Fig. 5.26 *Jointless ceilings*
(a) *Plasterboard on suspended metal frame*
(b) *Access to suspended plasterboard ceiling*
(c) *Expanded metal lathing*
(d) *Suspended plaster ceiling*

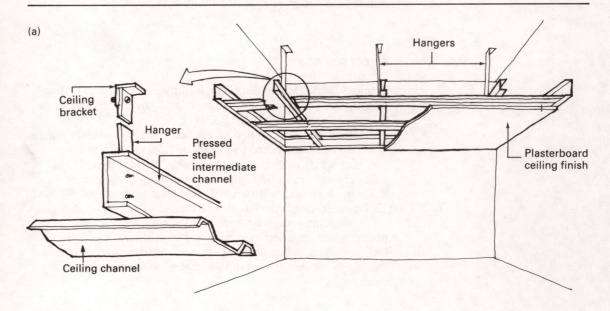

(a)

Hangers

Ceiling bracket

Hanger

Pressed steel intermediate channel

Plasterboard ceiling finish

Ceiling channel

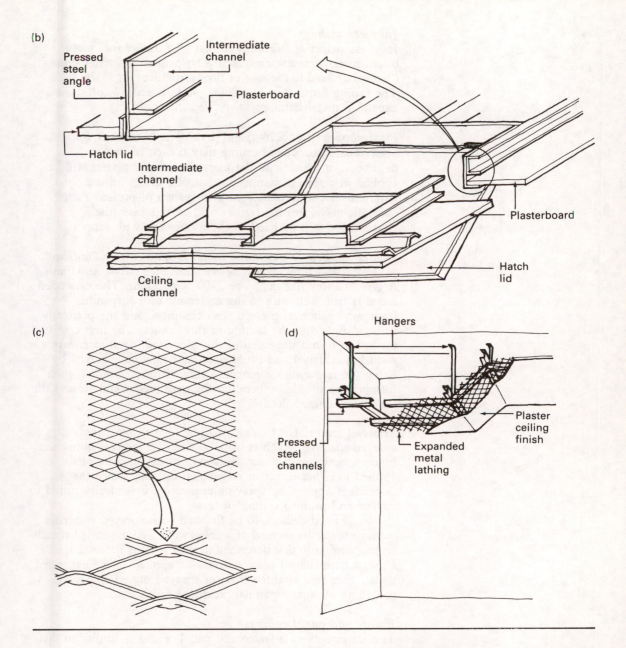

(b)

Pressed steel angle

Intermediate channel

Plasterboard

Hatch lid

Intermediate channel

Ceiling channel

Plasterboard

Hatch lid

(c)

(d)

Hangers

Pressed steel channels

Expanded metal lathing

Plaster ceiling finish

problem with framed or open ceilings, since with open ceilings there is no surface between the ceiling and services, and with framed ceilings the panels or strip can usually be easily removed. With jointless ceilings, however, it will be necessary to provide strategically placed hatches (Fig. 5.26(b)).

Jointless ceilings

Jointless suspended ceilings consist of a timber (or, more often, metal) framework which is suspended by metal supports ('hangers') fixed to the roof or floor structure (Fig. 5.26(a)). The ceiling finish may be of: (a) plasterboard; (b) plaster on expanded metal lathing; (c) sprayed materials.

Plasterboard (Figs 5.26(a) and (b))   The framework for a plasterboard suspended ceiling may consist of timber members, in which case the plasterboard and plaster will be applied in a similar manner to that of a direct ceiling. Alternatively, a prefabricated grid system of pressed metal channels may form the framework. In this case, the plasterboard will be fixed to the framework with screws.

Plaster applied to expanded metal lathing (Figs 5.26(c) and (d))   Expanded metal lathing is a sheet material of steel mesh. A typical sheet size might be 2400×675 mm. The expanded metal is tied with wire to the underside of a suspended framework (such as pressed steel channels) and the plaster is usually applied to the lathing in three coats. The first coat is pushed well into the mesh of the expanded metal to ensure a good key. Curved and profiled ceilings may be achieved by fixing the framework members at the required heights, then forming the profile by bending the expanded metal around the framework (Fig. 5.26(d)).

Sprayed materials   Ceilings may be formed by using a sprayed material which is usually based on mineral fibres and binders mixed with water. Sprayed finishes may be easily applied to either direct or suspended ceilings. Since the material is applied by spraying apparatus, it is ideally suited to curved and profiled ceiling surfaces.

Suspended ceilings to be finished with sprayed materials would usually be formed of a framework and expanded metal lathing similar to that described in the previous subsection. First, a base coat of plaster would be applied to the expanded metal, then two or three coats of sprayed material applied, providing a coarse, granular, textured finish.

Frame and panel ceilings

The framework of a frame and panel ceiling is similar to that of a metal jointless ceiling framework, consisting of a grid of steel or aluminium members suspended from the underside of the roof or ceiling by metal hangers. However, whereas jointless ceilings are formed by large sheets of plasterboard or expanded metal lathing covered with plaster, panels are small self-finished units. Typical panel sizes are 300 × 300 mm, 600 × 600 mm, 1200 × 600 mm, etc.

The framework of a frame and panel ceiling may be either exposed or concealed.

Exposed framework   Figure 5.27(a) illustrates a typical exposed framework ceiling, the components of which are:

(*a*) A grid of aluminium inverted 'T' sections forming the framework. The size of the grid will correspond with the size of panel to be used. The framework is suspended by hangers of 2 mm diameter steel wire which at the lower end is tied to the aluminium 'T' sections, and at the upper end is tied to a bracket secured to the structural ceiling.

(*b*) Standard-sized panels rest in the flanges of the framework and are secured by steel spring clips.

Fig. 5.27 *Frame and panel ceiling*
(a) *Exposed framework ceiling*
(b) *Concealed framework ceiling*

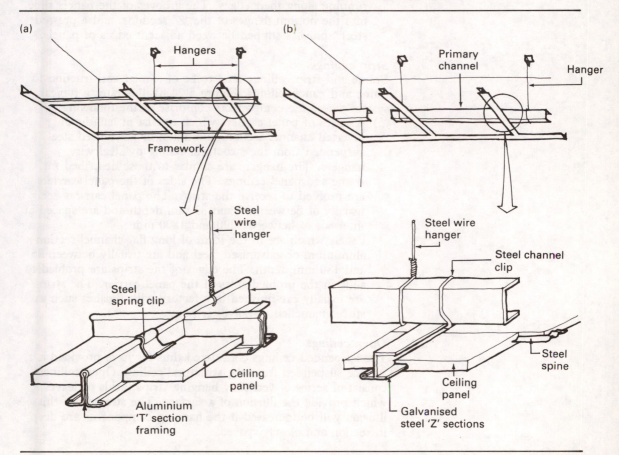

(a)

Hangers

Framework

Steel wire hanger

Steel spring clip

Ceiling panel

Aluminium 'T' section framing

(b)

Primary channel

Hanger

Steel wire hanger

Steel channel clip

Ceiling panel

Steel spine

Galvanised steel 'Z' sections

There are many different types of panel available including smooth or textured mineral fibre, smooth, textured or profiled glass-reinforced plastic and acrylic (a semi-transparent plastic material used for illuminated ceilings, where fluorescent tubes are installed above the framework). Figure 5.27(b) illustrates a concealed framework ceiling where the components comprise of the following:

(a) A framework of galvanised steel primary channels spaced at intervals of 1200 mm, and suspended by steel wire in a similar manner to an exposed framework ceiling.

(b) Galvanised steel 'Z' sections, fixed at intervals which correspond to the width of the panels (e.g. 300 mm, 600 mm). The 'Z' sections are fixed to the channels by means of steel wire channel clips.

(c) Panels (which may be of similar size and material as those used for exposed framework ceilings) which have grooves running along their edges. The grooves of the panels slide into the bottom flanges of the 'Z' sections, and a pressed steel 'spine' is slipped between adjacent edges of panels.

Strip ceilings

Frame and strip ceilings tend to be of similar construction to frame and panel ceilings. Figure 5.24(b) illustrates a typical system where the components comprise of the following:

(a) A series of panel carriers which consist of lengths of inverted channel section aluminium or galvanised steel suspended from the structural ceiling by steel wire hangers. The hangers are similar to those described for frame and panel ceilings. The sides of the panel carriers are profiled to receive the strips. The panel carriers are usually of between 25 and 33 mm depth and are spaced at intervals of between 1200 and 1800 mm.

(b) Panels which are in the form of long flat channel-section aluminium or galvanised steel and are usually between 85 and 180 mm depth. The edges of the strips are profiled to slide in the profiled sides of the panel carrier. The strips are usually pre-finished in a factory with finishes such as stove enamelled or polyester paint.

Open ceilings

The suspended ceilings considered thus far have provided a surface suspended from the structural ceiling. Open ceilings consist of series of vertically hanging strips, grids or leaves which provide the illusion of a surface. The strength of this illusion will be increased if the hanging components are deep in section and closely spaced.

The open ceiling illustrated in Fig. 5.24(d) closely resembles the strip ceiling described in the previous subsection. A series of profiled channel-section panel carriers typically of 90 mm depth are suspended from the structural ceiling by wire hangers. The strips, which may typically be of 100 or 200 mm depth, are profiled to slot into the profiled sides of the panel carriers.

Hanging grids may be constructed of either aluminium interconnecting strips (which may be of similar profile to the hanging strips illustrated in Fig. 5.24(d), or of thicker strips such as box-section metal or extruded chipboard (Fig. 5.28). The grid is hung from the structural ceiling by metal hangers.

Suspended leaves consist of:
- primary channels, spaced at intervals of typically 900 mm hung from the structural ceiling by steel hangers;
- small-section steel channels fixed at right angles underneath the primary channels; typical spacing of these channels might be 100 mm;
- suspended from the small channels are closely spaced leaves of either glass or polished steel either 150 × 75 mm or 200 × 75 mm in size.

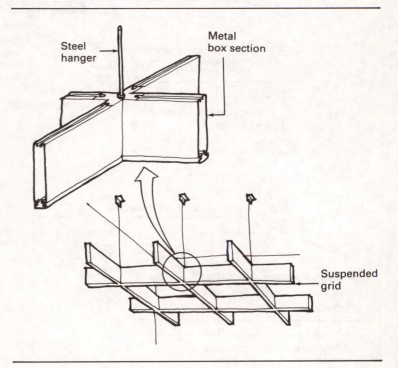

Fig. 5.28 *Open ceiling*

237

Raised floors

Raised or 'cavity' floors are useful for rooms such as laboratories or computer rooms where service cables, pipes and ducts might be housed beneath the floor finish. There are many prefabricated raised floor systems available which enable a floor finish to be placed at heights from 50 mm or less, to 500 mm or more above the sub-floor. The floor finishes for such systems are often in the form of panels which may be easily withdrawn for inspection, maintenance or rerouteing of services. Figure 5.29(a) illustrates a low-height, raised-floor system suitable for concealing cables or small-diameter pipework. The floor surface consists of 1200 × 1200 mm tongued and grooved chipboard panels which are laid on

Fig. 5.29 *Raised floors*
(a) *System using corrugated sheeting*
(b) *System using steel stools*

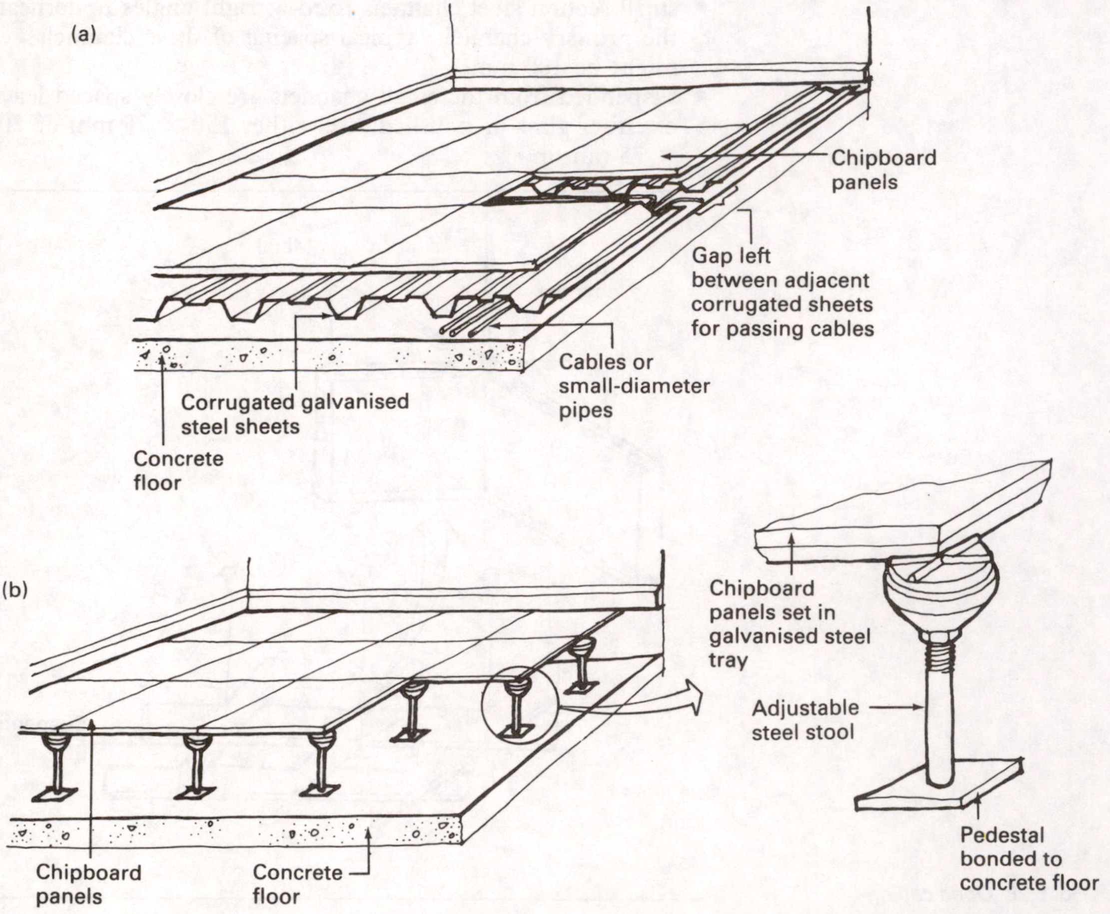

(a)

Chipboard panels

Gap left between adjacent corrugated sheets for passing cables

Cables or small-diameter pipes

Corrugated galvanised steel sheets

Concrete floor

(b)

Chipboard panels set in galvanised steel tray

Adjustable steel stool

Chipboard panels

Concrete floor

Pedestal bonded to concrete floor

galvanised steel corrugated sheets fixed to the sub-floor. Cables and pipes may be routed between the corrugations of the sheeting. Where services run at right angles to the corrugations, a gap of sufficient width is left between adjacent corrugated sheets. Chipboard panels of suitable width bridge these sheets and are screwed to them to enable easy withdrawal for access.

Figure 5.29(b) illustrates a typical raised-floor system used for greater cavity depths, where larger diameter pipes or services such as ventilation ducts might be concealed. The system comprises of:

- 600 × 600 × 38 mm deep chipboard floor panels each set in a galvanised steel tray;
- an adjustable steel stool located beneath the intersection of each panel which is capable of providing a cavity of up to 500 mm depth, the pedestal of the stool being bonded to the sub-floor with epoxy adhesive. The stool head, which is fitted with a p.v.c. bearing pad, may be adjusted for level by turning a levelling nut.

Sprung floors

Sprung floors are used for rooms such as gymnasia or dance-halls where a degree of resilience is required in the floor surface in order to reduce fatigue in the feet of the people using the floor.

Sophisticated fully sprung floor systems employ steel 'I' beams mounted on sets of coil springs. The floor finish is usually fixed to counter battens and joists which rest on the steel 'I' beams. A locking mechanism wedges the springs rigid for occasions when the floor is not being used for dancing or sports activities.

Figure 5.30 illustrates a much simpler semi-sprung floor. The floor finish, which might typically consist of hardwood strips, is fixed to 50 × 50 mm softwood battens. Rubber buffers, which are inset into the battens at 600 mm intervals, rest on the sub-floor.

Fig. 5.30 *Semi-sprung floor*

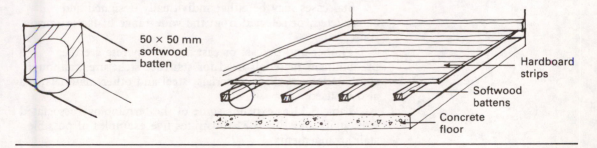

50 × 50 mm softwood batten

Hardboard strips

Softwood battens

Concrete floor

Table 5.11(a)   Stair terminology

1. Flight	A series of steps between different levels or storeys of a building
2. Landing	An area of floor at the top of, or between, flights of stairs. Functions of landings include: (a) providing the facility for changing direction in the stairs; (b) to provide a resting place between long flights of stairs
3. Half landing	A half landing is rectangular in plan shape, the length of which equals twice the width of the stairs (Fig. 5.31(b))
4. Quarter landing	A quarter landing is square in plan shape, the dimensions of which equal the width of the stair (Fig. 5.31(d))
5. Tread	The horizontal surface of a step
6. Riser	The vertical surface of a step
7. Nosing	The portion of a tread which overhangs the riser
8. Going	The horizontal distance between two consecutive nosings
9. Rise	The vertical distance between two consecutive treads
10. Total rise	The vertical distance between the floor levels at the top and bottom of a flight of stairs

STAIRS

Introduction

A stair is a system of steps by which people and objects may pass from one level or storey of a building to another.

Staircases may be either individually designed and constructed, or selected from the wide range of factory-made staircases.

Timber and precast or cast *in situ* concrete are the predominant materials used for staircase structures, although combinations of these materials, steel and other materials, may also be used.

Table 5.11(a) explains some of the terminology associated with stairs and Fig. 5.31 illustrates five examples of possible stair arrangements.

Table 11(a) continued

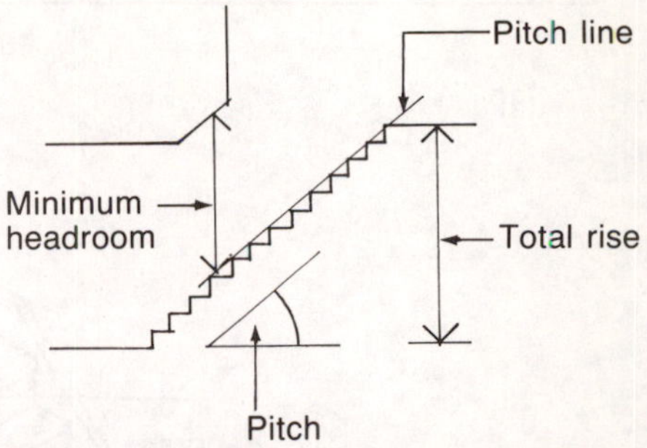

11. Pitch line	An imaginary line linking the nosings of a flight of stairs	
12. Pitch	The angle between the pitch line and the horizontal	
13. Headroom	The vertical distance between the pitch line and any surface (such as a celing or beam) above it	
14. String	A structural member at the side of a stair which provides support for the treads (Fig. 5.31(a))	
15. Wall string	A string which is fixed to the wall	
16. Outer string	A string which is independent of a wall	
17. Handrail	A rail fixed parallel above the pitch line at the sides of the stair (Fig. 5.31(a))	
18. Balusters	Vertical members which support a handrail	
19. Newel	A vertical post which might provide support for either (a) the handrail, at the top and bottom of the stairs (Fig. 5.31(a)), or (b) support for the upper end of an outer string	

Diagram labels: Pitch line, Minimum headroom, Total rise, Pitch

Straight flight (Fig. 5.31(a))
The simplest form of stair arrangement which consists of one straight flight of stairs linking two levels or storeys of a building.

Dog leg (Fig. 5.31(b))
An arrangement of two short flights of stairs with a half-space landing between them. The lower flight rises to the landing and the upper flight rises from the landing to the upper floor level. With a dog-leg stair, the outer strings of the two flights: (a) lie in the same vertical plane; (b) meet together at the landing.

Open well with two flights (Fig. 5.31(c))
An open-well stair is similar to a dog leg, only the half-space landing is longer, which causes a space between the lower and upper flights. This space is known as a 'well'.

241

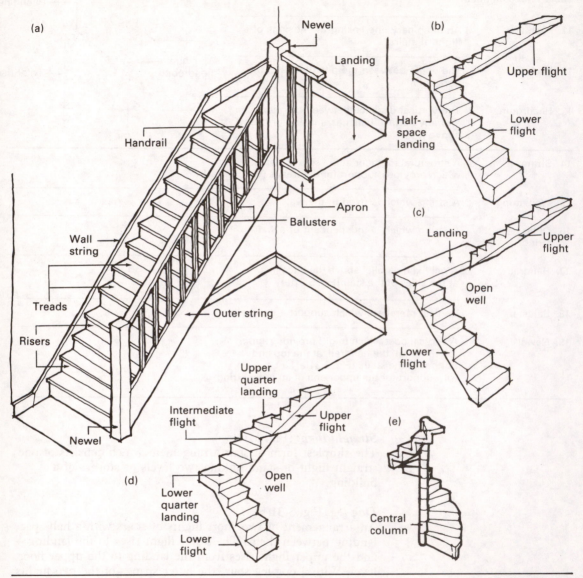

(a)

Newel

Landing

(b)

Upper flight

Handrail

Half-space landing

Lower flight

Wall string

Treads

Outer string

Risers

Apron

Balusters

(c)

Landing

Upper flight

Open well

Lower flight

Newel

Upper quarter landing

Intermediate flight

Upper flight

(e)

Open well

(d)

Lower quarter landing

Lower flight

Central column

Fig. 5.31 *Different stair arrangements*
(a) *Straight flight*
(b) *Dog leg*
(c) *Open well with two short flights*
(d) *Open well with three short flights*
(e) *Spiral*

Open well with three flights (Fig. 5.31(d))
In this stair, the well is of sufficient size to allow an intermediate flight of steps. The lower flight rises to a quarter landing, the intermediate flight rises from the quarter-space landing to a higher quarter-space landing and the upper flight rises from the higher landing to the upper floor level.

Spiral stair (Fig. 5.31(e))

The treads of a spiral stair turn and rise around a central column. Important factors to consider in the choice of a stair arrangement include:

(*a*) *The size and shape of the room*: straight flights require at most only half the width of other stair arrangements, but require at least twice the length.

(*b*) *The use of space around the stair*: a spiral staircase located in a restricted square area will result in the creation of spaces around and below the stair which could be awkward to use and difficult to clean. The space below stairs other than spiral stairs can be used for such purposes as storage, seating and circulation areas.

(*c*) *Use of stairs*: a straight flight provides the easiest means of passage for people, or for carrying bulky objects up and down the stairs. Manoeuvring arrangements (particularly on spiral stairs) can be difficult.

(*d*) *Economy*: although it is difficult to generalise, the simple construction and installation of a straight flight is likely to make it cheaper, in most instances, than other stair arrangements.

Functional requirements

Some of the major functional requirements of staircases are: (*a*) stability; (*b*) protection from fire; (*c*) suitable dimensions; (*d*) appearance.

Stability

Staircases are required to carry the loads of people and objects without excessive deflection either: (*a*) throughout the length of the staircase; (*b*) within the width of the tread.

Table 5.11(b) illustrates some of the bearing conditions that might apply to a straight-flight stair with the possible means of support. The means of support might be achieved by wall strings, stair strings, built-in stairs or by cantilevering the stairs. Similar conditions and means of support might also apply to dog-leg and open-well stairs, although a further means of support might include the use of newels to support the stair strings at junctions with landings (Fig. 5.32(a)). Spiral stairs are frequently supported by being cantilevered from the central column (Fig. 5.31(e)). Alternative methods include casting an *in situ* r.f.c. helical slab where the reinforcing bars of the stair link with the floor reinforcement at the top and bottom of the stair (Fig. 5.32(b)).

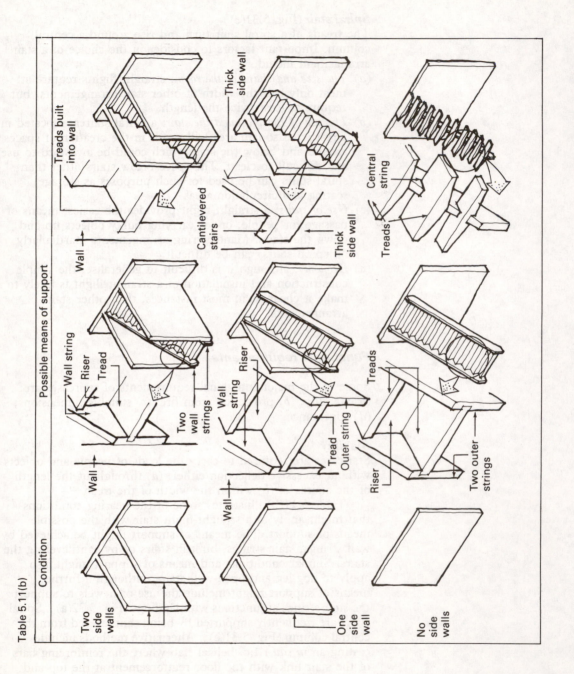

Table 5.11(b)

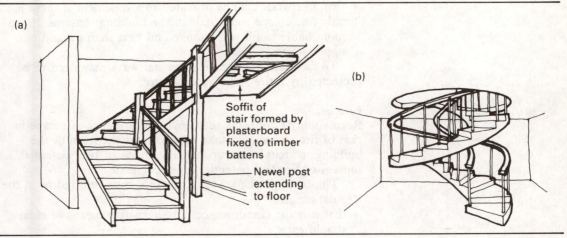

(a)

Soffit of stair formed by plasterboard fixed to timber battens

Newel post extending to floor

(b)

Fig. 5.32 *Means of support for stairs*
(a) *Stair with newels*
(b) *Helical stair*

Fig. 5.33 *Guide for measurement of tapered stairs (see Table 5.11(d))*

Protection from fire

Approved document B of the Building Regulations 1991 requires fire escape protected stairways and landings to be constructed with 'materials of limited combustibility'. Combustible materials, however, are permitted as a floor finish for landings and stairs since they are not significant in the early stages of fire spread.

Buildings requiring such stairway construction are:

- dwelling houses, flats and maisonettes other than those of less than three storeys;
- buildings other than dwellings.

Requirements for protected stairs include:

- that they are situated within a fire-resisting enclosure;
- that they are kept free of potential sources of fire (eg limitations might be made on the provision of lifts, reception desks and cupboards within the stairway);

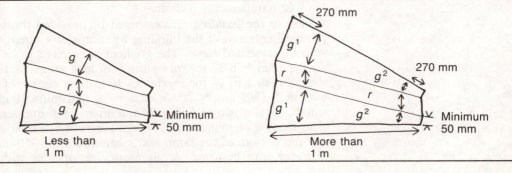

Less than 1 m

g

r

g

Minimum 50 mm

270 mm

g^1

r

g^2

270 mm

g^1

r

g^2

Minimum 50 mm

More than 1 m

- that a separate stair is provided to a basement if there is only one escape route/stair in the building: basements are more liable to fill with smoke and heat than ground or upper stories.

To facilitate fire escape some stairways may need to be protected by a fire-resisting enclosure.

Fire escape

Regulation B1 concerns the provision of means of escape in case of fire and should take into account the form of the building, activity of occupants, likelihood of fire, potential sources of fire and potential for fire spread.

The basic principles of means of escape referred to in the Regulations are:

- that in most circumstances an alternative means of escape should exist;
- to provide direct escape to a place of safety, or a place of 'relative safety' such as a protected stairway leading to an exit.

Regulation B involves:

- planning of escape routes: distance of travel, number and widths of routes;
- stairways and exits: number, siting and width of stairways/exits, enclosure and access lobbies to protected stairways and external escape stairways.

Suitability of sizes

The dimensions adopted for stairways affects their ease and safety in use. Approved document M of the Building Regulations 1991 concerns access for disabled people whilst document K concerns stairways, ramps and guards.

Access for disabled people

Approved document M requires the provision of access for disabled people who visit or work in buildings to which public are admitted.

The requirements include:

Access to the building Access must be provided through the principal entrance of the building by means of a ramp, with or without associated steps. The gradient is required to be not steeper than 1 in 12 for maximum 5 m long ramps and not steeper than 1 in 15 for maximum 10 m long ramps, of minimum 1200 mm width in sections of maximum 10 m length, separated by landings of minimum 1500 mm length. A clear landing of minimum 1200 mm length is required at the top and bottom of the ramp and a handrail of 45–50 mm diameter must be fitted. A minimum clear opening width of

Table 5.11(c)   Requirements for stairways (adapted from Building Regulations 1991, approved document K)

Relevant clause number	Summary of clause
Section 1: Stairways	

Rise and going

1.1	In a flight of stairs the steps should have the same rise and they should have the same going
1.4, 1.5	Twice the rise and one going (2R + G) should be between 550 and 700 mm

	Max. rise (mm)	Min. going (mm)
(1) Private stair	220	220
(2) Institutional and assembly stair	180	280
(3) Other stair	190	250

Steepness of stairs

1.4	Maximum pitch for a private stair = 42°

Construction of steps

1.8	Open riser treads (see Fig. 5.36) should overlap by min. 16 mm
1.9	Any stair with open risers and likely to be used by children under 5, should be constructed so that a 100 mm diameter sphere cannot pass between

Headroom

1.10	Minimum 2 m

Width of flights

1.11	No recommendations given, however, the following Approved Documents should be checked regarding widths of flights: ● B Fire Safety: for stairs forming part of a means of escape; ● M Access and facilities for disabled people: for stairs providing such access
1.12	Stairs in public buildings and which are more than 1800 mm wide should be divided into flights of less than 1800 mm width

Length of flights

1.13	Shop or assembly building: maximum 16 risers in a flight
	Stairways of more than 36 risers in consecutive flights to have at least one change in direction of 30° or more involving a landing. Landing length of not less than width of stair

Table 5.11(c) continued

Relevant clause number	Summary of clause
Section 1: Stairways	
Landings	
1.15	Landings should be provided at the top and bottom of every flight. Width and depth dimensions should be at least that of the stair width
1.16	For safety, landings should be kept clear of any permanent obstruction. A door may be positioned to swing close to the bottom of a staircase provided it leaves a space of at least 400 mm × width of stair
Tapered treads (refer to Fig. 5.33)	
1.18	(a) For stairways less than 1 m width the going should be measured in the centre of the tread
	(b) For stairs over 1 m width the going should be measured 270 mm from each side
	The rise and going should comply with 1.1, 1.4 and 1.5 (above)
	The tread should measure at least 50 mm at its narrowest end
1.19	Consecutive steps should have a uniform going
Handrails for stairs	
1.27	Handrails should be provided on at least one side of staircases less than 1 m wide and on both sides for wider staircases
	Handrail height: between 900 mm and 1 m above pitch line
Guarding of stairs	
1.28	Stairs and landings should be guarded at the sides • in dwellings where a drop of more than 600 mm occurs; • in other buildings where there are two or more risers
1.29	Apart from stairs unlikely to be used by children under 5 (eg in accommodation for old people) the stairs should be designed such that: • a 100 mm diameter sphere could not pass pass through any opening in the guarding; • children could not easily climb it
1.30	The guarding should be capable of resisting horizontal forces, and be of the following height:

Table 5.11(c) continued

Relevant clause number	Summary of clause

Section 1: Stairways

Building	Location	Horizontal force (kN/m)	Height (mm)
Single family dwellings	Stairs, landings, ramps, external balconies	0.36 0.74	900 1100
Factories and warehouses	Stairs, ramps, landings	0.36 0.36	900 1100
Residential, institutional, educational, office, and public buildings	All locations	0.74	900 (flight) 1100 elsewhere
Assembly	Within 530 mm in front of fixed seating	1.5	800
	All other locations	3.0	900 (flights) 1100 elsewhere
Retail	All locations	1.5	900 (flights) 1100 elsewhere

Section 2: Ramps

Steepness 2.1	Maximum 1 in 12
Headroom 2.2	Minimum 2 m
Width 2.3	No recommendations but check Approved Documents B and M (see 1.11, above)
Handrails 2.5	As for stairs, see 1.27, (above) Handrails not required if rise is less than 600 mm
Landings, guarding 2.6, 2.7	Landings should be provided (see 1.15, 1.16 above) Ramps and landings should be guarded (see 1.28–1.30, above)

Section 3: Guards and Barriers: Pedestrian guarding

Siting 3.1	Guarding should be provided to provide safety at the edges of floors, galleries, balconies, basements or sunken areas and vehicle parks to which people have access (unless only for maintenance and repair)

Table 5.11(c) continued

Relevant clause number	Summary of clause
Section 3: Guards and Barriers: Pedestrian guarding	
Design	
3.2	Walls, ballustrades and other obstructions may serve as guardings. They should be capable of resisting horizontal forces, and be of the height indicated in 1.30 (above)
3.3	Where the building is likely to be used by children under 5, the guarding should comply with the limitations indicated in 1.29 (above)

800 mm must be provided. (See 'Access' p. 70 and Figs. 2.25(a)(i) and (ii).)

Inside the building A minimum clear width of 750 mm for internal doors and certain minimum dimensions for internal lobbies are required (e.g. 1500 mm wide × 1700 mm or 1200 mm × 2000 mm). There are prescribed minimum sizes for lifts (e.g. 1100 mm wide × 1400 mm) and a clear door opening of minimum 800 mm must be provided. Internal ramps must comply with the requirements outlined in 'Access to the building' (above).

Building Regulations
Approved document K requires staircases and ramps to afford safe passage for the users of the building (K1).

Table 5.11(c) indicates some of the dimensional and other requirements in the regulations. Clause numbers adopted in the table refer to those used in document K. The table also indicates some of the dimensional and other requirements relating to the provision of guards and barriers for raised floors, balconies, etc., (K).

The Regulations apply to changes in level within the following buildings:
- dwellings: where the difference in level is 600 mm or more;
- other buildings: where the difference in level represents more than two risers, or 380 mm if stairs are not involved.

Appearance
Stairs are a complex item of building structure and very careful attention is essential to such details as:
- junctions between skirting board and wall strings;
- profiles and junctions of handrails, strings, newels and balusters;

- the detailing of finishes for treads, risers and nosings (particularly with concrete stairs).

Timber stairs

Timber stairs may be individually designed and purpose made, although it is often cheaper to use a joinery manufacturer's staircase where straight flights may be available as ready-made products, or special stairs such as dog legs may be made to order using standardised components and manufacturing techniques.

The basic components of most timber stairs consist of: (*a*) treads and risers supported by strings; (*b*) newels, landings, handrails and balusters.

Treads, risers and strings

Cut string (Fig. 5.34(a))   The upper surface of a cut string is profiled to the shape of the treads and risers. The treads rest on the horizontal surfaces of the string, and the risers are fixed to the vertical surfaces. Additional support for the treads is provided by timber 'carriages' which run beneath the entire length of the stair. One carriage is fixed against each string at the sides of the stair, and one is placed centrally. Timber brackets are fixed from the central carriage to the underside of each tread. Triangular section blocks fixed between the strings and treads and between the risers and treads provide rigidity to the steps (Fig. 5.34(c)). The nosings of the treads project beyond the vertical faces of the string and lengths of moulding of similar profile to the nosings are fixed against the end of the treads (Fig. 5.34(b)). Where a stair is fixed against one wall, a 'close' wall string (described below) is usually used. This is securely fixed to the wall.

Typical timber thicknesses for cut-string stairs are: cut outer string 46 mm; wall string 34 mm; riser 20 mm; tread 25 mm.

Because cut-string stair construction does not lend itself to mass-production techniques, it is used rarely today.

Close string (Fig. 5.35)   A close string is a deep section length of timber with a series of horizontal and vertical grooves or 'housings' cut into it which receive the treads and risers. The inside edges of the housings are tapered so that timber wedges may be driven against the inside of the treads and risers, in order to push them firmly against the outer edge of the housings (Fig. 5.35).

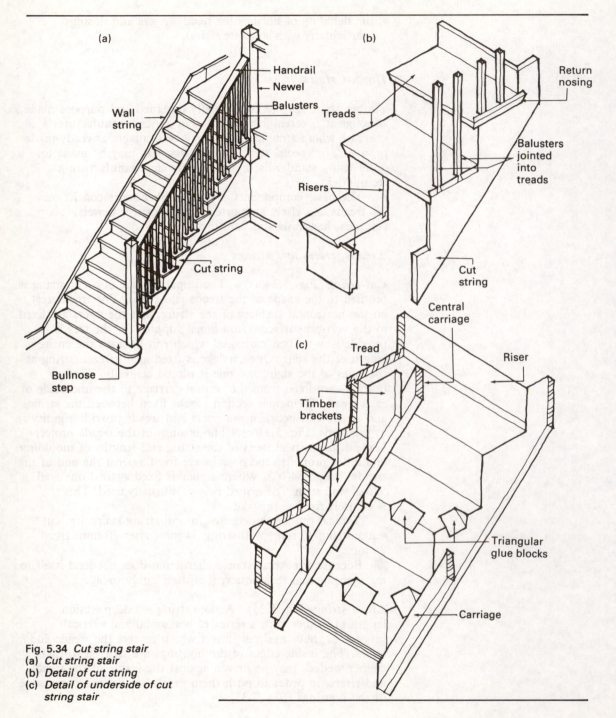

Fig. 5.34 *Cut string stair*
(a) *Cut string stair*
(b) *Detail of cut string*
(c) *Detail of underside of cut string stair*

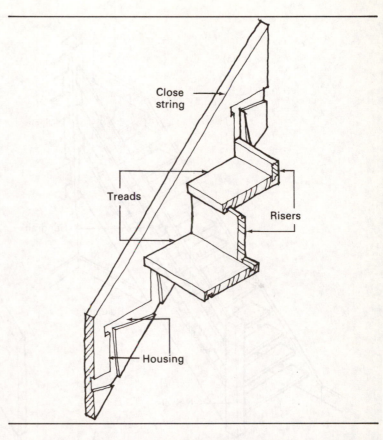

Close
string

Treads

Risers

Housing

Fig. 5.35 *Detail of close string*

If the stair is over 900 mm in width, a central carriage and brackets will be fixed beneath the length of the stair to provide additional support for the stair. Triangular-section blocks are fixed between the strings and treads and between the risers and treads to provide rigidity to the steps.

The treads are grooved below the front edge and above the back edge to receive the tongued top and bottom edges of the risers.

Wall strings are fixed securely to the wall. Typical timber thicknesses for close-string stairs are: close outer string 28 mm; close wall string 24 mm; riser 20 mm; tread 25 mm.

Close-string stairs are suitable for mass-production techniques, and consequently are commonly used for domestic buildings today. The underside or 'soffit' of a stair is usually covered in to conceal the blocks, carriages or wedges. This may be achieved by fixing timber nogging pieces betweeen the strings and nailing plasterboard to the noggings (see Fig. 5.32(a)).

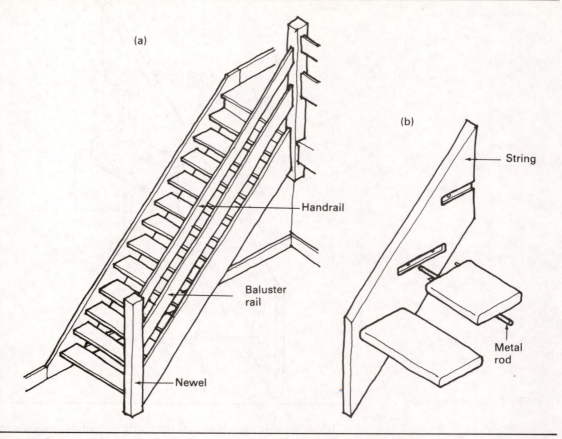

Fig. 5.36 *Open-riser stair*
(a) *Open-riser stair*
(b) *Detail of open risers and string*

Open-riser (Fig. 5.36(a)) In the construction of open-riser stairs, the risers are omitted. This results in a less bulky looking structure which may be an advantage in such instances as a small room where other forms of stair could visually reduce the volume of the room.

Since there is no riser to give support along the front and back edges of the tread, the thickness of tread is greater than that of other stairs, 34–40 mm being typical.

If a close string is used, the junction between the housing and tread might be insufficient to ensure that the strings are adequately tied together. Figure 5.36(b) shows a metal rod which is fixed below every fourth tread to provide additional restraint between the strings. The thickness of the string is likely to be 34–40 mm.

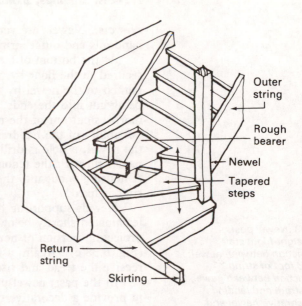

Outer string

Rough bearer

Newel

Tapered steps

Return string

Skirting

Fig. 5.37 *Tapered steps*

Tapered steps (Fig. 5.37)   Tapered steps may be used to save the amount of space taken up by other stair arrangements. In Fig. 5.37, the bottom of the stair terminates at the corner of two walls. If a straight flight were used in this instance, it would terminate at the floor level vertically below the back of the third tapered tread.

Since tapered steps are less easy to negotiate than parallel steps, they are usually located at the bottom of a stair, so that if an accident occurred on the stair, there would be less distance to fall.

In the past, stairs known as 'winders' have been used. These are steps which taper towards the centre-line of the newel, resulting in very small tread widths. Such stairs may not comply with the Building Regulation requirements for tapered steps (Table 5.11(c), 1.16).

Since the length of a tapered step is greater than the parallel steps, it may be necessary to provide additional support. This may be achieved by 'rough bearers' which are lengths of timber (typically 100 × 50 mm) spanning between the newel and the strings. These support the back of the tapered treads and the risers.

Newels, landings, handrails and balusters (Fig. 5.38(a))

Newels   Newels are vertical posts which provide support for handrails and outer strings.

At the bottom of a stair, the newel should be firmly secured to the floor by bolts. The string and handrail are jointed to the newel by mortise and tenon joints, the tenons being cut into the ends of the handrail and string, and the mortises cut out of the newel (Fig. 5.38(b)). To deter the possibility of the handrail and string being pulled away from the newel, a hole is drilled through the newel post which passes through the tenons, and a hardwood dowel is inserted. A housing is cut into the newel into which the lower tread and riser is fixed.

Newels supporting the top end of a string are usually terminated shortly below the level of the string. The projecting portion of newel is known as the 'newel drop' (Fig. 5.38(c)). Again, a housing is cut into the newel to receive the tread and riser.

In the past, newel posts were frequently turned on a lathe to provide a decorative treatment (Fig. 5.38(d)). Although turned newels are still available today, it is more common to use a simple 70 × 70 mm square-section newel.

Fig. 5.38 *Newel posts*
(a) *Straight-flight stair*
(b) *Junction between newel and foot of string*
(c) *Junction between newel, top of string and landing*
(d) *Decorative newel and balusters*
(e) *Quarter-space landing*

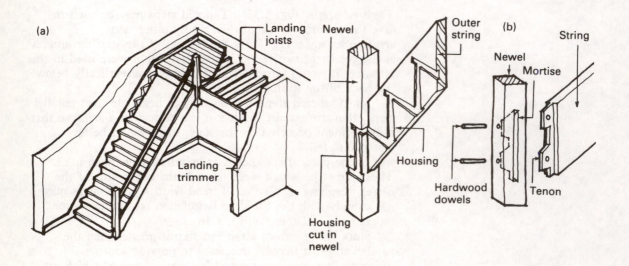

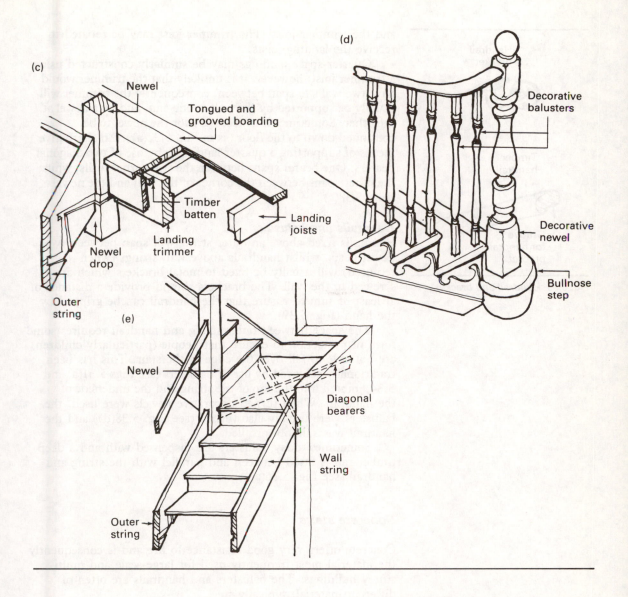

(c)

Newel

Tongued and
grooved boarding

Timber
batten

Landing
trimmer

Landing
joints

Newel
drop

Outer
string

(e)

Newel

Diagonal
bearers

Wall
string

Outer
string

(d)

Decorative
balusters

Decorative
newel

Bullnose
step

Landings Timber landings are, in effect, small suspended floors which require joists to support the floor finish (which usually consists of tongued and grooved timber boarding).

Figure 3.28(d) (Ch. 3) illustrates a system for forming an opening for a landing at the top of a stair.

Half-space landings are formed by a trimmer joist which is supported either by newel posts, or by being built into walls (Fig. 5.38(a)). A series of landing joists span between the wall

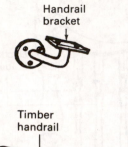

Fig. 5.39 *Handrail and bracket*

and the trimmer joist. The trimmer joist may be rebated to receive the landing joists.

Quarter-space landings may be similarly constructed using a trimmer joist; however, it is unlikely that the trimmer would have two walls to span between, consequently the trimmer will usually be supported by the wall at one end, and by a newel at the other. Sometimes it is necessary for the newel to be continued down to the floor (as in Fig. 5.32(a)). An alternative means of supporting a quarter landing is by use of two diagonal bearers. One bearer spans between the two corner walls, whilst the other spans between the corner of the wall and the newel (Fig. 5.38(c)).

Handrails and balusters

Handrails fixed above an outer string will span between two newel posts, whilst handrails above wall strings (when required) will usually be fixed to metal brackets which are screwed to the wall. The brackets should provide a distance of at least 35 mm to ensure that the handrail can be gripped by the hand (Fig. 5.39).

The space between outer string and handrail requires some form of protection to ensure that people (particularly children) are not able to fall over the edge of the stair. This has been traditionally achieved by use of balusters (see Figs 5.31(a) and 5.34) jointed into the top of the string and the underside of the handrail. When turned decorative newels were used, the balusters were usually also turned (see Fig. 5.38(d)) and the handrail was suitably profiled.

Sometimes today balusters are dispensed with and a deep timber rail is fitted between and parallel with the string and handrail (see Fig. 5.36(a)).

Concrete stairs

Concrete offers very good resistance to fire and is consequently the material most frequently used for large-scale and multi-storey buildings. The balusters and handrails are often of different material, typically steel.

Precast concrete stairs

Precast concrete may be used in stair construction either for: (*a*) treads; (*b*) steps; (*c*) complete flights of stairs.

Precast concrete treads Figure 5.40(a) illustrates a series of simple cantilever treads that would be suitable for forming a stair for a small-scale building. One end of each tread is built

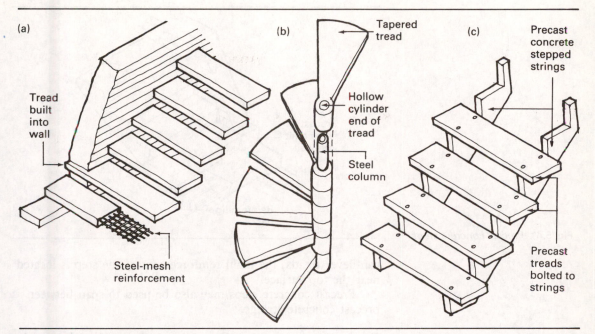

Fig. 5.40 *Precast concrete treads*
(a) *Cantilever treads*
(b) *Precast concrete spiral stair*
(c) *Precast concrete open stair*

Labels in figure:
(a)
Tread built into wall
Steel-mesh reinforcement

(b)
Tapered tread
Hollow cylinder end of tread
Steel column

(c)
Precast concrete stepped strings
Precast treads bolted to strings

into a wall which should be of sufficient thickness and weight to ensure that the treads remain firmly held in the wall. Since tension occurs in the top surface of a cantilever the reinforcement, which consists of steel mesh, is placed towards the top surface of the step. Typical dimensions of such a stair might be as follows: depth of tread 75 mm; width of stair 1.000 m, depth of tread built into the wall 200 mm.

Other examples of precast concrete tread stairs include:

(a) Spiral stairs (Fig. 5.40(b)) where each tread is tapered with a hollow cylinder formed at the narrow end. The cylinders, which are the same depth as the dimension of one riser, interlock and are fitted over a central steel column.

(b) Open-riser stairs (Fig. 5.40(c)) where precast concrete stepped strings span between the floors or landings and precast concrete treads are bolted between the horizontal surfaces of the strings.

Precast concrete steps   Figure 5.41 illustrates a system of precast concrete step units which are cantilevered from a wall. The section of each step is triangular which provides a suitable soffit for the application of plaster. The portion of step which is built into the wall is rectangular in section so that brickwork may be easily built around it. Like precast concrete

259

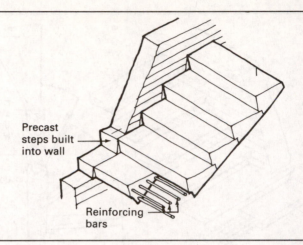

Fig. 5.41 *Precast concrete steps*

cantilever treads, the main reinforcement of the step is located near the top surface.

Precast concrete steps may also be used to span between precast concrete strings.

Precast concrete stairs   Complete concrete stair units may be precast, comprising of straight flights of stairs with landings at the head and foot of the stairs. The units may be either built into the walls or located in the framework of the building.

Cast in situ ***concrete stairs***   *In situ* concrete, being a fluid material when it is cast, is suitable for forming stairs of complex curved shapes such as the helical stair illustrated in Fig. 5.32(b). However, the complexity and quantity of

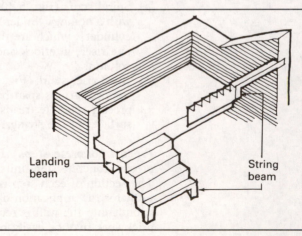

Fig. 5.42 *Cast* in situ *stair*

reinforcement and formwork necessary tends to make such stairs expensive. Figure 5.42 shows a typical example of a cast *in situ* concrete stair which incorporates cast *in situ* landings. Beams are formed into:

- the outer edges of the landings to form trimmers;
- the outer edges of the stairs to form strings.

The inner edges of the landings and stairs are built into the surrounding walls.

The landings and stairs are reinforced in a similar manner to the reinforcement of slabs, and the trimmers and strings are reinforced in a similar manner to r.f.c. beams.

Handrails, balusters and nosings for concrete (or other) stairs

Handrails and balusters   Balusters and handrails for concrete stairs may be individually designed and purpose made, or selected from the wide range of proprietary systems available. The systems described may be also used in conjunction with timber or metal stairs.

Figure 5.43 illustrates different baluster and handrail arrangements:

(*a*) The balustrades are of square- or round-section steel tube (typically of 15–18 mm width or diameter). Each baluster is fixed to the stair and provides both support for the handrail and a means of guarding people from falling between the handrail and stairs.

(*b*) Substantial main balusters of square- or round-section steel tube (25 mm or more in width or diameter) are fixed to the stair at wide intervals (perhaps in excess of 1 m). Two steel flat rails of 6 mm thickness are welded between the balusters, one at the top of the balusters, and one above

Fig. 5.43 *Handrails and balusters for concrete stairs*
(a) *Each baluster fixed to tread*
(b) *Main and intermediate balusters*

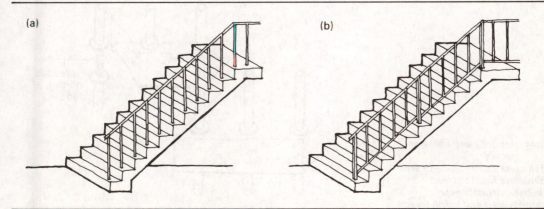

(a)   (b)

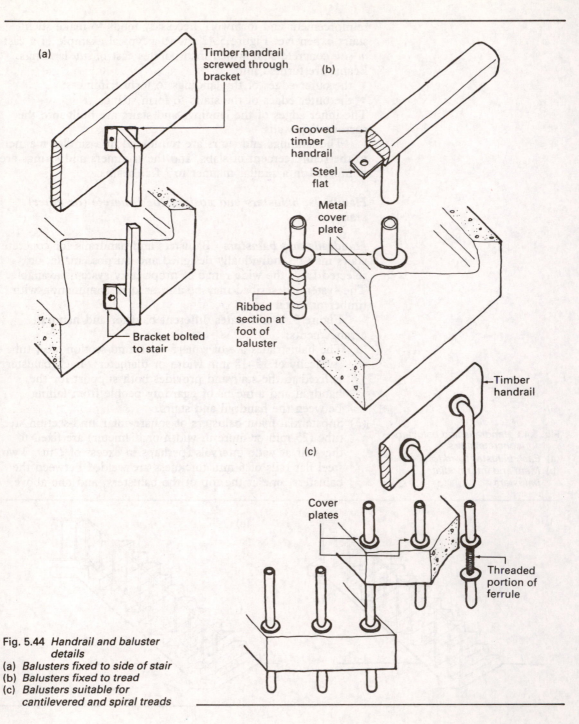

(a)

Timber handrail
screwed through
bracket

(b)

Grooved
timber
handrail

Steel flat

Metal
cover
plate

Ribbed
section at
foot of
baluster

Bracket bolted
to stair

Timber
handrail

(c)

Cover
plates

Threaded
portion of
ferrule

Fig. 5.44 *Handrail and baluster
details*
(a) *Balusters fixed to side of stair*
(b) *Balusters fixed to tread*
(c) *Balusters suitable for
cantilevered and spiral treads*

the level of the nosings. A series of smaller-section vertical steel rods are welded between the top and bottom rails to provide a guard for the stairs. Alternative means of guard provision include fixing panels such as acrylic or deep rails such as timber between the main balusters.

Figure 5.44 (a,b,c) illustrates three different means of fixing the balusters to the stairs:

(a) Steel brackets are formed at the top and bottom of each baluster. The balusters are bolted through the bottom brackets to the side of the stair using expandable bolts (see Ch. 6, p. 284). A timber handrail is fixed by being screwed through the top brackets.

(b) A length of 6 mm thick steel flat is welded to the tops of the balusters which acts as a core for a grooved timber handrail which sits over the steel flat and is fixed by being screwed through the underside of the steel flat. The foot of each baluster is ribbed and set in a hole in the concrete stair tread. The space between the edge of the hole and baluster is grouted (filled with fine aggregate concrete) which forms a key around the ribs. A metal cover plate conceals the junction between the baluster and the tread.

(c) This arrangement is suitable for cantilevered or spiral stairs where each baluster is threaded at its foot and screwed over a threaded metal ferrule which passes through a hole formed in the tread. A metal cover plate conceals the junction between the baluster and the tread. The baluster is angled at the top and welded to a steel bracket through which the handrail may be screwed.

Nosings The nosings of stairs, and concrete stairs in particular, are vulnerable to damage due to being kicked.

Special nosing trims are manufactured, which both protect the nosing of the tread and provide a non-slip surface. A wide range of profiles and sizes are available of which Fig. 5.45 illustrates a typical example. The trim consists of an angle

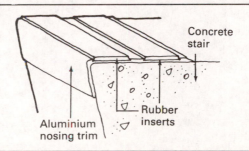

Fig. 5.45 *Nosing trim*

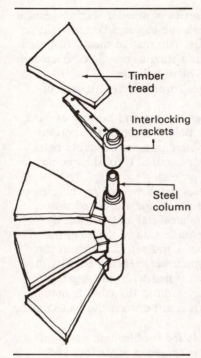

Timber tread

Interlocking brackets

Steel column

Fig. 5.46 *Interlocking steel spiral staircase*

profile aluminium section which is either screwed or bonded by adhesive to the step. A rubber insert fits into a shallow channel in the upper surface of the trim to provide a non-slip surface.

In order to obtain a neat junction at the edge of the stair, trim strips are available which fit to the edges of the risers and treads.

Metal stairs

Although metal stairs are often associated with cast-iron industrial and fire-escape stairs, elegant metal stairs may be either purpose made or selected from the range of proprietary metal stair manufacturers who may use such metals as stainless steel or aluminium.

Figure 5.46 illustrates a metal spiral stair of the type that might be obtained from manufacturers. The stair comprises of a series of interlocking aluminium tread brackets. At the nosing end of each bracket a hollow cylinder is formed which passes over a central steel column in a similar manner to the precast concrete spiral stair previously described. The tread brackets are drilled in order that tapered timber treads may be screwed to their top surface. The balusters and handrail may be of a similar nature to that shown in Fig. 5.44(c).

CHAPTER 6

Internal furnishings and fittings

BUILT-IN FURNITURE: FIXINGS AND IRONMONGERY

Introduction

Parts 6.1 and 6.2 of this chapter deal mainly with 'built-in' furniture. Built-in furniture derives some of its structural support from the internal structure of a building, by being attached to the walls and/or floor and ceiling. 'Free-standing' furniture, however, is structurally independent of the building.

Free-standing furniture allows freedom of room layout, since the furniture may be simply picked up and rearranged. Built-in furniture restricts freedom of room layout since it is fixed in one position. Some built-in furniture is designed to be relatively easily unfixed, and relocated elsewhere.

Purpose-built and mass-produced built-in furniture

'Purpose-built' built-in furniture is designed and constructed to suit the particular dimensions and visual characteristics of a particular room. Also, specific functional requirements may be accommodated in the design, such as specific volumes of storage space, or spaces designed to house particular items.

'Mass-produced' built-in furniture is made in large quantities in a factory and is available in a standard range of dimensions (often modular) and with a standard range of finishes and accessories. Usually, adaptable features, such as 'filler pieces' are available which enable the furniture to be fitted to the precise dimensions of an individual room.

Although the range of options offered in mass-produced built-in furniture may be considerable, the result is likely to compromise the visual and functional requirements. Provided an acceptable compromise is possible, mass production offers the advantage of low cost.

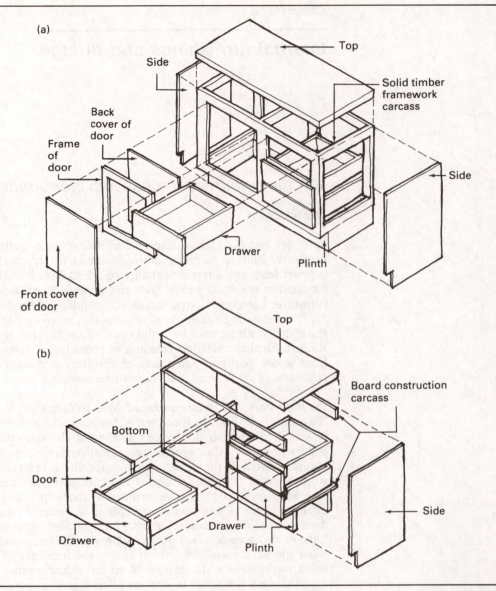

Fig. 6.1 *Carcass construction*
(a) *Traditional framed construction*
(b) *Modern construction using board material*

Traditional and modern construction

Traditional built-in furniture construction (say prior to 1914) tended to rely on a timber framework carcass fitted with timber or plywood panels. The introduction of timber-based boards (particularly chipboard) has caused a shift away from framed carcass construction, towards board carcass construction. Plastic laminate or veneer-faced timber-based boards offer the advantages of low cost, dimensional stability and simplification of joints and construction.

Figure 6.1(a) illustrates the traditional approach towards construction of a floor unit cupboard. The unit comprises of a structural framework of timber members, covered with a skin of plywood. The unit illustrated in Fig. 6.1(b) is a similar cupboard which uses a more modern approach to the construction. Here, the timber framework is dispensed with, since the board material acts as both structure and covering skin of the unit.

Different conditions of structural support for built-in units

Figure 6.2 (a,b,c) illustrates the extent to which built-in furniture might derive structural support from an interior structure.

Fig. 6.2(a). – The units are simply fitted against a flat wall. The floor unit, sometimes described as a 'base unit', is supported primarily by the floor, but is also fixed to the wall to prevent the unit from 'racking'. The wall unit is fixed to the wall and, in effect, is suspended from it. The wall unit could be raised to ceiling level, and derive some additional support from the ceiling.

The floor unit could, if required, be designed without a back or bottom, since the wall could act as the back, and the floor as the bottom, of the unit. Similarly, the wall unit could

Fig. 6.2 *Means of support for built-in furniture*
(a) *Unit fixed against wall*
(b) *Unit fixed in corner*
(c) *Unit fixed in recess*

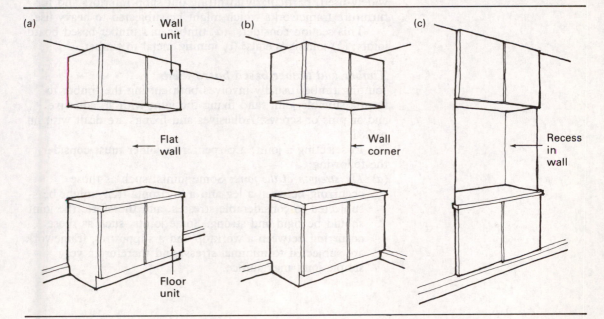

(a)   Wall unit

Flat wall

Floor unit

(b)   Wall corner

(c)   Recess in wall

be designed without a back and (if fixed to the ceiling) without a top.

Fig. 6.2(b). – The units are fitted into the corner of a wall. The means of support are much the same as the units illustrated in Fig. 6.2(a), but take advantage of deriving support from two walls. Such units could, if required, be designed without a back, a base (in the case of the floor unit) or a top (in the case of the wall unit). Also, one side of the unit could be omitted.

Fig. 6.2(c). – The units are fitted into a recess in a wall. In this case, both the wall and floor units might derive much of their support from the walls.

The design of the floor unit could be reduced to little more than a furniture top and front, and the wall unit, a furniture front and bottom.

In practice, mass-produced units are often designed to fit into any of these conditions of structural support, consequently the omission of sides, tops and bottoms of units is not as common as in the design and construction of purpose-built built-in furniture.

Joints

Timber and timber-based boards are very commonly used materials for built-in furniture construction. Metal is also widely used, particularly in fitting out shop interiors and for furniture frameworks which might be subjected to heavy use.

This section considers: (*a*) timber and timber-based board joints; (*b*) joining metals; (*c*) joining metal to timber.

Timber and timber-based board joints

Jointing timber usually involves both cutting the timber to form a suitable joint, and fixing the joint with an adhesive, and/or pins or screws. Adhesives and fixings are dealt with on p. 281.

In selecting a joint, a carpenter or joiner must consider the following:

(*a*) *The strength of the joint*: Some joints, such as those occurring between a leg and a horizontal rail, might be subjected to considerable stresses, and therefore the joint should be rigid and strong. Some joints, such as those occurring between a worktop and a supporting framework, are subjected to minimal stress, and therefore a very simple joint may suffice.

Table 6.1 Timber joints

Joints	Type of joint and illustration ref. (E = three way joint)	Description and use
Butt joints and dowelled joints	L (Figs. 6.3(a)(i) and (ii)) T (Figs. 6.3(a)(i), (ii) and (iv)) X (Figs. 6.3(a)(iii))	Butt joints are the simplest forms of joint, relying entirely on the glue and/or screws for their strength. The joints may be strengthened by use of dowels, which are inserted and glued into holes drilled in the members. The end grain of one of the joined members is exposed, which may be unsightly if visible. Butt and dowelled joints may be used for timber framing or boards
Mitred joint	L (Fig. 6.3(b))	The joint relies on the glue and/or screws for its strength, but may be strengthened by use of dowels. The 45° cut of the joint means that no end grain is visible, resulting in a neat appearance. The joint is suitable for timber framing or boards
Other simple L joints suitable for boards	Angle connecting plate (Fig. 6.3(c)(i)) Plastic corner jointing block (Fig. 6.3(c)(ii)) Chipboard joint (Fig. 6.3(c)(iii))	A metal bracket with slotted screw holes is fitted at the intersection of two boards. This also facilitates rapid removal of the joint for 'knock-down' furniture A two-part plastic block which is held together with a locking screw. One portion of the block is screwed to one board, and the other portion to the other board. The boards are brought together and the locking screw tightened to hold the boards together Since chipboard has a poor capacity to hold nails or screws, either: (a) The screws or nails may be coated with adhesive before being driven; (b) a length of dowel may be inserted and glued to act as a plug capable of receiving the screws
Lap joints and halving joints	T (Figs. 6.3(d)(i) and (ii)) X (Fig. 6.3(d)(iii))	Lap 'T' joints are much stronger than butt joints, since they restrain any lateral movement in the 'limb' of the 'T'. The full lap joint is used for forming a 'T' joint from timbers of different depths, whilst the half lap joint (also known as the 'halving' joint) is used for joining timbers of similar depth. The cross-halving joint is used for forming an 'X' joint from either timber framing or boards
Rebated and rebated and grooved joint	L (Figs. 6.3(f)(i) and (ii)) T (Fig. 6.3(f)(ii))	Rebated, and rebated and grooved joints might be typically used for shelving. The rebated joint is a simple 'L' joint relying on the glue and/or screws for its strength. The rebated and grooved joint (also known as the 'rebate housing' joint) is stronger since it resists lateral movement in the 'limb' of the 'T'. These joints are used principally for boards

Table 6.1 continued

Joints	Type of joint and illustration ref. (E = three way joint)	Description and use
Housing joints	T (Figs. 6.3(e)(i) and (ii))	The housing joint is a 'T' joint used typically for shelving. The housing of the through housing joint is visible, whereas the stopped housing is neater in appearance. Housing joints are used principally for boards
Mortise and tenon joints	L (Fig. 6.3(g)(iii)) T (Figs. 6.3(g)(i), (ii), (iv) and (v)) E (Fig. 6.3(g)(vi))	The mortise and tenon joints are a family of strong 'T' joints which are suitable for heavier framing, or where the joint is subjected to considerable stresses (such as the legs of chairs). The tenon is formed at the end of the 'limb' of the 'T', by reducing the thickness of the timber. The mortise is simply a slot cut in the opposite member to receive the tenon. The thickness of the tenon is not usually less than one-third of the thickness of the timber, otherwise the strength of the tenon would be impaired. The joints may be further strengthened by: (a) driving wedges between the mortise and tenon, which stiffens the joint; (b) inserting a dowel through the joint which prevents the tenon from being withdrawn
	Fig. 6.3(g)(i)	The 'through' mortise and tenon is the joint from which the others are derived. The tenon is visible on the outside
	Fig. 6.3(g)(ii)	The 'stub' mortise and tenon conceals the tenon by making both the mortise and tenon shorter than the depth of the timber
	Fig. 6.3(g)(iii)	The 'haunched' mortise and tenon is used for either 'L' joints, or 'T' joints where a minimal thickness of timber exists above the mortise. Tenons of larger dimension could severely weaken the mortised timber, hence
	Fig. 6.3(g)(iv)	'twin' tenons are used for deep-section timbers such as middle rail to style joint in doors, and
	Fig. 6.3(g)(v)	'double' tenons are used for wide section timbers such as window or door jambs to sill Mortise and tenon joints are usually not used for joining boards
Dovetail joints	L (Figs. 6.3(h)(i), (ii) and (iii))	Dovetails are strong joints used typically for drawer construction. The handmade versions are time-consuming and expensive to make, but the machine version is frequently used for drawers. The 'through dovetail' can be very attractive if well made, alternatively the 'lapped' dovetail may be used which reduces the amount of visible end grain

(b) *The size and type of material used in the joint*: Some joints, such as the through mortise and tenon, are ideal for strong junctions between timbers of relatively small cross-section. However, such a joint used with timbers of substantial size would be structurally inefficient and weak.

Some materials, such as chipboard, may become radically weakened by the cutting of an intricate joint. A simple joint such as those using plastic jointing blocks might be more appropriate (see Fig. 6.3(c)(ii)).

(c) *The appearance of the joint*: Some joints are practical and strong but might be unsightly. Consequently, such joints might be more suitable for inconspicuous locations. Other joints are neat in appearance, and more suitable for visible locations.

(d) *The type of joint required*: Joints may be used for forming:
- a simple 'L', such as at the corner junction of a horizontal and vertical board;
- a 'T', such as at the intersection of a leg into the middle of a horizontal rail;
- an 'X' joint, where two timbers or boards cross over each other;
- a 'three-way joint' (indicated by 'E' in Table 6.1) such as the joint between a leg and the corner of two horizontal rails.

Fig. 6.3 *Timber and timber-based joints*
(a) *Butt joints*
 (i) *Plain butt joint*
 (ii) *Dowelled joint*
(iii) *Dowelled 'X' joint*
(iv) *Metal dowel joint*

Timber frame and board joints Table 6.1 describes some of the commonly used timber and timber-based board joints, and Fig. 6.3 illustrates them. Reference to 'boards' in Table 6.1 includes timber-based boards and solid timber boards.

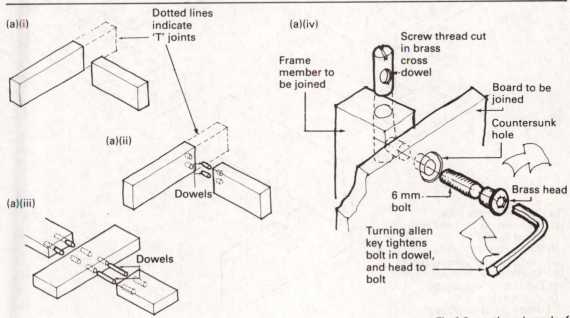

(a)(i)

Dotted lines indicate 'T' joints

(a)(iv)

Screw thread cut in brass cross dowel

Frame member to be joined

Board to be joined

(a)(ii)

Countersunk hole

Dowels

6 mm bolt

Brass head

(a)(iii)

Dowels

Turning allen key tightens bolt in dowel, and head to bolt

Fig 6.3 continued overleaf

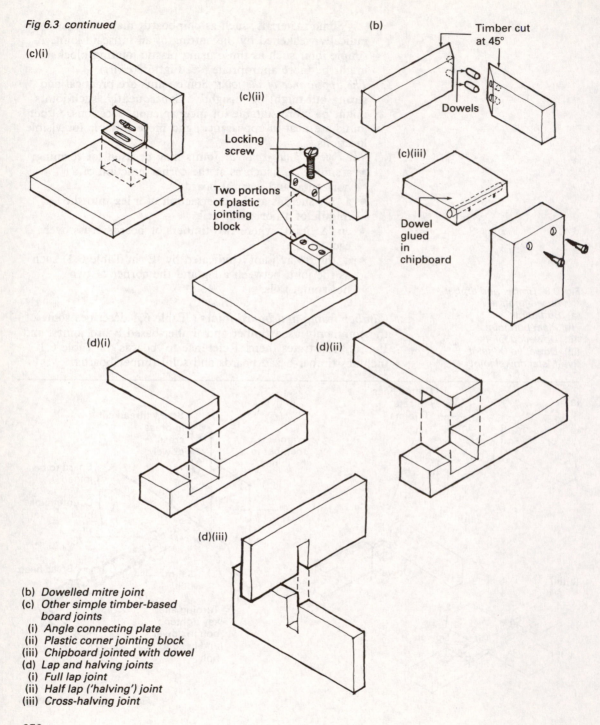

Fig 6.3 continued

(b)

Timber cut at 45°

Dowels

(c)(i)

(c)(ii)

Locking screw

Two portions of plastic jointing block

(c)(iii)

Dowel glued in chipboard

(d)(i)

(d)(ii)

(d)(iii)

(b) *Dowelled mitre joint*
(c) *Other simple timber-based board joints*
 (i) *Angle connecting plate*
 (ii) *Plastic corner jointing block*
 (iii) *Chipboard jointed with dowel*
(d) *Lap and halving joints*
 (i) *Full lap joint*
 (ii) *Half lap ('halving') joint*
 (iii) *Cross-halving joint*

272

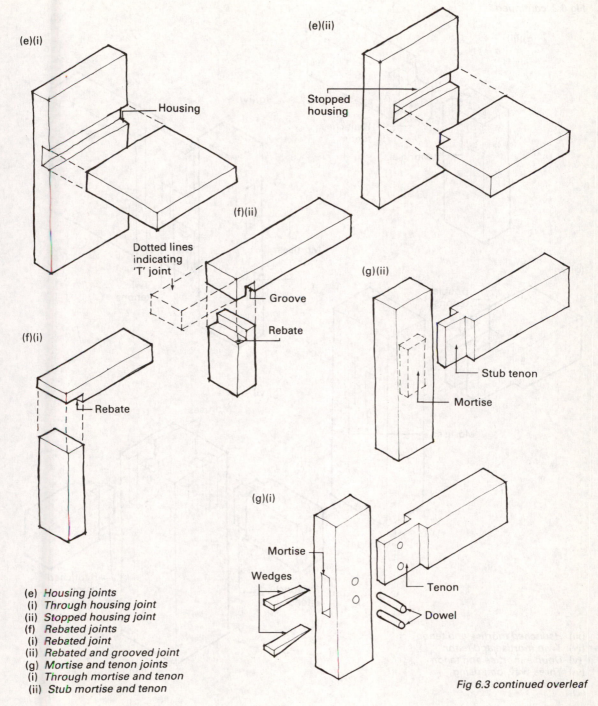

(e)(i)

Housing

(e)(ii)

Stopped
housing

(f)(ii)

Dotted lines
indicating
'T' joint

Groove

Rebate

(g)(ii)

Stub tenon

Mortise

(f)(i)

Rebate

(g)(i)

Mortise

Wedges

Tenon

Dowel

(e) Housing joints
(i) Through housing joint
(ii) Stopped housing joint
(f) Rebated joints
(i) Rebated joint
(ii) Rebated and grooved joint
(g) Mortise and tenon joints
(i) Through mortise and tenon
(ii) Stub mortise and tenon

Fig 6.3 continued overleaf

Fig 6.3 continued

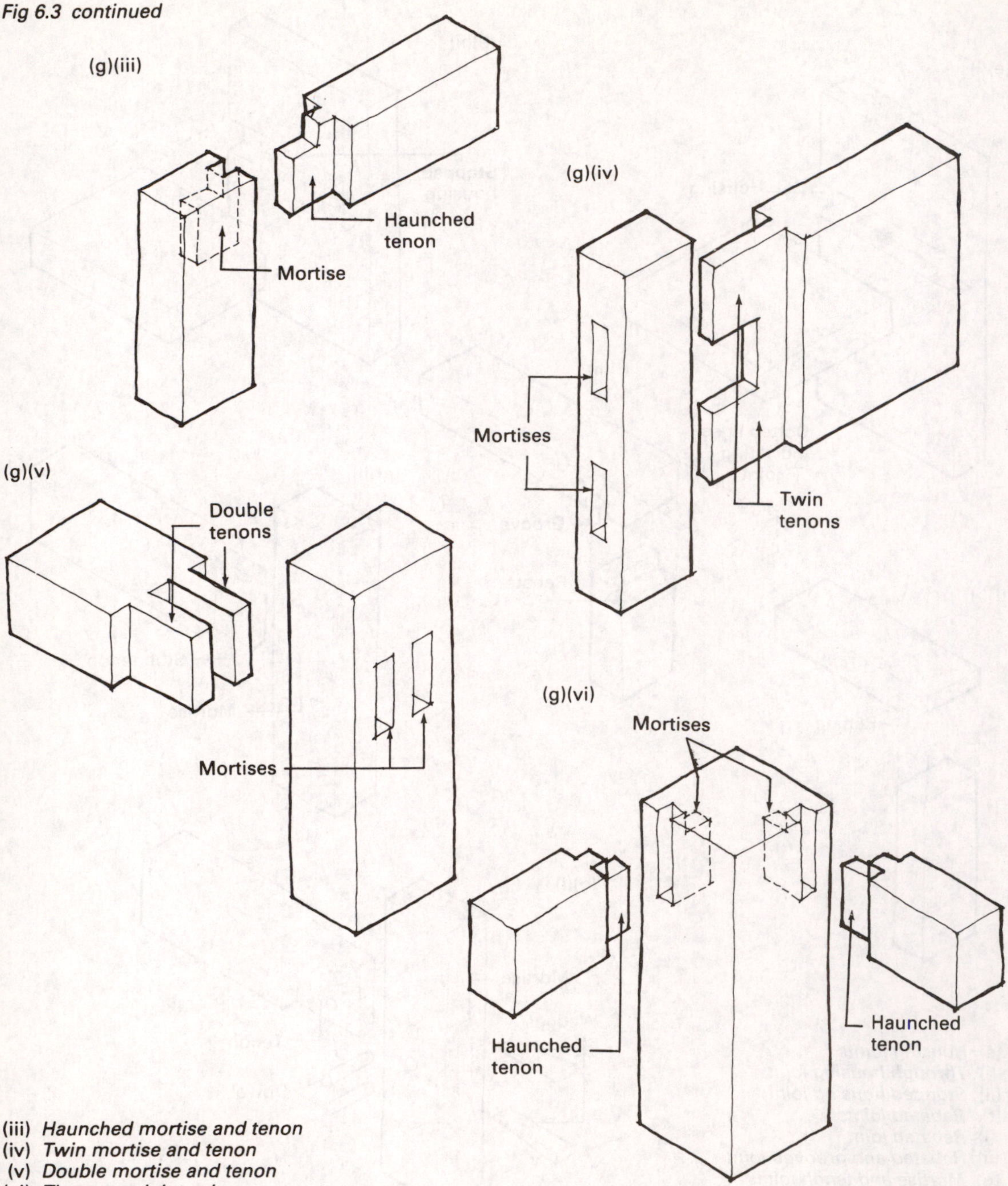

(g)(iii)

Haunched
tenon

Mortise

(g)(iv)

Mortises

Twin
tenons

(g)(v)

Double
tenons

Mortises

(g)(vi)

Mortises

Haunched
tenon

Haunched
tenon

(iii) *Haunched mortise and tenon*
(iv) *Twin mortise and tenon*
(v) *Double mortise and tenon*
(vi) *Three-way joint using mortises and tenons*

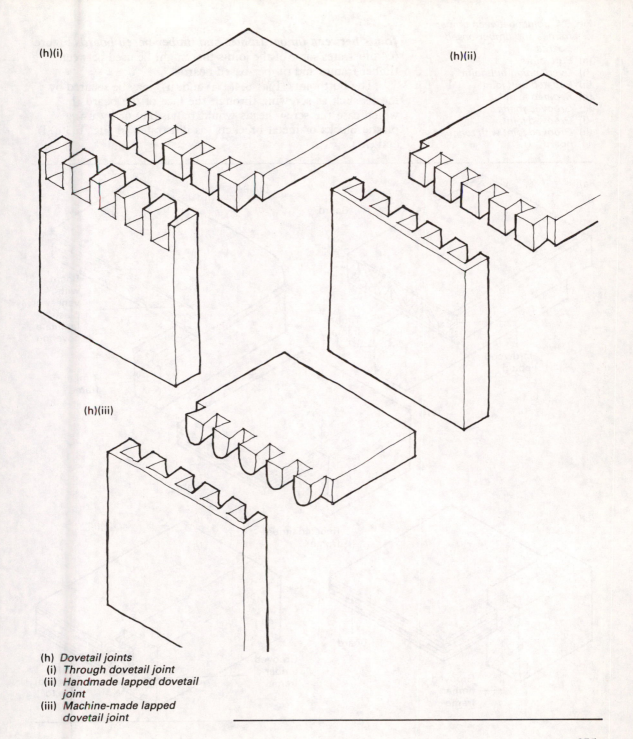

(h) *Dovetail joints*
 (i) *Through dovetail joint*
(ii) *Handmade lapped dovetail
    joint*
(iii) *Machine-made lapped
    dovetail joint*

Fig. 6.4 *Joints between timber frames and timber-based boards*
(a) *Butt joints*
 (i) *Overlapping butt joint*
(ii) *Rebated butt joint*
(b) *Rebated joint*
(c) *Grooved joints*
 (i) *Grooved joint*
(ii) *Grooved joint with tongued board*

Joints between timber frames and timber-based boards Figure 6.4 illustrates some of the joints that might be used between timber frames and timber-based boards.

The butt joints (Figs 6.4(a)(i) and (ii)) may be secured by means such as screwing through the face of the board (in which case the screw heads would be visible) or by using plastic blocks or metal brackets (as illustrated in Figs 6.3(c)(i) and (ii)).

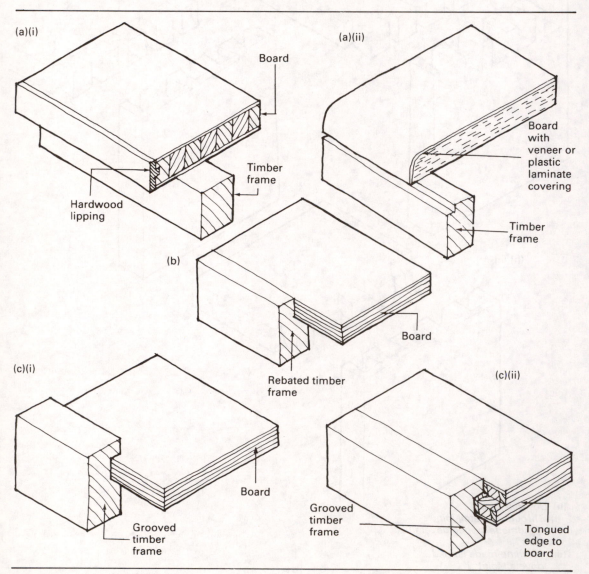

Two problems that may occur with such butt joints are:
- the problem of moisture movement in the timber causing an unsightly gap to occur between the frame and board;
- the problem of dealing with the exposed edge face of the board which is vulnerable to damage and visually unsatisfactory.

The overhanging butt joint (Fig. 6.4(a)(i)) overcomes these problems by:
- concealing the frame/board joint under the overhang of the board;
- using a hardwood lipping to conceal the edge face of the board.

The rebated butt joint (Fig. 6.4(a)(ii)) overcomes these problems by:
- the use of a rebate which forms a deep gap such that any moisture movement would be visually insignificant;
- the edge face of the board is covered with either veneer or plastic laminate.

The rebated joint (Fig. 6.4(b)) has the advantage of concealing and protecting the edge face of the board material, although moisture movement could cause a gap to occur between the frame and the board. The board may be fixed to the frame by similar means as those described for butt joints.

The grooved joint (Fig. 6.4(c)(i)) both conceals the end face of the board and conceals any gaps that may occur due to moisture movement. The tongued and grooved joint (Fig. 6.4(c)(ii)) has the advantage of a flush edge detail. A disadvantage of both grooved joints is that should the board need to be removed, the grooved frame would have to be dismantled first.

Joining metals

Methods of joining metals include: (*a*) mechanical connections; (*b*) soldering and brazing; (*c*) welding. These methods vary in their appearance, their degree of permanence and their strength.

Mechanical connections   Mechanical connections are made relatively simply, by drilling a hole in the pieces of metal to be joined, then inserting and fastening either nuts and bolts, rivets or screws.

Nuts and bolts   Nuts and bolts may be used to form either a permanent connection or one which occasionally needs to be undone (in which case washers are used, see Fig. 6.5(a)(ii)). If the connection is to be frequently undone, a 'wing nut' may be used which may be tightened by hand.

In some locations, bolted connections could cause injury by people knocking themselves against the exposed nut or bolt head. The nut and bolt heads may also be visually unacceptable. A possible solution is to cover the nut or bolt head with plastic cap.

The simplicity of bolted connections makes them suitable for joining metal items on the site. Typical bolt sizes are: diameter 6–12 mm, and lengths 20–120 mm.

Fig. 6.5 *Joining metal*
(a) *Mechanical joints*
 (i) *Nut and bolt joint*
(ii) *Wing nut and bolt joint*
(b) *Riveted joints*
 (i) *Holes drilled in metal*
(ii) *Rivet inserted*
(iii) *Head formed at end of rivet*
(c) *Metal to timber connections*
 (i) *Screw connection (viewed from below)*
(ii) *Recessed bolted connection*

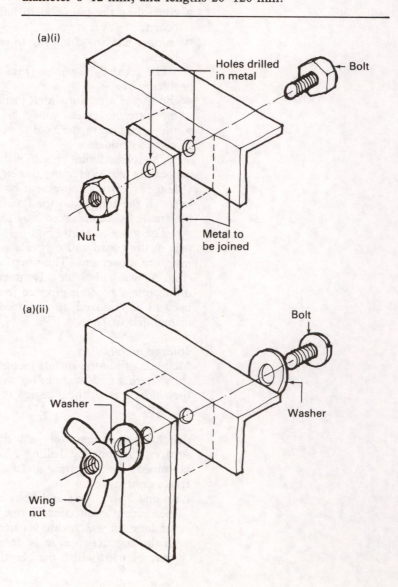

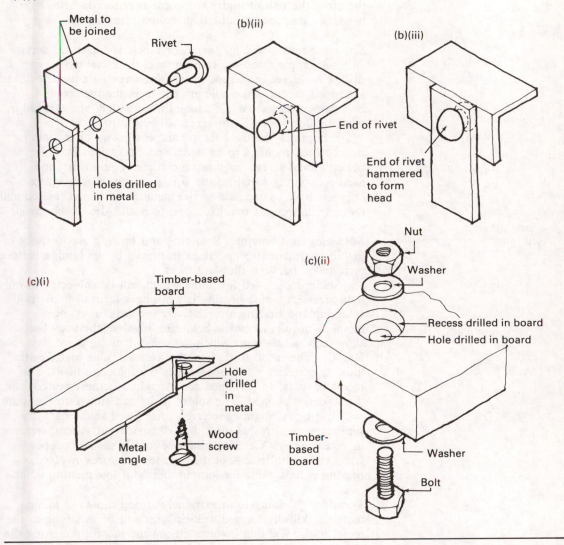

(b)(i)

Metal to be joined

Rivet

Holes drilled in metal

(b)(ii)

End of rivet

(b)(iii)

End of rivet hammered to form head

(c)(i)

Timber-based board

Hole drilled in metal

Wood screw

Metal angle

(c)(ii)

Nut

Washer

Recess drilled in board

Hole drilled in board

Timber-based board

Washer

Bolt

Rivets Rivets are suitable for forming permanent metal connections. Figure 6.5(b) illustrates the stages involved in forming such a connection:

- a hole is drilled in the pieces of metal to be jointed;
- the rivet is inserted;
- the rivet end is hammered to form a head.

Riveted connections are less obtrusive than bolted joints, and therefore the risk of injury to people is reduced. Also, rivet heads are less conspicuous than bolted connections.

Screws   Screws used for metalwork include 'machine' screws and 'self-tapping' screws. In both cases the metal to be joined is drilled to receive the screw. If machine screws are used, a thread is 'tapped' or cut in the hole prior to inserting the screw. Self-tapping screws have a strong thread which taps a thread in the wall of the hole, as the screw is being turned.

Screws may be used for permanent fixing, or for joining metal which requires to be taken apart occasionally. Like bolted joints, screws require no complex equipment, and are ideal for joining metal on the site, particularly where access is only available to one side of the metal to be joined (bolted and riveted connections require access to both sides of the metal).

Soldering and brazing   Soldering and brazing are methods of making permanent connections in metals by applying a molten metal alloy between the joint faces.

Soldering is used for joints which will be subjected to only light stresses, whilst brazing is a stronger form of joint. Both soldering and brazing are relatively inconspicuous, neat methods of joining metal. Soldering involves the use of an alloy such as tin—lead, which has a low melting point (less than 300 °C). The metal to be joined is cleaned, and coated with 'flux' (a material which encourages the solder to flow). The pieces of metal to be joined are brought together, heated, and the solder is applied. The solder melts and runs into the joint.

Brazing (sometimes referred to as 'hard soldering') is a similar process to soldering, but is carried out at temperatures in excess of 600 °C, using alloys of metals such as copper, silver and zinc. Because of the high temperatures involved, brazing is unsuitable for joining metals of low melting point.

Welding   Welding is an extremely strong means of joining metal: a skilfully formed welded joint will be as strong as the metal itself. Welding involves fusing the metal by melting the areas of metal to be joined. Sometimes a 'filler rod' of similar metal is also melted into the joint. Welding tends to be carried out in the factory, rather than on the site, by either electric arc, or oxyacetylene methods.

Electric-arc welding involves equipment which produces high-voltage current. A connection is made from the equipment to the metal to be joined (which is also earthed), and a second connection is made from the equipment to the filler rod. The weld is formed by touching the filler rod to the metal to be joined (which completes the circuit), then immediately withdrawing the rod a short distance. This causes the current to jump the gap (or 'arc') between the filler rod and the metal. This produces sufficient heat to melt and fuse the metal. Arc welding is suitable for thick or thin metal.

Oxyacetylene welding is carried out by igniting a mixture of acetylene (a hydrocarbon gas) and oxygen. The flame produced is capable of melting the metal. Oxyacetylene welding is used principally for sheet metal.

Metal to timber connections
Connections between metal and timber are usually made with either screws or bolts.

Screw connections may be made by drilling a hole in the metal and screwing through the metal into the timber (Fig. 6.5(c)(i)).

In conditions where the timber is liable to tend to pull away from the metal, a bolted connection might be better. To avoid the danger of injury to people knocking against an exposed bolt head or nut, the bolt could be recessed into the timber (Fig. 6.5(c)(ii)).

Adhesives and fixings

This section considers: (*a*) adhesives suitable for interior joinery; (*b*) nails and screws used in interior joinery; (*c*) methods of fixing interior joinery to walls.

Adhesives
Table 6.2 describes some adhesives commonly used for interior joinery purposes. Although all the adhesives are strong, the traditional animal glues have been largely superseded by other adhesives which have better moisture resistance, heat resistance and other properties.

Nails and screws
Tables 6.3 and 6.4 respectively outline the sizes and uses of nails and screws used in joinery.

Generally, nails are used for holding joints in locations which will not be visible, such as *in situ* frameworks covered by board material. Pins are fine nails which are often used in conjunction with adhesives for fixing some of the simple joints, such as butt and rebated joints.

Screws have more resistance against being withdrawn due to the friction of their thread. Consequently, screws may be used:

Table 6.2 Adhesives used for interior joinery

Adhesive	Description and use	Properties Colour when dry	Moisture resistance	Mould resistance	Heat resistance
Animal glues	Traditional glues derived from bones and skins of animals and fish. Now largely superseded by p.v.a. and other adhesives. Available in tube form for small quantities, or powder or cake form for large quantities	Opaque light brown	Poor	Poor	Poor
Casein glues	Derived from milk protein. Usually supplied in powder form to be mixed with water	Transparent yellow	Fair	Fair	Good
Contact adhesives	Not suitable for wood-to-wood joints, but is ideal for bonding plastic laminate and plastic foams to wood. Provides an instant bond	Clear or yellow	Good	Good	Good
Epoxide resin adhesives	Costly adhesive which is capable of bonding almost any material including metal. Available as two-part adhesive: resin and hardener	Yellow	Very good	Very good	Very good
Polyvinyl acetate (p.v.a.)	Relatively cheap general-purpose joinery adhesive, suitable for all joints not subjected to excessive stresses. Requires no mixing	Transparent	Fair	Good	Poor
Urea formaldehyde (u.f.) adhesives	Relatively cheap adhesive used widely in plywood manufacture. Available in two-part form (resin and hardener) or in powdered form for mixing with water	Opaque white	Good	Very good	Good

- for joining timber which is required to be periodically taken apart;
- as a device for drawing together two pieces of wood to be joined.

Where screws or nails are used for visible joints which are not required to be taken apart, their heads should be driven below the surface of the timber, and a filler material or small piece of matching veneer used to provide a smooth surface finish.

Table 6.3   Nails used for interior joinery

Nail	Typical length (mm)	Uses
Round wire nail	15–200	For rough joinery or carpentry. Unsuitable for visible locations. The thick, round cross-section of the nail could cause splitting in thin timber members
Oval wire nail	20–150	As above, but since head may be driven below the surface and filled over, it could be used for visible locations. The oval cross-section is less liable to split timber
Lost head wire nail	15–75	As above, but nail has round cross-section
Panel pin	10–75	Often used in conjunction with glued joints. Also used for fixing plywood and hardboard, the pinhead being easily driven below the timber or board surface

Table 6.4   Screws used in interior joinery

Screw	Typical sizes (mm)	Uses
Countersunk	Length: 6–100. Diameter: 1.8–7 or screw gauge No. 1–16	The tapered or 'countersunk' head is suitable for either finishing flush with the surface if not visible, or for driving below the surface and filling, if visible
Raised head	Length: 12–38. Diameter: 2.6–4 or screw gauge No. 4–8	For locations where screws will be required to be withdrawn occasionally. Because they are usually used in visible locations, they are often used with 'cups' (see below) and made of non-ferrous metals
Round head	Length: 12–60. Diameter: 2.6–5 or screw gauge No. 4–10	For fixing materials such as hardboard, which are too thin for countersinking

Cups and screws: used to provide a neat finish for raised-head screws which are required to be occasionally withdrawn

Raised cup   Countersunk cup

See also 'Dome-headed screws', Fig. 8.8 p. 403.

Whilst most nails and screws are made of steel, screws which are to be used in visible locations may be made of materials such as brass or bronze, or finished with chromium or nickel plating.

Fixings to walls
A wide variety of proprietary fixing devices are available, which enable timber or metal members to be fixed to solid or hollow walls.

Fig. 6.6 *Fixings to solid walls*
(a) *Masonry bolt*
 (i) *Holes drilled*
(ii) *Bolt installed*
(b) *Plugs*
 (i) *Holes drilled*
(ii) *Plug installed*

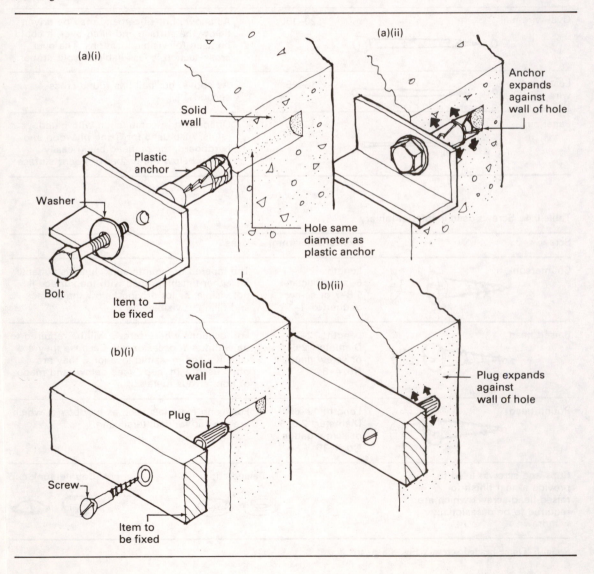

Solid walls For fixing heavy items to solid walls, a 'masonry bolt' might be used (Fig. 6.6(a)). The bolt is used as follows:

(*a*) A hole is drilled in the wall to fit the plastic anchor of the bolt. A hole is also drilled in the item to be fixed: the diameter corresponding to the diameter of the bolt thread. The bolt is inserted in the hole and the item to be fixed is held in place.

(*b*) The nut is turned, which both tightens the item against the wall, and causes the plastic anchor to expand, jambing the anchor against the wall of the hole.

Typical sizes of masonry bolts are: length 50–300 mm; diameter 6–25 mm.

For fixing lighter objects, 'plugs' may be used (Fig. 6.6(b)). These are cylinders of fibre, metal or plastic, which are made in sizes to suit standard screw sizes. The plugs are used as follows:

- a hole is drilled in the wall to accept the plug (the item to be fixed is also drilled, the hole diameter corresponding to the diameter of the screw shank);
- the plug is inserted into the hole, and the item to be fixed is held in place;

as the screw is turned, the plug expands and jams tight against the wall of the hole.

Hollow walls Devices for fixing items to hollow walls (such as plasterboard and stud walls) work on the principle of some form of metal or plastic sleeve or toggle, which is inserted in a hole drilled in one skin of the wall. As the screw is tightened, so the sleeve or toggle is tightened against the back surface of the wall skin. Figure 6.7 (a,b,c) illustrates three of these devices:

Fig. 6.7 *Fixings to hollow walls*
(a) *Steel gravity toggle*
 (i) *Holes drilled*
(ii) *Toggle installed*
(b) *Steel spring toggle*
 (i) *Holes drilled*
(ii) *Toggle installed*
(c) *Rubber-sleeved bolt*
 (i) *Hole drilled*
(ii) *Bolt inserted*

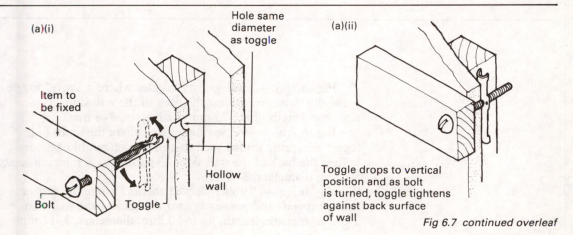

(a)(i)

Item to be fixed

Bolt

Toggle

Hole same diameter as toggle

Hollow wall

(a)(ii)

Toggle drops to vertical position and as bolt is turned, toggle tightens against back surface of wall

Fig 6.7 continued overleaf

Fig 6.7 continued

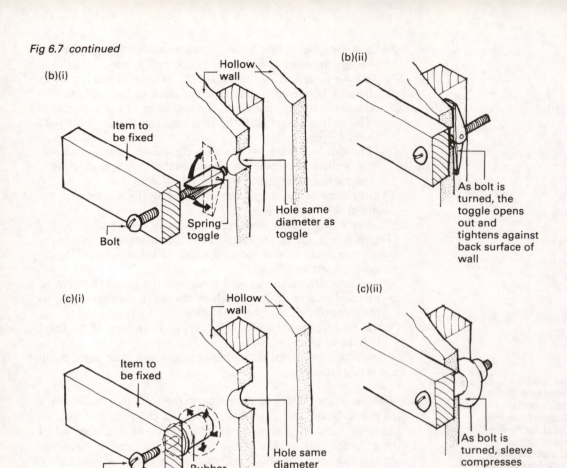

(b)(i)

Item to be fixed

Bolt

Spring toggle

Hollow wall

Hole same diameter as toggle

(b)(ii)

As bolt is turned, the toggle opens out and tightens against back surface of wall

(c)(i)

Item to be fixed

Bolt

Rubber sleeve

Hollow wall

Hole same diameter as rubber sleeve

(c)(ii)

As bolt is turned, sleeve compresses against back surface of wall

Fig. 6.7(a). – Steel gravity toggle, where a swivel toggle is gradually tightened against the back of the wall skin. Typical sizes are: length, up to 75 mm; diameters, 3–6 mm.

Fig. 6.7(b). – Steel spring toggle, where the limbs of the toggle are spring loaded such that they open out like scissors against the back of the wall skin. Typical sizes are: length, up to 50 mm; diameters, 3–6 mm.

Fig. 6.7(c). – Rubber-sleeved anchor, where the rubber sleeve expands and presses against the back of the wall skin. Typical sizes are: length, up to 50 mm; diameters, 3–12 mm.

Ironmongery for furniture

Furniture ironmongery fulfils the same basic purpose as ironmongery for internal and external doors: to provide the means and mechanism for opening, closing and locking doors. However, furniture ironmongery also includes the means and mechanism for opening, closing and locking drawers. Some furniture ironmongery resembles a scaled-down version of internal and external door ironmongery – for example hinges, sliding-door mechanisms and locking devices.

Furniture ironmongery is dealt with in this section as follows: (*a*) furniture hinges; (*b*) sliding mechanisms for doors and drawers; (*c*) catches, locks, handles and pulls for doors and drawers.

Furniture hinges

Hinged cupboard doors and flaps may be side, top or bottom hung (Fig. 6.8(a)). The doors may be either hung in 'lay on' position (hinged on the front edge of the cupboard) or in 'inset' position (hinged on the inside of the front edge of the cupboard). These positions are illustrated in Fig. 6.8(b).

Fig. 6.8 *Cupboard door hinge arrangements*
(a) *Hanging of doors*
 (i) *Side hung doors*
 (ii) *Top hung door or flap*
 (iii) *Bottom hung door or flap*
(b) *Positioning of doors*
 (i) *Lay on door*
 (ii) *Inset door*

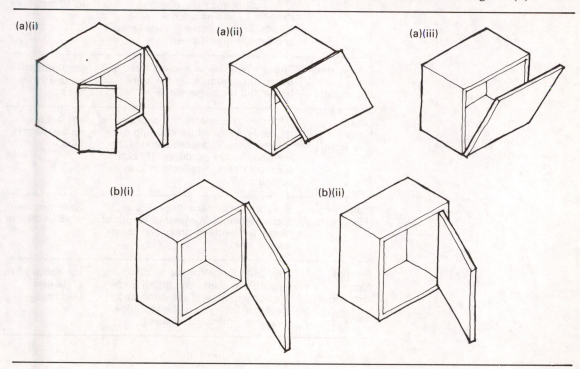

(a)(i)   (a)(ii)   (a)(iii)

(b)(i)   (b)(ii)

Table 6.5 and Fig. 6.9(a)–(g) describe and illustrate some of the many different types of hinges suitable for cupboard doors and flaps. Cupboard hinges are also available for automatic closing, and are frequently used for kitchen unit doors.

Table 6.5 Furniture hinges

Hinge	Description and use	Typical sizes
Back flap hinge (Fig. 6.9(a))	Similar to butt hinges, but with the knuckle formed on the opposite side to the countersinks of the screw holes. Used mainly for lids and flaps for furniture such as bureaux and cabinets. Available in steel and brass	18 and 38 mm (depth of hinge)
Concealed cylinder hinge (Fig. 6.9(b))	Suitable for lay-on or inset doors, or for lids and flaps. The cylinders are hidden in holes, one in the door/flap, and one in the cupboard side	For doors of 16 and 25 mm thickness
Cranked hinge (Fig. 6.9(c))	The cranked leaves of the hinge enable the door to be opened clear of the cupboard side, enabling drawers or sliding shelves to be withdrawn. Available in nickel-plated and bronze-plated steel	60 mm deep for doors of up to 10 mm thick
Double door hinge (Fig. 6.9(d))	These hinges allow two adjacent doors to be hung from the same frame and the same hinge	For doors of up to 18 mm thick
Lift-off hinge (Fig. 6.9(e))	Lift off hinges enable the doors to be easily removed for cleaning or maintenance. Available either: (a) cranked, for lay-on doors; (b) butt, for inset doors. Available in zinc-coated steel	For doors of up to 18 mm thick
Piano hinge (Fig. 6.9(f))	Suitable for lids and flaps. Cut to required length from a 2 m length of hinge. Available in brass and brass-coated steel	2000 mm long × 25 or 30 mm
Pivoted hinge (Fig. 6.9(g))	Easy-to-fit hinge for lay-on doors. Available in pairs, one fitting under the bottom edge of the door, and the other fitting over the top edge. Available in zinc-plated steel	For doors of 15 or 18 mm thickness

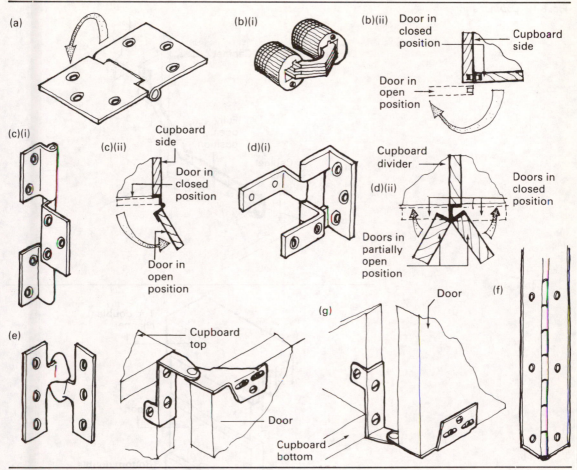

Fig. 6.9 *Furniture hinges*
(a) *Back flap hinge*
(b) *Concealed cylinder hinge*
 (i) *Pictorial view*
(ii) *Plan of cupboard with concealed hinge*
(c) *Cranked hinge*
 (i) *Pictorial view*
(ii) *Plan of cupboard with cranked hinge*
(d) *Double door hinge*
 (i) *Pictorial view*
(ii) *Plan of cupboards with cranked hinge*
(e) *Lift-off Hinge*
(f) *Piano hinge*
(g) *Pivoted hinge*

Very large hinged cupboard doors, such as wardrobe doors, will often be hung on ordinary butt hinges.

Where bottom-hung doors or flaps are used, it will often be necessary to install a 'stay' which holds the door or flap in the horizontal position. Figure 6.10 illustrates a typical stay which is fixed to the side of the cupboard and top surface of the flap, and has an elbow half-way along its length. Such a stay would be typically made of brass and be of 150–300 mm in length when closed. Other types of stay include friction stays which allow the door or flap to be held open at any angle, and pneumatic stays which automatically close the lid.

289

Fig. 6.10 *Cabinet stay*

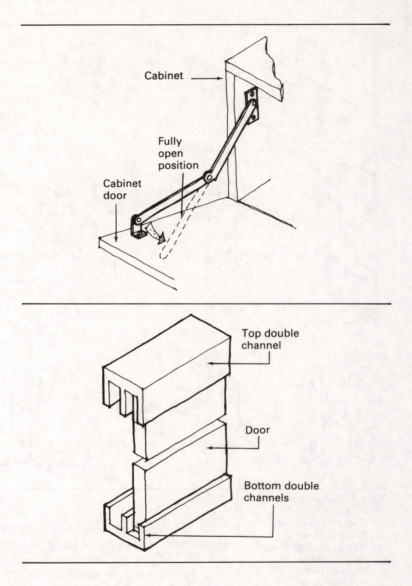

Cabinet

Fully
open
position

Cabinet
door

Top double
channel

Door

Bottom double
channels

Fig. 6.11 *Channels for small
sliding doors*

Sliding methods for doors and drawers

Doors Cabinet and cupboard doors may open by a simple
sliding mechanism, or by a sliding/folding mechanism.

Figure 6.11 illustrates a typical device for small cupboard
and cabinet sliding doors. The top and bottom edges of the
doors locate in the grooves of double channels. The channels,
which are usually made of plastic or metal, are either fixed to

the surface, or recessed into the top and bottom of the front of the cupboard. The top grooves of the top channel are twice the depth of the lower channel; this enables the doors to be lifted in and out of the grooves. The sliding action of the doors may be eased by either fitting sections of smooth material (such as nylon or p.t.f.e. – 'Teflon') to the bottom edge of the door, or by fitting a metal strip which incorporates small nylon wheels.

Heavier and larger cupboard doors may employ a sliding mechanism similar to those used for sliding internal doors. Figure 6.12 illustrates such a mechanism, which consists of:

- bottom rollers which are fixed to the bottom rails of the doors, and locate over a bottom track;
- top guides which are fixed to the top rail of the door, and locate in the grooves of top channels.

Sliding/folding doors are sometimes used for larger items of furniture such as wardrobes. Figure 6.13 illustrates a typical sliding/folding mechanism suitable for such doors. The

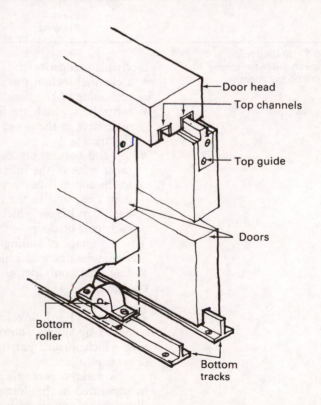

Fig. 6.12 *Sliding mechanism for large sliding doors*

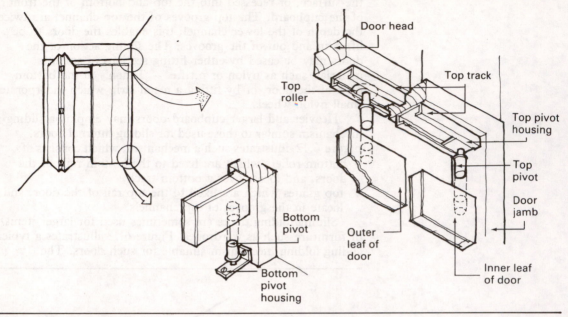

Fig. 6.13 *Sliding/folding mechanism for large cupboard doors*

Labels on figure: Door head; Top track; Top roller; Top pivot housing; Top pivot; Door jamb; Bottom pivot; Outer leaf of door; Inner leaf of door; Bottom pivot housing

mechanism consists of:

- a channel-section top track which is screwed to the underside of the door head;
- top rollers which are fixed to the top outer edge of the outer leaf of the door, the wheels of the rollers locating in the track;
- top and bottom pivots fixed to the top and bottom of the inner edge of the inner leaf of the door, the pivots locating in the top and bottom pivot housings which are fixed to the door head and to the floor, at the jamb of the opening;
- back flap hinges which are fitted to the meeting stiles of each pair of doors.

The advantage of sliding/folding doors is that the entire door area may be opened, which provides better access. Simple sliding doors only permit half of the door area to be opened at one time.

Drawers The sliding action of traditional drawers was often achieved by simply allowing the drawer bottom to slide over rails which formed part of the carcass of the furniture (Fig. 6.14(a)).

Nowadays, some form of sliding device is usually incorporated in the drawer construction. Figure 6.14(b) illustrates a drawer with a grooved side. The groove runs over

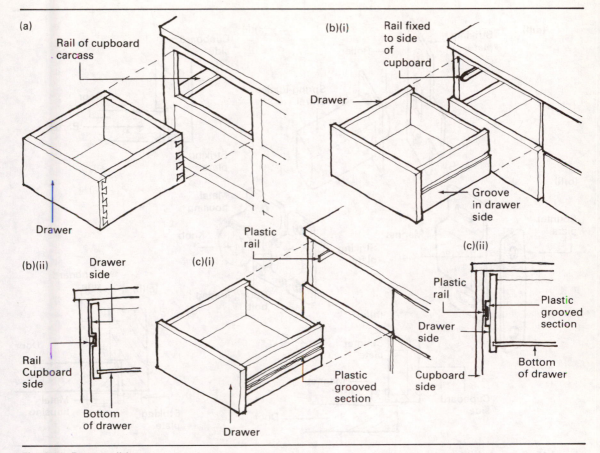

(a)

Rail of cupboard carcass

Drawer

(b)(i)

Rail fixed to side of cupboard

Drawer

Groove in drawer side

(b)(ii)

Drawer side

Rail Cupboard side

Bottom of drawer

Plastic rail

(c)(i)

Plastic grooved section

Drawer

(c)(ii)

Plastic rail

Plastic grooved section

Drawer side

Cupboard side

Bottom of drawer

Fig. 6.14 *Drawer slides*
(a) *Traditional drawer*
(b) *Grooved drawer side*
(i) *Pictorial view*
(ii) *Section through drawer side*
(c) *Plastic slide*
(i) *Pictorial view*
(ii) *Section through drawer side*

a rail attached to the side of the carcass of the furniture. In better-quality drawers, and drawers required to contain heavy items, the drawer–slide system might incorporate roller wheels.

Proprietary plastic drawer slides are available consisting of a grooved section which is fixed to the carcass of the furniture, and a slide section which is attached to the drawer side (Fig. 6.14(c)).

Catches, locks, handles and pulls for cupboard doors and drawers

Catches Catches are used for holding cabinet and cupboard doors in the closed position, although if a lock is fitted, a catch will usually be unnecessary. Figure 6.15 illustrates three types of cupboard catch.

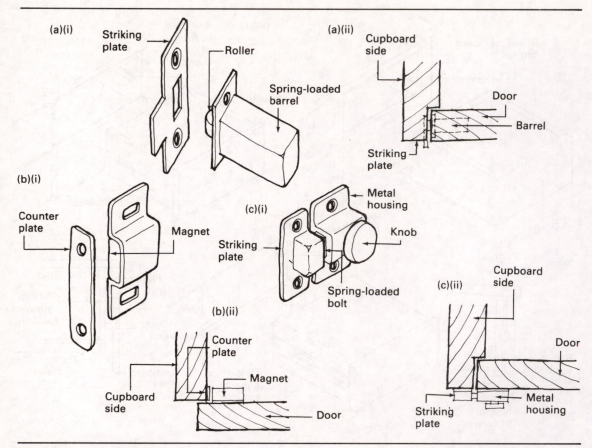

Fig. 6.15 *Cupboard catches*
(a) *Roller catch*
 (i) *Pictorial view*
 (ii) *Plan*
(b) *Magnetic catch*
 (i) *Pictorial view*
 (ii) *Plan*
(c) *Mechanical catch*
 (i) *Pictorial view*
 (ii) *Plan*

Roller catch (Fig. 6.15(a)) These catches consist of a spring-loaded barrel which activates a roller. The barrel is mortised into the stile of the door, and when the door is closed, the roller rests in a slot in the striking plate. The striking plate is fixed to the side of the cupboard. Roller catches are typically of 18–25 mm in depth and made of steel and/or nylon.

Magnetic catch (Fig. 6.15(b)) These very simple catches comprise of a plastic block which is fixed to the inside surface of the door, and a counter plate which is fixed to the side of the cupboard. Both the block and counter plate have magnets attached to them, which hold the door in the closed position. A typical size for such a catch might be 40 × 10 × 10 mm.

Mechanical catch (Fig. 6.15(c))   These catches consist of a spring-loaded bolt contained in a metal housing, which is fixed on the outside surface of the door. The bolt is activated by a twist or press-action knob. With the door in a closed position, the bolt rests in the raised portion of the striking plate which is fitted to the side of the cupboard front. Mechanical catches might be typically of 60 × 40 mm in size, made of zinc alloy, or aluminium.

Locks   Furniture locks resemble scaled-down versions of internal door locks. Figure 6.16 (a,b,c) illustrates three kinds of locks:

Fig. 6.16(a). – A typical cupboard lock which is fixed to the inside surface of the door. The locks are usually of 60–70 mm depth and made of brass.

Fig. 6.16(b). – A typical drawer lock which is partially recessed in the back surface of the drawer front. The locks are usually of 50–60 mm depth and made of brass.

Fig. 6.16(c). – A cylinder lock which is fixed to the inner surface of a door or drawer front. The cylinder is fitted into a hole drilled through the drawer or cupboard. These locks are suitable for doors or drawers of typically 16–25 mm and are made of brass.

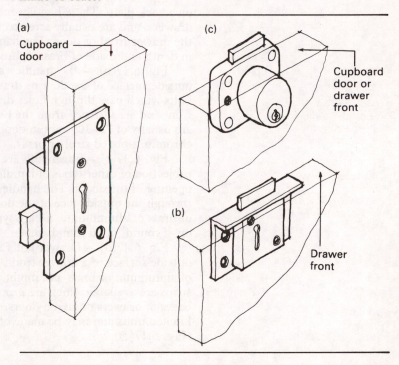

Fig. 6.16 *Furniture locks*
(a) *Cupboard lock*
(b) *Drawer lock*
(c) *Cylinder lock*

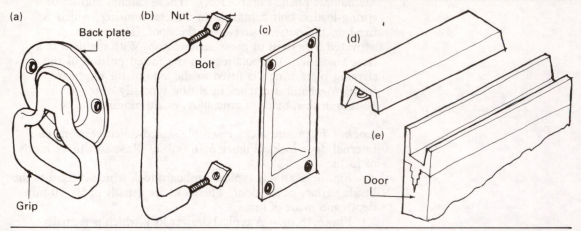

Fig. 6.17 *Furniture handles and pulls*
(a) *Drop handle*
(b) *'D' handle*
(c) *Flush pull*
(d) *Drawer pull*
(e) *Pull strip*

Handles and pulls Handles and pulls are devices for gripping a door or drawer in order to open and close them. Figure 6.17 (a,b,c,d) illustrates some of the commonly used types, of which there are many variations.

Fig. 6.17(a). – Drop handles which consist of a grip pivoted on a back plate. These handles may be used for doors or drawers, and are usually screwed through the outside surface of the drawer front or door. The handles are usually of 38–75 mm in depth and made of brass or bronze.

Fig. 6.17(b). – 'D' handles are either screwed through the outside surface of the door or drawer front, or may be fixed by bolts which pass through holes drilled in the door or drawer front and are fastened from the back with nuts. These handles are usually of 75–100 mm in depth, and made of aluminium, chromium-plated steel or brass.

Fig. 6.17(c). – Flush pulls are used for sliding doors, since the projection of other types of handle would interfere with the opening of the doors. The handles are recessed and screwed through the outside face of the door. The handles may be made of brass or aluminium, and are typically of 75–100 mm in depth or, if round, of 40 mm diameter.

Fig. 6.17(d). – Drawer pulls are usually screwed through the outside surface of a drawer front. These pulls are typically made of aluminium or brass and might be 70–80 mm in width. Pull strips are available which are grooved into the top or bottom edges of drawers or lay-on doors. These are frequently used for kitchen units and may be made of aluminium or plastic (Fig. 6.17(e)).

BUILT-IN FURNITURE CONSTRUCTION

Much of the built-in furniture seen in modern interiors (particularly of domestic buildings) is of the mass-produced variety. Although an interior designer might be called upon to design items of built-in furniture, this part deals mainly with the factory-made, mass-produced built-in furniture.

Mass-produced built-in furniture

An enormously wide variety of mass-produced built-in furniture systems are available on the market for fitting out kitchens, bathrooms and bedrooms. The main reasons for their popularity are:

(a) *Reduced design time*: since the construction methods and sizes of units are predetermined, the designer is required to consider only the kitchen plan layout and location of units. Often, all that is necessary, once the layout has been decided, is a drawing showing the layout and location of the built-in units, a list indicating which units and which accessories are to be installed and details of which of the range of finishes is to be used.

(b) *Cheapness*: factory-made, mass-produced furniture will almost inevitably be less expensive than equivalent furniture, which is individually designed and purpose built. However, the use of standardised units is likely to result in a compromise, since the units are not designed to the specific requirements of either a particular client, or a particular room.

(c) *Simplicity of construction*: many of the mass-produced systems available are based on 'knock-down' construction principles. This means that the components of the furniture are assembled by the use of very simple fixing techniques. The items of furniture arrive on the site in flat packs, and are very speedily erected, which radically reduces the time spent on site by skilled craftsmen.

Cupboards and drawers

Figure 6.1 compared the contrasting nature of traditional and modern techniques of carcass construction. The ranges of mass-produced kitchen bedroom and bathroom built-in furniture are mostly based on the principles illustrated in Fig. 6.1(b), where a series of timber-based boards are connected by simple joint fixings.

Figure 6.18(a) illustrates a kitchen fitted out with mass-produced units selected from a range offered by a manufacturer. For comparison, it might be useful to refer to manufacturers' catalogues.

Cupboards Figure 6.18(b) indicates a typical range of kitchen cupboards available from a manufacturer. Most units today are based on a plan module of 600× 600 mm or 500 × 500 mm. The range consists of:

- base (or 'floor') units, some of which incorporate sink tops, some of which are able to house cookers/hob units, refrigerators and washing machines, and some of which are fitted with combinations of drawers and cupboards;
- tall units, some of which are able to house cookers and freezers;
- worktops;
- wall units (usually 300 mm in depth), some of which are able to accommodate a cooker hood.

Most manufacturers also offer a range of accessories, such as:

Fig. 6.18 *Built-in kitchen units*
(a) *View of kitchen*

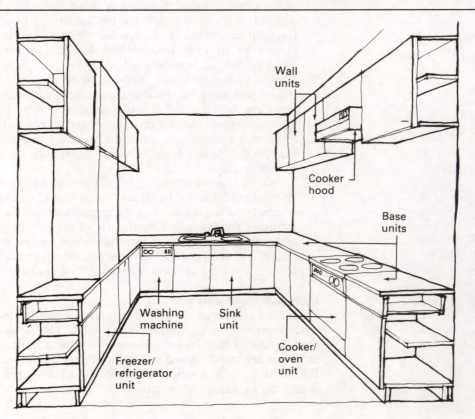

Wall units

Cooker hood

Base units

Washing machine

Sink unit

Cooker/ oven unit

Freezer/ refrigerator unit

298

(b) Typical range of units

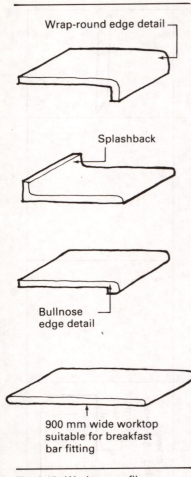

Wrap-round edge detail

Splashback

Bullnose edge detail

900 mm wide worktop suitable for breakfast bar fitting

Fig. 6.19 *Worktop profiles*

- pull-out table-tops which are drawn out on a telescopic sliding mechanism;
- cooker hoods, often with a small cupboard above;
- swing-out waste bins;
- bins and trays which fit in the cupboards;
- adjustable feet which screw to the underside of the base units.

The adjustment allows the height of the base units to be either adjusted to compensate for irregularities in the floor level, or to provide a slightly higher worktop level.

Most of the units are constructed from a carcass of chipboard faced on both sides with plastic laminate (total thickness typically of 15–19 mm). The boards will usually be butt jointed and fixed by such means as plastic or wooden dowels and screws. The worktops will usually be of similar material, but of 20–30 mm thicknesss. The worktops will usually be fixed to the top of the floor units by fixings such as angled connecting plates, which will be screwed to the top of the inside of the floor unit and to the underside of the worktop (see Fig. 6.3(c)(i)).

'Postformed' plastic laminate surfaced worktops are available with shaped profiles (Fig. 6.19).

The cupboard doors may be of similar material to the carcass, although some manufacturers offer timber-framed doors with plywood or timber veneered panels. The doors are hung from the cupboard side or end by hinges such as either those illustrated in Figs 6.9(c) and (d) (for less expensive units) or Fig. 6.9(b) (for better-quality units). The doors may be opened by use of either knobs, 'D' handles or pull strips (see Fig. 6.17). Typical closing devices for the doors include magnetic catches (see Fig. 6.15(b)) and moulded nylon catches. Some manufacturers offer 'decor panels' which fix to the front of fridges, freezers, etc. in order to maintain co-ordination in appearance.

The units will be fixed back to the wall by either being plugged and screwed to solid walls or by use of toggle fixings to hollow walls (Figs 6.6(b) and 6.7 respectively).

Cupboard fronts Ranges of bedroom and bathroom furniture are also available in mass-produced form, employing similar construction techniques to those of mass-produced kitchen furniture. Wardrobe and cupboard fronts are also manufactured. These consist of a softwood framework comprising of jambs, top rail and bottom rail. Sometimes an intermediate rail is used to split the cupboards into top and bottom storage areas. The fronts are delivered to the site with the doors already fitted. The doors are usually either:

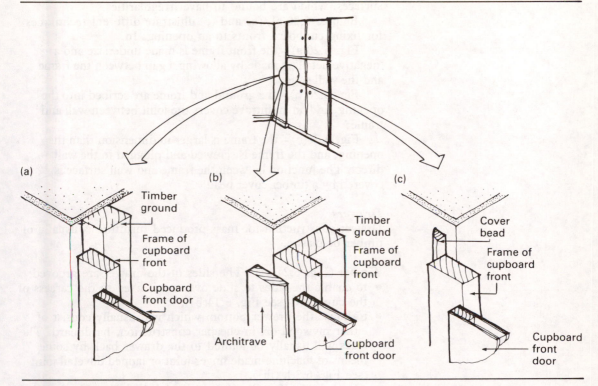

Fig. 6.20 *Built-in cupboard fronts*
(a) *Recessed ground*
(b) *Use of architrave*
(c) *Fixed direct to wall*

- plywood- or hardboard-faced flush doors;
- louvred timber doors;
- sliding or sliding/folding doors (sometimes available with mirror surfaces).

The fronts may be used to either fit into:
- a recess (such as at the sides of a chimney breast);
- for fitting into the corner of a room, where a side panel will be necessary (which some manufacturers supply).

A typical range of cupboard fronts might be of 2300 mm in height, and between 600 and 1200 mm in width. In order to fit a specific opening size, the cupboard front will often require 'packing out' and 'scribing in'. 'Packing out' involves the fixing of timber grounds to the opening to which the frame of the cupboard may be screwed. The grounds may be fixed to the wall by use of masonry bolts or plugs, for example (see Figs 6.6(a) and (b)). Packing out has the added advantage of providing a means of fixing the front frame true in the vertical.

'Scribing in' involves shaping the frame of the front with a plane, in order to fit the frame exactly to the wall/ceiling surfaces, which are bound to have irregularities.

Figures 6.20(a), (b) and (c) illustrate different techniques for fixing cupboard fronts to an opening. In:

Fig. 6.20(a) – the front frame is made undersize and a 'negative detail' is made by allowing a gap between the frame and the wall edge;

Fig. 6.20(b) – the ground and frame are scribed into the opening, and an architrave covers the joint between wall and frame;

Fig. 6.20(c) – the frame is larger in dimension than the opening, and the frame is screwed and plugged to the wall direct. The junction between the frame and wall surface is covered by a timber cover bead.

Drawers

Drawer construction for mass-produced furniture is usually of timber or plastic.

Timber (Fig. 6.21(a)) The sides of the drawer are grooved:
- to enable the sides to slide over a rail fixed to the carcass of the cupboard (see Fig. 6.14(b));
- to house the drawer bottom which will usually consist of either plywood or, in cheaper construction, hardboard. The sides will usually be jointed to the drawer back by some means of machine-made finger joint or lapped dovetail joint (see Fig. 6.3(h)(iii)).

In the case of a drawer front which is flush with the sides (such as that illustrated in Fig. 6.14(a)) a machine-made lapped dovetail joint might be employed. In the case of a drawer front which overlaps the sides, some form of dowel joint might be used (Fig. 6.3(a)(ii)). An alternative form of construction (Fig. 6.21(a)(ii)) involves the sides and back of the drawer being made in one continuous piece of plywood. The plywood is grooved for receiving the drawer base, and for the drawer to slide. In cheaper drawer construction, the front and back might be made of veneer or plastic laminate-faced chipboard.

Plastic (Fig. 6.21(b)) Many of the cheaper drawers are made with plastic moulded sides. As with timber construction, the sides are grooved to:
- locate in a slider rail fixed to the carcass of the cupboard;
- to house the drawer bottom (which will usually be of hardboard).

The detail illustration indicates the jointing system for fixing the sides to the back of the drawer by means of moulded corner blocks with projecting lugs (Fig. 6.21(b)). The lugs

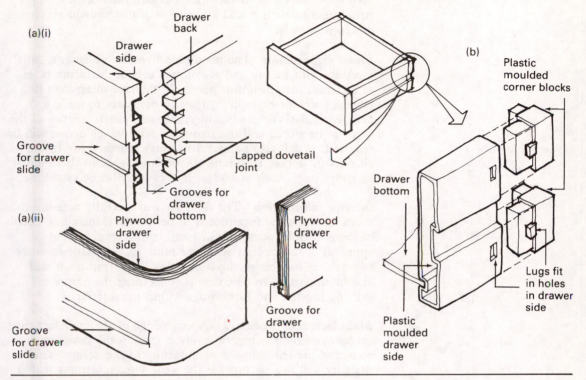

Fig. 6.21 *Drawer construction*
(a)
 (i) *Timber*
(ii) *Plywood*
(b) *Plastic*

locate in holes in the drawer side. A similar system is used to fix the sides to the drawer front, where plastic lugs are screwed to the inside face of the drawer front in order to locate in holes positioned at the front of the drawer sides. The drawer front is usually made from plastic laminate-faced chipboard. An alternative form of plastic drawer comprises of a complete moulded drawer – sides, bottom and back attached to a chipboard drawer front.

Purpose-built, built-in furniture

Design of purpose-built built-in furniture
The design of purpose-built built-in furniture requires both an appreciation of the physical requirements of the furniture, and an understanding of the manufacturing processes involved.

Physical requirements Some of the factors involved in the design of purpose-built built-in furniture are unique in that they pertain to a specific client and a specific interior. Having deduced the nature of these unique factors, the designer may then apply expertise in such fields as anthropometrics, structure/construction and knowledge of materials to evolve design proposals.

Client information   The main objective of a designer's initial meetings with a client and/or eventual user of the furniture, is to ascertain the client/user needs. In the case of an item like a storage unit, for example, it may be necessary to make a detailed analysis of the quantity, size and characteristics of the items to be stored and the frequency with which access will be required. It will also be useful to gauge whether any future changes in storage requirements are envisaged. In these initial meetings items such as budget figures will also be established.

Interior information   The designer will carefully measure the room in which the furniture is to be installed, including details of items such as skirting board and cornice profiles. An appraisal of the interior structure must also be made in order to establish suitable methods of fixing the furniture. It will also be important to note details concerning the nature of existing finishes and furnishings within the interior.

Manufacturing processes   A knowledge of the manufacturing process and the machinery involved enables the designer to recognise the feasibility of manufacture of the design. The designer will usually provide the joinery manufacturer with a set of drawings, showing overall plans, elevations and sections to a scale of usually 1 : 5 or 1 : 10, and details to a scale of usually 1 : 2, or full size. These drawings will enable the joinery manufacturer both to estimate a price for the work, and also later to produce a 'workshop rod' which is a full-size drawing, usually made on a plywood or hardboard. The workshop rod (or 'setting-out rod') is used throughout the manufacturing process as a pattern for setting out the joints, profiles, etc. on to the timber. A 'cutting list' lists the quantity, size, shape, material and finish for each of the components of the furniture.

Manufacturing purpose-built built-in furniture
In most furniture joinery workshops, there will be four distinct areas of activity: (*a*) preparation; (*b*) machining; (*c*) assembly; (*d*) finishing.
   Usually the joinery manufacturer will aim to prepare as much of the furniture as is possible in the workshop, thus reducing the amount of time spent on the site. For small joinery items made largely on the site, hand tools and power hand tools might be used.

Preparation   The joiner will take timber from the stock, and saw it into suitable sections and lengths of approximately the correct dimensions. The sawing will usually be carried out by some form of 'circular saw' such as the table saw illustrated in Fig. 6.22(a). The saw blade consists of a steel-toothed circular disc which is rotated by motor. The blade is housed in a table to enable the timber to be fed through the saw. The saw may be used for 'cross-cutting' (cutting timbers to length) or 'ripping' (cutting timber longitudinally). A 'bandsaw' (Fig. 6.22(b)) may be used for cutting shaped pieces. Here, the blade consists of a continuous toothed steel band which travels around two rotating wheels (one top and one bottom). The blade passes through a table along which the timber is fed. The table surface of most bandsaws is adjustable in angle to enable bevelled cutting to be carried out.

After cutting, the timber is usually planed in order to:

- provide the timber with flat surfaces with faces and edges at true right angles to each other;
- to reduce the timber to the correct dimension of cross-section.

A 'surface planer' is a machine which enables a face and adjacent edge to be planed to a true right angle. A 'thicknesser' is used to both produce a true right angle to the other face and adjacent side, and to reduce the timber to a cross-section of specific dimensions. Frequently the two operations are carried out on one machine, a 'combined surface planer and thicknesser' (Fig. 6.22(c)(i)) which incorporates facilities for both operations.

The surface planer (Fig. 6.22(c)(ii)) consists of a cylindrical cutter block which houses two cutters and is rotated by motor. The cutter block is recessed between two tables; a front table from which the timber is fed, and a back table. The angle of the front table is adjustable in height in order to vary the depth of cut. The timber is fed from the front table, over the cutters and through to the back table. A thicknesser (Fig. 6.22(c)(iii)) consists of a table which is adjustable in height and a cutter block with cutters installed above the table. The depth of cut is varied by adjusting the height of the table. The timber is fed, good face down, from the front table, through the infeed rollers, under the cutting block, through the outfeed rollers, on to the back table.

Machining   After sawing and planing the timbers, the joiner 'sets out' the work. This involves drawing full-size sections of the timbers on a piece of board known as a 'setting-out rod' which will be used as a pattern to mark out the sections, shapes and joints on the pieces of timber. The timbers may then be machined to provide the necessary joints, profiles and sections.

Spindle moulder (Fig. 6.22(d))   A spindle moulder is used to cut profiled sections in a length of timber for purposes such as making moulded beadings, architraves and lippings. The machine consists of a vertical rotating spindle to which shaped cutter blades are fixed. The spindle is located above a table which enables the timber to be fed against the cutter blades. The cutters may be either chosen from the wide range of factory-made profiles, or purpose made for a particular profile. The timber is fed:

• from the front of the table alongside a fence, the distance between the fence and cutter blades determining the depth of cut;

• past the cutter blades which reduce the timber to the required profile;

• on to the back of the table.

Figure 6.22(d)(ii) illustrates some typical profiles which may be achieved on a spindle moulder.

Lathe (Fig. 6.22(c))   A lathe is used to produce timbers of circular cross-section; however, the diameter of the cross-section may be varied throughout the length of the timber in order to produce items such as shaped rails and legs. For 'turning' (forming an item on a lathe) an item such as a rail, a square section of timber is fixed to the spindles of the headstock at one end, and the tailstock at the other.

Fig. 6.22 *Joinery machinery*
(a) *Table saw*
(b) *Bandsaw*
(c)
 (i) *Combined surface planer and thicknesser*
 (ii) *Detail of surface planer*
 (iii) *Detail of thicknesser*
(d)
 (i) *Spindle moulder*
 (ii) *Shapes possible from a spindle moulder*
(e) *Lathe*

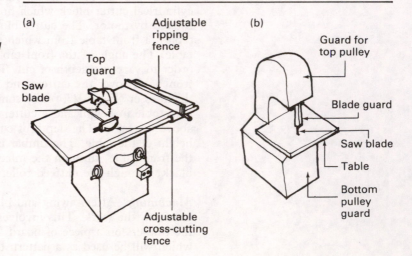

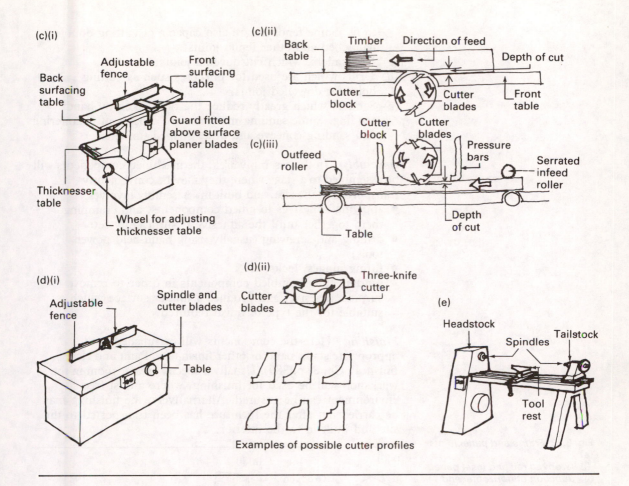

(c)(i)

Back surfacing table

Adjustable fence

Front surfacing table

Guard fitted above surface planer blades

Thicknesser table

Wheel for adjusting thicknesser table

(c)(ii)

Back table

Timber

Direction of feed

Depth of cut

Cutter block

Cutter blades

Front table

(c)(iii)

Cutter block

Cutter blades

Pressure bars

Serrated infeed roller

Outfeed roller

Depth of cut

Table

(d)(i)

Adjustable fence

Spindle and cutter blades

Table

(d)(ii)

Cutter blades

Three-knife cutter

Examples of possible cutter profiles

(e)

Headstock

Spindles

Tailstock

Tool rest

The motor drives the spindle of the headstock which rotates the timber. A gouge, chisel or scraper is held carefully against the toolrest, and the cutting edge of the gouge, chisel or scraper is gradually moved towards the revolving timber. The timber is gradually reduced to the required circular profile.

Other machines Other machines include:

(a) Mortisers, which are used for cutting rectangular-shaped holes for mortise joints. The machine acts rather like a drill where the drill bit is housed in a hollow square-shaped chisel. As the chisel is brought down, a square hole is formed in the timber.

(b) Tenoners, which are used for cutting tenon joints. The mortises are cut first on the mortiser, and the tenon is cut

307

to fit. Some tenoners are also capable of cutting double, haunched and other tenon joints.

(c) Dovetailers, which cut dovetail joints.

(d) Drills which are used for purposes such as cutting holes for dowelled joints.

(e) Sanders which greatly reduce the time taken by hand sanding. Some sanding machines are capable of chamfering and sanding concave and other shaped timbers.

Assembly   In the assembly area, the machined components will be assembled to a stage where they may be conveniently transported to the site, and built in. Assembly involves:

- applying adhesives to glued components, and cramping them together until the adhesive is sufficiently set;
- drilling and screwing (usually using hand-held power tools);
- fitting hinges, locks, etc.;
- sanding the assembled components in order to remove surplus adhesive, and produce a timber surface which is suitable for the type of finish required.

Finishing   Here the components will be finished with appropriate stain, paint or other finish (see 'Paint and clear finishes', Ch. 8, p. 389). Usually a special area or room in the workshop will be used for finishing, where a dust-free environment can be ensured. Alternatively, the finishing may be carried out after the furniture has been transported to the site and built in to the interior.

Fig. 6.23 *Frame and panel joints*
(a)
 (i) *Grooved rail and inset panel*
(ii) *Junction of horizontal and vertical grooved rails*
(b)
 (i) *Rebated rail with inset panel*
(ii) *Rebated rail with flush panel*
(c) *Frame and flush panel*

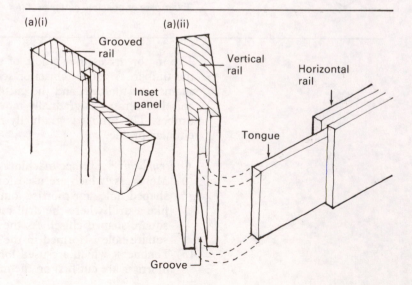

308

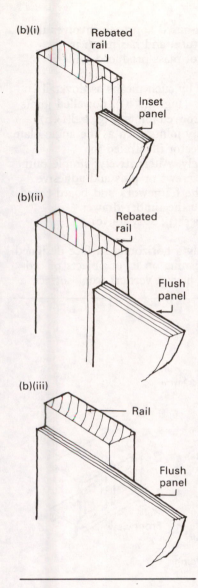

(b)(i)

Rebated rail

Inset panel

(b)(ii)

Rebated rail

Flush panel

(b)(iii)

Rail

Flush panel

Construction methods for purpose-built built-in furniture

Cupboards Purpose-built built-in cupboards are usually based on either a timber framework carcass (Fig. 6.1(a)) or a board-based carcass (Fig. 6.1(b)).

Timber framework carcass This traditional approach to built-in furniture construction is based on a skeleton framework infilled or covered by panels of board material such as plywood. A framework carcass involves the forming of many more joints than that of a board-based carcass, and consequently is more time consuming to manufacture.

Traditionally, the framework rails would have used mortise and tenon joints such as the 'L' and three-way haunched mortised and tenoned joints (Figs. 6.3(g)(iii) and (iv)). Today, dowelled joints might be used as an alternative (Figs. 6.3(a)(ii), (a)(iv) and (b)).

The frame rails housing panels would have traditionally been grooved, resulting in panels being inset from the face of the frame (Fig. 6.23(a)(i)). The junction between two grooved rails might be made with a tongued and grooved joint (Fig. 6.23(a)(ii)). Simpler forms of panelling may be achieved by use of:

- a rebated rail, with either a flush or an inset panel (Figs. 6.23(b)(i) and (ii) respectively);
- by applying the panel to the face of the framework (Fig. 6.23(c)). This arrangement would be unsuitable for instances where the edge is visible, due to the unsightly nature of the edge junction.

The top of the cabinet might be designed to either overlap or fit flush with the side. If fitted flush, the junction between the top and the panel rail might be formed by such means as:

- fitting the top into a rebate cut into the rail (Fig. 6.4(b));
- use of a dovetail joint, which provides considerable strength to the junction (Fig. 6.3(h));
- by use of a dowelled joint (Fig. 6.3(c)(iii)), although where the panel rail is slender, insufficient strength may be derived from such a joint.

If fitted with an overlapping top, the junction between the top and the panel rail might be formed by:

(*a*) a dowelled joint; (*b*) a stopped housing joint (Fig. 6.3(e)(ii)).

Both flush and overlapping joints could be fitted with either angle connecting plates or plastic jointing blocks (Figs 6.3(c)(i) and (ii)) or other means which enable rapid and simple site assembly.

Board-based carcass The board-based approach involves the construction of a box-like structure, and has much in common with the construction of mass-produced built-in furniture.

The boards may be jointed by such means as dovetail joints for rigid, strong junctions (Fig. 6.3(h)), dowelled joints (Fig. 6.3(c)(iii)), tongued and grooved joints (Fig. 6.3(f)(ii)) or, for quick *in situ* assembly, by joints such as the angle plate connector or plastic block connector illustrated in Figs. 6.3(c)(i) and (ii) respectively. Alternatively, simple butt joints may be used, fixed with screws or pins and adhesive. Drawer construction for both the framework and board-based carcass will usually be similar to the timber drawer construction described in 'Timber' (p. 302, above).

Fig. 6.24(a) *Counter unit – framed carcass*
(b) *Counter unit – board-based carcass*

Counters A counter is basically a horizontal surface designed to support the load of people leaning on it, or objects placed on it. A counter will often also have a vertical front surface

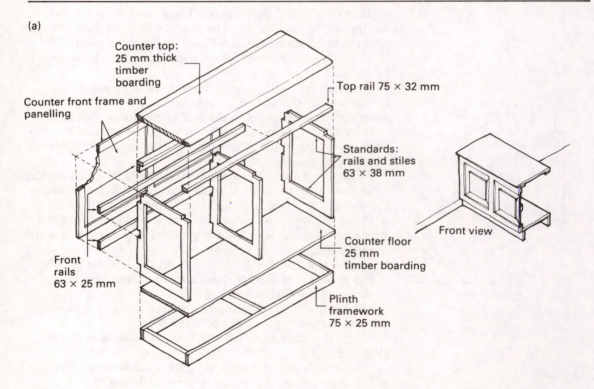

(a)

Counter top: 25 mm thick timber boarding

Top rail 75 × 32 mm

Counter front frame and panelling

Standards: rails and stiles 63 × 38 mm

Front rails 63 × 25 mm

Counter floor 25 mm timber boarding

Front view

Plinth framework 75 × 25 mm

6.24a

(b)

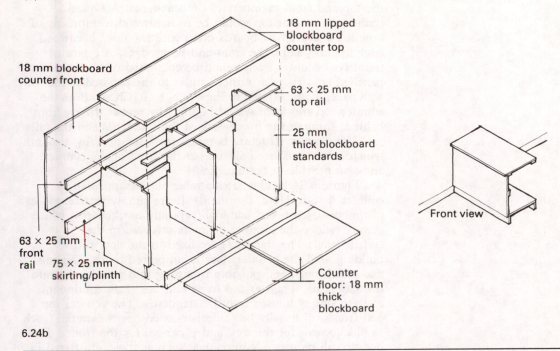

18 mm lipped
blockboard
counter top

18 mm blockboard
counter front

63 × 25 mm
top rail

25 mm
thick blockboard
standards

Front view

63 × 25 mm
front
rail

75 × 25 mm
skirting/plinth

Counter
floor: 18 mm
thick
blockboard

6.24b

designed to act as a physical and/or psychological barrier.

The counter top and front will be supported by a structure based on a series of vertical 'standards'. The spacing of the standards will be determined by:

- the loads which the counter is designed to carry;
- the nature and thickness of material used for the counter top.

The standards are linked together by rails, which both:

- provide rigidity to the supporting structure;
- provide suitable support for the front and back edges of the counter top and the top and bottom edges of the counter front.

Like the floor unit cupboards described in 'Mass-produced built-in furniture' (p. 297, above), counters may be built on either a framed carcass, or a board-based carcass.

Figure 6.24(a) illustrates a traditional framed carcass structure, where the standards consist of frameworks of timber fixed together, for example, with haunched mortise and tenon joints.

The top and front rails are continuous and are housed into the top and front members of the standards. A typical traditional counter top would be of hardwood, comprising of tongued and grooved boards fitted with moulded hardwood edge lippings along the front and back edges. A typical traditional counter front would be composed of a framed and panelled construction, rather similar to the exposed framework wall panelling illustrated in Figs. 8.9–8.10 (Ch. 8). Both the counter top and front would be fixed by screws driven from behind, through the framework of the standards and the rails.

The counter structure is raised from the floor by a 'plinth' which forms a recessed skirting at the front of the counter, and also provides a bottom shelf.

Figure 6.24(b) illustrates a more simple approach to counter design. Here, the standards are simply vertical pieces of timber-based board material (typically blockboard), which greatly reduce the number of joints involved in the construction. The standards are fixed to the floor of the building, and the counter floor is supported by the skirting and plinth rails, and possibly by fitting into grooves cut into the standards. The top and front rails are again continuous and are simply housed into the standards. The counter top and front will usually be of timber-based board material (such as blockboard, for the top, and plywood, for the front). These boards will require edge lippings, such as those illustrated in Figs. 6.4(a), (b) and (c).

Shelves Like counters, shelves are horizontal surfaces, designed to support loads. This section looks at examples of fixed and adjustable shelving as applied to: (*a*) walls; (*b*) cupboards or other furniture.

Wall shelving – fixed shelves Perhaps the simplest and cheapest means of fixed shelf provision is by use of pressed steel brackets (Fig. 6.25(a)(i)), which are either screwed and plugged to the wall direct, or screwed to timber battens fixed to the wall. The range of bracket sizes is in the region of 125 × 100 mm to 355 × 300 mm.

Shelves built into a wall recess could be simply fixed, by screwing and plugging horizontal bearers to the back and sides of the recess, and screwing the shelves to the bearers.

Wall shelving – adjustable shelves Figure 6.25(a)(ii) illustrates an aluminium adjustable shelving system comprising of upright supports and brackets. The brackets have lugs formed at the back, and these locate and lock into slots cut in the upright supports. The upright supports are screwed and plugged to the wall at intervals suited to the thickness and expected loading of the shelves. These shelving systems often offer a range of accessories such as bookshelf ends, brackets fitted with rubber stoppers suitable for glass shelving and sloping brackets for magazine display shelves.

Cupboard shelving – fixed shelves Fixed shelving in a cupboard may be simply achieved by such means as:
- housing the shelf in a groove cut in the cupboard end (see Figs. 6.3(e)(i) and (e)(ii));
- by use of dowelled joints;
- by fixing battens to the back and sides of the cupboard, fixing a cross rail under the front edge and pinning a thin plywood shelf to the battens and cross rail.

Cupboard shelving – adjustable shelves Figure 6.25(b) indicates two forms of adjustable cupboard shelving. The adjustable

Fig. 6.25
(a)
(i) *Steel shelf bracket*
(ii) *Adjustable aluminium shelving system*
(b)
(i) *Adjustable aluminium cupboard shelving system*
(ii) *Adjustable shelf pins*

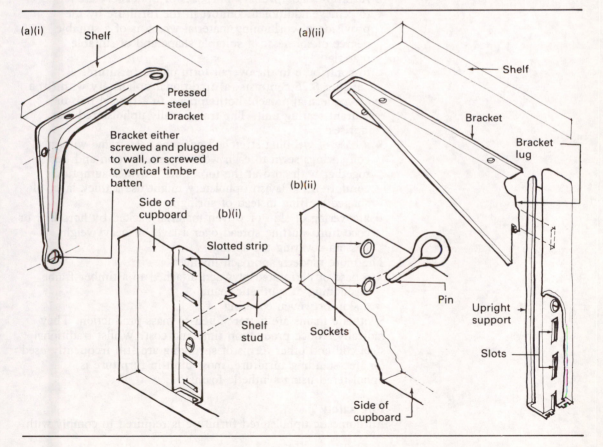

313

aluminium strip system (Fig. 6.25(b)(i)) comprises of pairs of slotted strips which are screwed to the cupboard sides, and aluminium shelf studs with lugs which locate in the strip slots. The shelves rest on the lugs. Figure 6.25(b)(ii) illustrates a system comprising of brass sockets which are pushed into holes drilled in the cupboard sides. Brass pins (on which the shelves rest) are pressed into the sockets. Similar fittings are also available in plastic.

UPHOLSTERY, CURTAINS AND BLINDS

Upholstery

Introduction

Upholstery consists of the cushioning and covering of an item of furniture. The primary purposes of upholstery are:

- to achieve additional comfort in the furniture by the provision of cushioning material which is of a suitable degree of softness, of suitable shape and of suitable material;
- to contribute to the overall form of the furniture.

Figure 6.26 contrasts (a) traditional upholstery with (b) a more modern approach; both applied to a typical pub or restaurant seating unit. The traditionally upholstered seat comprises:

- a base of webbing stretched over the seat frame with steel coil springs sewn to each webbing intersection and tied together with cord at the top. This provides sumptuous comfort. Less lavish upholstery might use a thick layer of shaped stuffing instead of springing;
- a covering of skin or woven material backed by horsehair or coco-fibre stuffing spread over a layer of heavyweight hessian covering the springs.

The more modern approach involves

- a base of perforated plywood attached to a timber frame cushioning of synthetic foam;
- a skin or woven covering.

Synthetic foams are better suited to mass-production. They radically reduce production time and cost. Whilst traditional steel coil and other forms of springing are still frequently used for free-standing furniture, most built-in furniture is upholstered using synthetic foams.

Fire safety

All domestic upholstered furniture is required to comply with

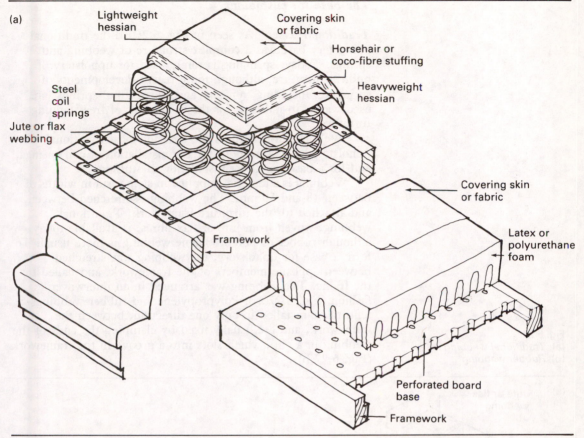

(a)

Lightweight hessian

Covering skin or fabric

Horsehair or coco-fibre stuffing

Heavyweight hessian

Steel coil springs

Jute or flax webbing

Framework

Covering skin or fabric

Latex or polyurethane foam

Perforated board base

Framework

Fig. 6.26
(a) *Upholstered seating – traditional approach*
(b) *Upholstered seating – modern approach*

the Furniture and Furnishing (Fire Safety) Regulations, 1988, which make reference to British Standard 5852, an assembly test where a sample chair is set alight. The test is in two parts: part 1 ignition source 0 (burning cigarette) and source 1 (butane flame/match) and part 2 ignition sources 2–7.

Domestic upholstery fabrics must pass the match test, although certain other fabrics containing at least 75 per cent of cotton, linen, silk or wool are permitted if used with an interliner or barrier fabric which passes part 2 ignition source 5.

Foam fillings must be of CMHR composition (see polyurethane foam pp. 318–19) passing part 2 ignition source 2.

The following section looks firstly at the methods used for forming a base for cushioning, and secondly at the cushioning itself.

The base for cushioning

Traditional bases   As seen in Fig. 6.26(a), the traditional upholstery base was a complex structure of webbing and springs. Today, springing is rarely used for upholstery of built-in furniture, although more recent developments of springing, including 'tension' and 'serpentine' springs, are used widely in industry for the upholstery of free-standing furniture.

'Webbing' was used as the base for most traditional upholstered furniture both for the installation of coil springs, and for a base for direct installation of stuffing.

Webbing is a material supplied in strip form in widths of between 18 and 50 mm. The webbing is stretched between, and attached to, the furniture framework. Traditional webbing, woven from jute or flax fibres, is available today, although rubber or polypropylene webbing is more usual. To form a base for upholstery, the webbing was stretched between opposite members of the framework, and nailed to the frame. The webbing was arranged in an interwoven fashion (Fig. 6.27(a)). Polypropylene and rubber webbings tend to be installed in only one direction, between the framework, and are usually fixed by clamping the ends of the webbing in a clip, which slots into a groove in the framework (Fig. 6.27(b)).

Fig. 6.27
(a) *Traditional webbing*
(b) *Rubber webbing*

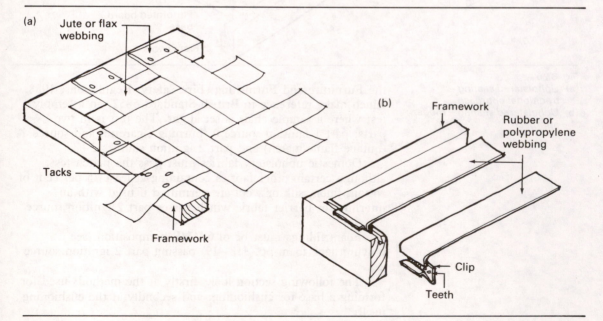

(a) Jute or flax webbing

Tacks

Framework

(b) Framework

Rubber or polypropylene webbing

Clip

Teeth

Modern bases The simplest base for a modern upholstered built-in unit consists of a board material such as plywood or blockboard fixed to, or recessed into, the furniture framework. The board should be perforated to allow for ventilation when plastic or rubber-based foams are used for the upholstery cushioning, since unlike traditional upholstery bases, modern foams do not 'breathe'. Alternatively, a base of timber slats could be used.

Cushioning

Traditionally, fibrous materials such as horsehair or coco fibre were used for stuffings. The loose fibre would be packed and shaped over a layer of heavy hessian covering either coil springs or webbing. A later development was the use of 'stuffing pads' which comprised of fibres woven into a hessian backing by a needle-point process. The pads are available either in rolls, or in cut lengths.

Today, latex and polyurethane foams are almost always used for the cushioning of upholstery for built-in furniture. Fibre stuffings might still be used in exceptional circumstances, such as the restoration of furniture in a historic interior.

Latex foams Latex foams are made from either naturally occurring latex obtained from rubber trees, from synthetic sources, or from a combination of both. Latex foams are very durable and resilient, but are expensive in comparison with polyurethane foams.

The softness or hardness of latex foams are directly related to their density; seven grades of density are available, varying from 'super soft' to 'extra firm'.

Some of the available forms of latex foam are:
- 'plain sheet' which is available in typical sizes of 1800 × 1400 mm and 12.5 or 25 mm thickness;
- 'cavity sheeting' (Fig. 6.28(a)) has a series of circular cavities on one side, which results in a thicker, less dense material. Cavity sheeting is available in typical sizes of 1800 × 1400 mm and 25–100 mm in thickness;
- 'pincore foam' (Fig. 6.28(b)) which has a series of small-diameter holes formed in both surfaces of the sheet.

Pincore foam has the advantages of:
- being reversible (cavity foam is not);
- being more robust both in handling and in its installed condition, than cavity foam;
- being, to an extent, self-ventilating.

Figure 6.29 illustrates the ways in which an edge detail might be achieved in both cavity and pinhole latex foams.

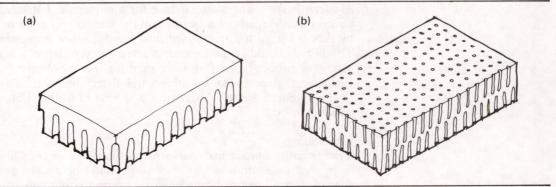

Fig. 6.28
(a) *Latex cavity foam*
(b) *Latex pincore foam*

When cavity foam is used to create a square edge (Fig. 6.29(a)), a 'side wall' is necessary in order to provide an even density at the edge. Tape is applied to the vertical surface of the foam, and the tape is fixed to the board base with tacks. The 'feathered edge' (Fig. 6.29(b)) provides a gentle rounded edge, and is achieved by:

- cutting the edge of the foam to a 45° taper;
- applying tape above the tapered edge;
- pulling the tape and foam down to the base, and taking the tape to the base.

The 'cushioned edge' provides a firmer edge and is formed in a similar manner to the feather edge, except the edge is left square prior to being pulled down to the base (Fig. 6.29(c)).

Polyurethane foam   Polyurethane foam has become perhaps the most widely used material for upholstery cushioning, largely due to its cheapness and simplicity in handling and fabrication.

Whereas the relative softness or hardness of latex foams are directly proportional to their density, polyurethane foams are available in differing combinations of hardness and density, such as soft high-density foam, or hard low-density foam. Typical densities of foam available are:

- 16–24 kg/m^3 for light upholstery work, such as headrests and arm cushions;
- 24–29 kg/m^3 for back cushions and seats for chairs;
- 29–48 kg/m^3 for seat cushions and hard-wearing upholstery.

Typical thicknesses of foam used for a seating unit might be 90 mm of firm grade foam, or 115 mm of softer grade. Forming edge details are much the same for polyurethane foam as for latex square, feathered and cushioned edges.

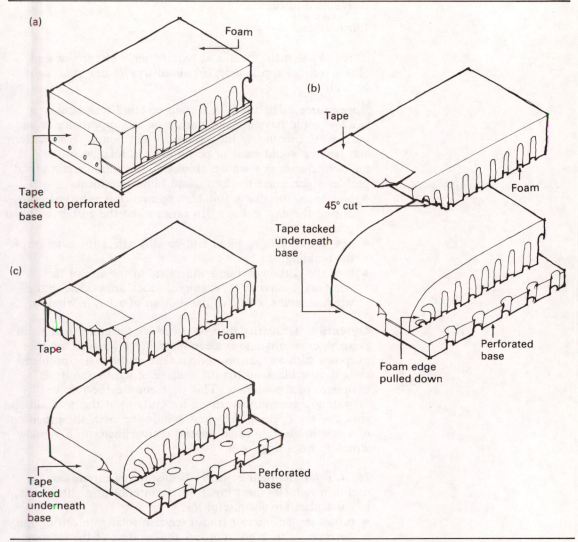

Fig. 6.29
(a) *Square edge*
(b) *Feather edge*
(c) *Cushion edge*

CMHR foams Standard polyurethane foam is a fire hazard since when alight it emits toxic smoke and fumes.

Foam used for upholstered domestic furniture is required to be of CMHR (combustion modified high resilience) composition (see 'Fire safety' p. 314). These are polyurethane foams modified by graphite and melanine additives which retard the time before combustion occurs.

Curtains and blinds

Introduction

Three of the main functional requirements of curtains and blinds are: (*a*) appearance; (*b*) obscurity; (*c*) heat/solar/light control.

Appearance   The use of a curtain or blind treatment to a window could have a major impact on the appearance of an interior. Consequently the nature of the colours, style, pattern and texture would need to be resolved in relation to the interior scheme as a whole. However, there are certain visual problems particular to curtain and blind treatments:

- how to ensure that a suitable appearance is achieved both during the day and at night time, when the curtain or blind is drawn;
- how the curtain or blind will visually affect the exterior of the building;
- how the curtain or blind will relate to the size of the window. A curtain, for example, which draws clear of the window jambs, can give the illusion of a larger window.

Obscurity   Obscurity may be necessary for either reasons of privacy, or in order to achieve black-out for rooms used for purposes such as film projection. One of the major problems in achieving black-out, is the spillage of light around the perimeter of the window. This problem may be resolved by providing a generous overhang for curtains at the head sill and jambs of the opening, or where blinds are used, the provision of a channel section fitted around the perimeter of a window which forms a light trap.

Heat, solar and light control   Whilst heavy curtains are useful in reducing heat loss from an interior at night, blinds, in particular, are also useful for:

- protecting an interior from excessive solar gain during the daytime – this both improves the comfort of the interior, and reduces the deteriorating effect which ultraviolet light has on certain finishes;
- enabling glare from sunlight to be deflected.

   In designing either a curtain or a blind treatment, it is important to consider the type of opening window. Pivot windows and casement stays might intrude into interior space when the window is open, which could interfere with the fabric of a drawn curtain or blind.

Curtains

Curtaining is a system for suspending fabric on a 'track' or 'rod' such that it may be pulled or 'drawn' to one side of an opening. This section looks at: (*a*) forms of curtain treatment; (*b*) headings, pelmets and valances; (*c*) tracks and rods.

Forms of curtain treatment   Figure 6.30 (a, b, c, d, e) illustrates some of the decorative treatments applied to curtains:

Fig. 6.30(a). – Café curtains are short curtains suspended from rod or pole which is often attached to a window transom. The form of curtain is often of the 'scalloped and strap' variety consisting of straps which hook over the rod, and rounded scallops cut between the straps. Alternatively, the curtain could be hung by a series of wooden or metal rings which are sewn to the top of the curtain and fit over the rod.

Fig. 6.30(b). – Pelmets are stiff lengths of material which span the head of the curtain in order to form a decorative treatment, and to conceal the curtain track or rod. Figure 6.30(b)(i) illustrates a plain pelmet, whilst (b)(ii) illustrates a decorative form with 'festoons' (hanging loop-shaped folds of fabric) and 'cascades' (falling folds of fabric at the sides of the pelmet).

Fig. 6.30(c)(i) – Tie-back curtains are designed to be gathered at the side and held by such means as a decorative tie, sleeve or cord. Figure 6.30(c)(ii) illustrates a decorative treatment known as a 'puff'. Here a secondary curtain is held back permanently against the side of the window, whilst an ordinary drawn curtain is fixed behind it.

Fig. 6.30(d) – Tiered curtains consist of two or more curtains hung one over the other. The top tier will usually be hung over the window head, whilst lower tiers will be hung from transoms (Fig. 6.30(d)).

Fig. 6.30(e). – A valance is a means of concealing the curtain track or rod by means of a length of fabric hung above the curtain (Fig. 6.30(e))

Curtains may be made either unlined, lined or interlined. Unlined curtains are usually used in locations such as kitchens or bathrooms, where the curtains may require frequent cleaning. 'Sheer' curtains (used mainly for daytime privacy, sometimes referred to as 'net' curtains) are unlined.

Lined curtains have a lining fabric (usually cotton) sewn to the back. The lining both improves the hanging characteristics of the curtain and provides a measure of protection to the curtain fabric from the effects of sunlight. A further advantage is that from the exterior of a building, the

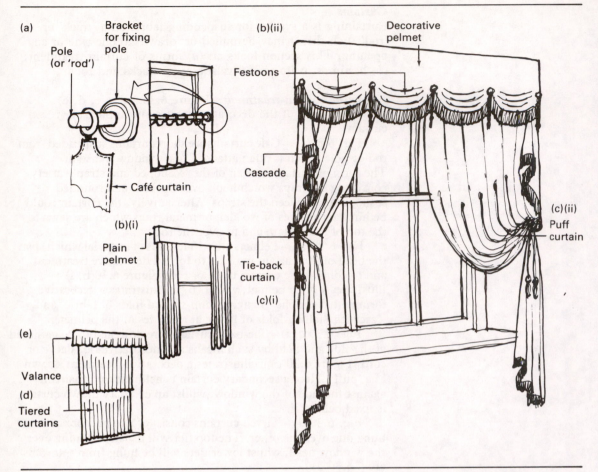

Fig. 6.30 *Decorative forms of curtain*
(a) *Café curtains*
(b) *Pelmets*
 (i) *Decorative*
(ii) *Plain*
(c)
 (i) *Tie-back curtain*
(ii) *Puff curtain*
(d) *Tiered curtains*
(e) *Valance*

effect of different fabrics seen in different windows will be eliminated. Detachable linings may also be made which are useful for instances where the lining fabric and curtain fabric require different washing treatments.

Interlined curtains are lined with two layers of lining fabric. This provides a luxurious appearance to the curtains, improves the hanging characteristics and increases the degree of obscuration and thermal insulation and so aids insulation.

Headings, pelmets and valances Headings, pelmets and valances are methods of forming a neat and acceptable appearance at the top or 'head' of a curtain.

Headings   The 'heading' is the area at the top of a curtain which incorporates the system of curtain suspension from a track. Figure 6.31(a) illustrates a typical method of curtain suspension, where the basic elements comprise of:

- a track which is fixed, by means of brackets, above the head of the window opening;
- a series of 'runners' which are capable of sliding freely along the track, and which enable the attachment of curtain hooks;
- curtain hooks are fitted into the runners and are also inserted into the pockets of the 'heading tape'. In some curtain suspension systems, the hook and runner are incorporated into one fitment (see Fig. 6.31(b)(i)).

The heading tape is a strip of fabric which is sewn to the back of the curtain, and has a series of 'pockets' into which the hooks may be inserted. A typical construction of heading tape is 15 mm width of cotton, nylon or polyester fabric, which has two parallel lines of slots cut into it, and two lines of cord threaded through. When the cords are pulled, the heading tape bunches, which causes the slots to form into pockets.

Different combinations of heading tape and hook enable different styles of heading to be achieved. Figure 6.31(b) illustrates some of these different styles of heading:

Fig. 6.31(b)(i) – The use of 'standard' tape (typically 15 mm deep) produces a random, ruffled effect. Headings of this nature are often covered by a pelmet or valance. When the tape cords are tightened, the curtain will 'gather'.

Consequently, the width of fabric for the curtain should be in the order of $1\frac{1}{2}$ times the width of the window opening.

Fig. 6.31(b)(ii) – Deep triple pleats are achieved by the use of four-pronged hooks and deep pleating heading tape. To allow for the gathering, the fabric should be $2\frac{1}{4}$–$2\frac{1}{2}$ times the width of the opening.

Fig. 6.31(b)(iii) – 'Pencil pleats' are formed by use of a deep heading tape (often 75 mm in depth) which incorporates three sets of cords. To allow for gathering, the curtain fabric should be $2\frac{1}{4}$–$2\frac{1}{2}$ times the width of the window opening.

Pelmets and valances   Pelmets and valances are means of concealing the suspension system of a curtain.

Pelmets are made of either solid materials such as wood, plywood or hardboard, or stiff fabric such as 'buckram' (a coarse fabric stiffened by gum or paste).

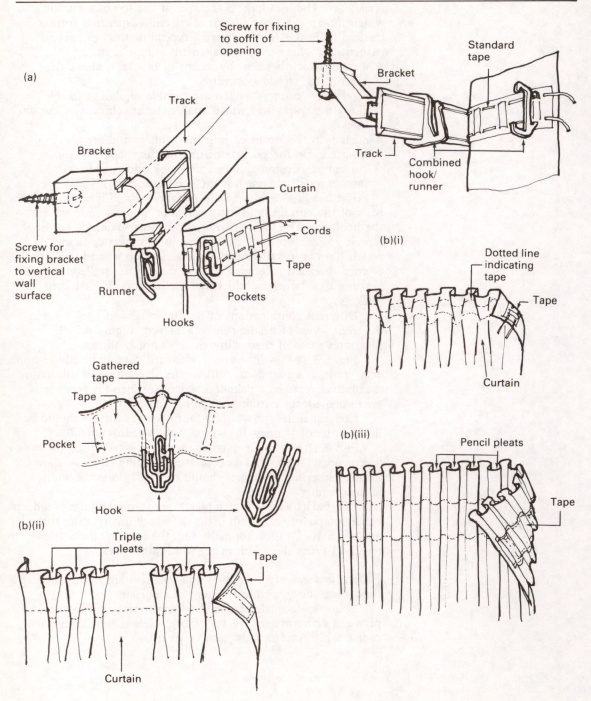

(a)

Screw for fixing to soffit of opening

Bracket

Standard tape

Track

Combined hook/runner

Bracket

Track

Screw for fixing bracket to vertical wall surface

Curtain

Cords

Tape

Runner

Hooks

Pockets

(b)(i)

Dotted line indicating tape

Tape

Curtain

Gathered tape

Tape

Pocket

Hook

(b)(iii)

Pencil pleats

Tape

(b)(ii)

Triple pleats

Tape

Curtain

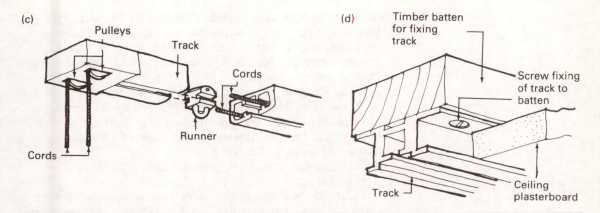

(c)

Pulleys

Track

Cords

Runner

Cords

(d)

Timber batten for fixing track

Screw fixing of track to batten

Ceiling plasterboard

Track

Fig. 6.31
(a) *Typical track system*
(b) *Curtain hanging details*
 (i) *Combined hook runner*
 (ii) *Deep triple pleats*
(iii) *Pencil pleats*
(c) *Cord-operated track*
(d) *Flush-fitting track*

Pelmets may be plain such as that illustrated in Figs. 5.10 and 5.12(c) (Ch. 5), where the pelmet forms a continuation of the cornice detail of the room, or decorated (such as that illustrated in Fig. 6.30(b)(ii)).

Wood or board pelmets are usually finished with paint, whilst buckram pelmets are usually covered with the same fabric used for the curtains. The pelmet is usually fixed to either a batten fixed to the ceiling, or to a simple wooden frame fastened to the wall with steel angle brackets. Valances are unstiffened lengths of fabric, usually of similar material to the curtain (see Fig. 6.30(e)). The valance is usually gathered by some form of pleating, and is suspended by a valance rail which is usually attached to the curtain track.

Both valances and pelmets may be embellished with braid, fringes, tassels, etc.

Tracks and rods   Tracks at one time were commonly made of brass; however, these have been largely superseded by plastic and aluminium tracks which are cheaper, and quieter in operation. Galvanised steel tracks may be used for very heavy installations.

An enormous range of styles and profiles of curtain track is available; however, in selecting a track, it is essential to ensure that:
● the track is capable of carrying the weight of the curtains;
● that the track is suitably supported.

For most domestic installations, a hand-operated curtain will suffice. For large domestic and commercial buildings, cord-operated curtains may be more convenient. Here, a system of cords operate on the runners, in conjunction with either pulleys or ball-bearing carriers installed at each end of the track. The cords themselves are usually housed within the track, sometimes in a separate channel (see Fig. 6.31(c)). Electrically operated curtains may be used for domestic purposes, but are particularly useful for interiors such as lecture theatres, where one person may operate all the curtains from a central console.

The fixing of the track may be by such means as:
- by use of brackets screwed and plugged to the vertical wall surface above the window opening (Fig. 6.31(a));
- by use of brackets screwed to the soffit of a window opening (Fig. 6.31(b)(i));
- by fixing the track flush with the ceiling finish (Fig. 6.31(d)).

Curtain rods (or 'cornice poles') are a simple alternative to curtain tracks. A pole is fixed by brackets to either the reveals of a window opening (such as the café curtains – Fig. 6.30(a)) or to the wall surface above the head of the window opening (in a similar manner to the hanging of the festoon blind in Fig. 6.32(a)).

Materials used for rods include timber, brass and aluminium anodised with bronze, brass or gold. When the rod is fixed above the window opening a finial (which is usually of an ornamental character) is fitted to each end of the rod.

The curtains are fitted with heading tape and hooks, usually to produce some form of deep-pleated heading. A series of wooden or metal rings fit over the pole, and each ring has a smaller ring fitted at the bottom. The curtain hooks fit into these small rings. Curtain rods are usually restricted to window openings of 3 m or less in width.

Blinds

Blinds are made in many different styles and materials and with different modes of operation. Festoon, roller, Roman, venetian and vertical louvre blinds are well-known examples. Most blinds may be fitted either to the soffit of the window opening, or to the wall above the head of the opening.

Festoon blind (Fig. 6.32(a)) The festoon blind provides an opulent effect due to the large amount of ruched material which is used (twice the width and three times the length of the window opening).

The blind operates by means of a cord which hangs from one side of the blind in a similar manner to the Roman blind. When the cord is pulled, a series of cords are raised, which are threaded into ringlets sewn to the back of the blind fabric.

Roller blinds (Fig. 6.32(b))   Roller blinds are perhaps the simplest form of blind, comprising of a length of fabric attached to a wooden or metal tube which is fixed at either end to brackets screwed and plugged above the window opening. The tube usually contains a spring-loaded mechanism, which enables the blind fabric to be rolled over the tube.

Roller blinds are economical in fabric, since they use little more fabric than the size of the window opening. They are often used in conjunction with curtains, and rails are available which combine curtain-rail and roller-blind facility.

'Pinoleum' blinds are operated in a similar fashion to roller blinds, but instead of fabric, a series of thin wooden strips are used, which are bound together with cotton.

Roman blinds (Fig. 6.32(c))   Roman blinds are similar to roller blinds, but differ in means of operation. Tapes with eyelets sewn in are sewn to the back of the blind fabric. Cords pass through the eyelets and are attached to a cord, which hangs at the side of the blind. Pulling the cord causes the blind to be raised in a concertina fashion, causing a horizontal pleated effect. When the blind is raised, the gathered pleats provide a pelmet effect. The head of the blind illustrated in Fig. 6.32(c) is concealed by a plain pelmet.

Like the roller blind, the Roman blind uses fabric which is of similar dimension to the window opening.

Venetian blinds (Fig. 6.32(d))   Venetian blinds consist of a series of thin horizontal slats made of materials such as plastic, metal or wood, which are able to be raised and bunched together at the window head, or lowered to the window sill. When in a lowered position, the slats may be tilted in order to either diffuse or deflect sunlight. The slats are usually 50 or 80 mm in width. Figure 6.32(a) shows a blind which uses a combined tilt and raise/lower mechanism operated by a cord. Sometimes these mechanisms are separated. Venetian blinds may also be operated by a pole with folding/cranked handle or electrically by remote control. Special venetian blinds include:
- external blinds operable from inside;
- blackout blinds which fit in a perimeter light-trap channel.

Micro blinds   Neat, slim-fitting venetian blinds are available in widths of 35, 25 and 16 mm. These provide good obscurity, but due to their slim size, a high degree of transparency when open. The smaller-width blinds may be installed within the air-space of double-glazing units and are dust and dirt free.

Vertical louvre blinds (Fig. 6.32(e))   Vertical louvre blinds are comprised of a series of vertical vanes made of materials such as fibreglass, fabric or aluminium. The vanes, which are usually between 90 and 120 mm in width, are suspended from a channel fixed at the head of the window opening. The vanes may be drawn across the opening, then angled in order to deflect or diffuse sunlight.

The vanes are usually weighted at the bottom and linked together by lightweight chains. The louvres are usually operated by either cord, pole with cranked handle, or electrical remote control, all of which usually operate both the drawing/withdrawing and angling of the vanes.

Some vertical louvre blind systems may be installed between the glass of a double-glazed window, and some are capable of installation with curved window openings.

Both venetian and vertical louvre blinds are made to order from blind manufacturers/suppliers, who will require accurate measurements and details of the window opening.

Fig. 6.32 *Blinds*
(a) *Festoon blind*
(b) *Roller blind*
(c) *Roman blind*
(d) *Venetian blind*
(e) *Vertical louvred blind*

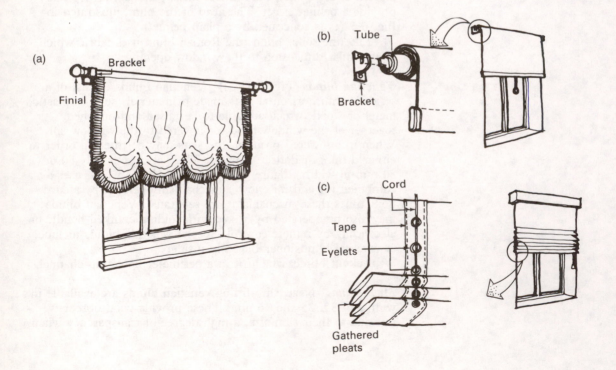

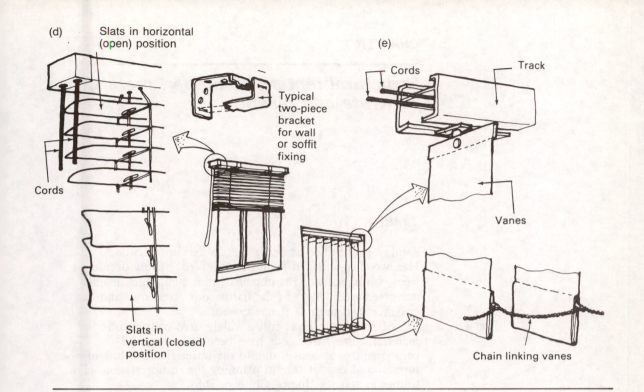

(d)

Slats in horizontal (open) position

Cords

Typical two-piece bracket for wall or soffit fixing

Slats in vertical (closed) position

(e)

Cords

Track

Vanes

Chain linking vanes

CHAPTER 7

Structural materials: timber, metals, concrete and plastics

TIMBER

Timber is classified as either 'hardwood' or 'softwood'. Hardwood is obtained from broad-leaved, mainly deciduous trees, whilst softwood is obtained from coniferous, mainly evergreen trees. Most of the hardwoods are denser and stronger than most of the softwoods.

Although timber is still a widely used construction material, some hardwoods have become scarce, and consequently expensive, due to the unrestricted felling of forests, and lack of careful planning for reafforestation. (For timber joints, see 'Joints' Ch. 6 p. 268)

Growth of timber

A tree consists of crown, trunk and roots. The crown is the branch system which supports a wide surface area of leaf growth for the collection of sunlight and air. The trunk supports the crown and provides the means of passage of food between the leaves and roots of the tree. The root system spreads out into the soil, both to provide stability for the tree and to absorb moisture and certain mineral salts from the soil.

Figure 7.1(a) illustrates a typical cross-section through a tree-trunk showing its composition, consisting of a series of consecutive growth rings or 'annual rings'. Each growth season, a new ring will grow around the tree, beneath the bark. The basic composition of the growth ring is a series of squarish tube-shaped cells which run parallel to the length of the trunk and branches. The inner part of the growth ring is known as 'early wood' or 'spring wood' since it grows in the early part of the growth season, and the outer part is known as the 'late wood' or 'summer wood' since it grows in the later part of the season. The tubular cells of the early wood tend to be larger in area and the thickness of the cell walls less than

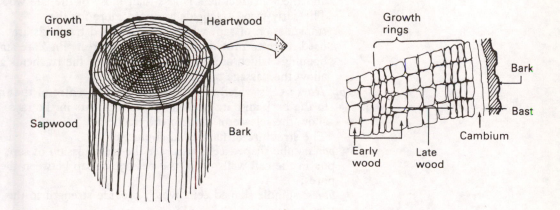

(a)

Growth rings

Heartwood

Growth rings

Bark

Bast

Cambium

Sapwood

Bark

Early wood

Late wood

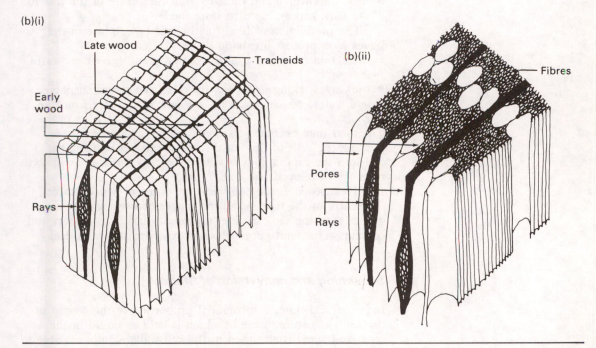

(b)(i)

Late wood

Tracheids

Early wood

Rays

(b)(ii)

Fibres

Pores

Rays

Fig. 7.1
(a) *Cross-section through a tree-trunk*
(b) *Cell structure of wood*
(i) *Softwood*
(ii) *Hardwood*

those of the late wood. The late wood grows more slowly, and is consequently rather more dense and darker in colour.

Figs 7.1(b)(i) and (ii) show highly magnified diagrams comparing the structure of softwood and hardwood. The structure of softwood comprises:

(*a*) '*Tracheids*': the hollow thin cells which form the bulk of the timber. The tracheids in early wood are larger in area and the wall thickness is less than those of the late wood. The early-wood tracheids are used more for the transmission of sap, whilst the late-wood tracheids are used more to provide strength for the tree. 'Pits' are small openings which are found in the walls of the tracheids and allow the passage of sap between them.

(*b*) '*Rays*': cells which run radially from the centre of the tree to the bark and are used to store food. Pits in the rays allow the release of food to the tracheids.

The structure of hardwood comprises:

- *pores*: tube-shaped cells used for the transmission of sap; pits in the cell walls allow the passage of sap between the pores;
- *fibres*: spindle-shaped cells which provide strength to the tree;
- *rays*: cells which run radially from the centre of the tree to the bark and are used to store food.

The production of food for the tree takes place in the leaves by a process involving both:

- the action of sunlight on the 'chlorophyll' (green matter of leaves);
- the leaves taking in carbon dioxide from the atmosphere, and water, which passes up the cells of the tree from the roots.

The food then becomes dissolved in the sap water, passes through the cells of the 'bast' (cells found beneath the bark) and into the cambium cells where the cells for the new growth rings are developed (see Fig. 7.1(a)).

'Heartwood' is the non-living or 'dormant' wood nearer the centre of the tree, which no longer conveys sap. It is denser than the sap-carrying 'sapwood', darker in colour, and contains resins and gums which provide resistance to decay.

Seasoning and conversion of timber

In its natural state, a substantial proportion of the weight of timber is moisture, some of which is held as liquid in the cells, and some is absorbed in the cell walls. Timber is able to give off or absorb moisture in relation to the humidity of its surroundings. When a piece of timber is dried, the moisture contained in the cells will first be drawn out and evaporate. 'Fibre saturation point' is reached when the cell walls are saturated, but no liquid is left within the cells. After further

Table 7.1 Moisture contents for timbers

Location of timber	Moisture content of timber (%)
External joinery	16
Internal joinery (a) With intermittent heating conditions	15
(b) Continuous heating (12–18 °C)	12
(c) Continuous heating (20–24 °C)	10
(d) Close to heat source	8

drying, the timber will lose moisture from the cell walls, and the timber will shrink.

A piece of freshly felled timber placed in a dry internal environment will dry out and shrink whilst, conversely, dry timber placed in a humid environment will take in moisture and swell.

It is important that timber used for a particular location should have a moisture content which is compatible with the humidity of that location. This moisture content is known as the 'equilibrium moisture content'. As a guide, Table 7.1 shows equilibrium moisture contents for timber in various locations.

The movement in timber due to shrinkage does not occur uniformly, and when timber is dried beyond the fibre saturation point, distortion may occur. Figure 7.2 illustrates the three types of movement that might occur in timber:

(a) Longitudinal, which takes place along the length of the trunk, but causes minimal dimensional change.

(b) Tangential, which takes place parallel to the direction of the growth rings. In some timbers tangential movement

Fig. 7.2 *Movement in timber*

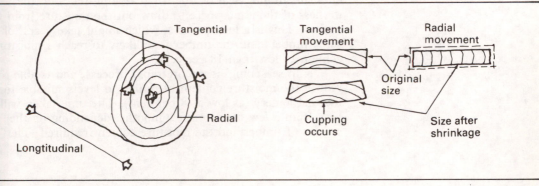

333

can cause a 3 per cent difference in the dimension of the timber. The shrinkage causes the timber to bend in the opposite direction to the growth rings; this is known as 'cupping'.

(c) Radial, which takes place perpendicular to the growth rings. For most timbers, the amount of radial movement is only about half of that caused by tangential movement.

Conversion and seasoning

Once a tree has been felled, it is transported to a sawmill where it is sawn (or 'converted') into boards. Figure 7.3 (a, b, c) illustrates three of the methods used for converting timber. The method of conversion will vary in accordance to the species of timber being converted, and the eventual use of the timber.

Fig. 7.3(a) – *Tangential*: here the long faces of the boards are tangential to the line of the growth rings.

Fig. 7.3(b) – *Quarter sawn*: here the growth rings run parallel to the short face of the board. Consequently, the problem of shrinkage is minimised, and the boards are suitable for interior joinery.

Fig. 7.3(c) – *Through and through*: whilst the boards furthest from the centre have growth rings arranged similarly to tangential sawn timber, the boards nearer the centre are similar to quarter-sawn boards.

A felled tree rapidly loses its moisture, which can lead to splits in the end and the surface of the timber. The timber is converted and dried ('seasoned') as soon as possible after felling to avoid the likelihood of splitting. The timber is seasoned in order to reduce its moisture content to a level compatible with the humidity level of its eventual location.

The traditional process for drying timber is 'air seasoning'. The boards are carefully stacked and lengths of timber batten are placed between the boards to ensure that air can freely circulate around each board. Air seasoning relies on the dryness of the atmosphere to draw out the moisture from the timber. This is a long process, which might take years for some timbers and the timber is unlikely to reach a moisture content of less than 18 per cent.

'Kiln seasoning' is a more rapid process, and is able to reduce the moisture content of timber to levels suitable for internal joinery (as low as 8 per cent). The time taken will vary from a few days to a few months, depending on the species of timber and the moisture content required. The

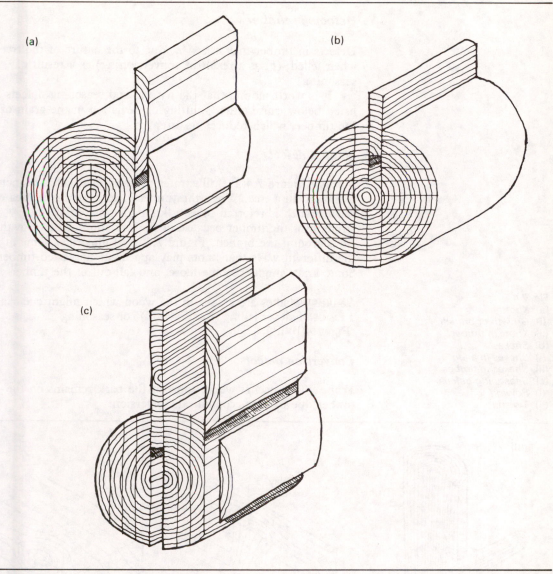

Fig. 7.3
(a) *Tangential conversion*
(b) *Quarter-sawn conversion*
(c) *Through-and-through conversion*

timber is stacked, then passed into a kiln building where the temperature is low, and the humidity high. The temperature is gradually increased, and the humidity reduced, until the moisture content is reached.

Defects in timber

Defects in timber may arise: (*a*) due to the nature of the tree when felled; (*b*) as a result of conversion; (*c*) as a result of seasoning.

It is worth noting that the natural and seasoning defects listed below cause either splitting or deviation in the grain of the timber, which reduces its strength.

Natural defects

Knots   Figure 7.4(a)(i) illustrates the way in which a branch is formed in a tree. When the timber is converted, the area of sawn branch is referred to as a 'knot'. Knots reduce the strength of the timber because of the deviation of the growth rings around the branch. Figure 7.4(a)(ii) illustrates some of the different ways that knots may appear in converted timber. Some knots might become loose, and fall out of the timber.

Shakes   Shakes are cracks in the wood which might occur as a result of either felling (Fig. 7.4(b)(i)) or seasoning (Fig. 7.4(b)(ii)).

Conversion defects

Wane   This occurs where some of the bark remains on the timber due to over-economical conversion.

Fig. 7.4
(a) *Knots*
 (i) *Growth of branch*
(ii) *Knots in timber*
(b) *Shakes*
 (i) *Shakes in a log*
(ii) *Shakes in timber*
(c) *Seasoning defects*
 (i) *Bowing*
(ii) *Twisting*

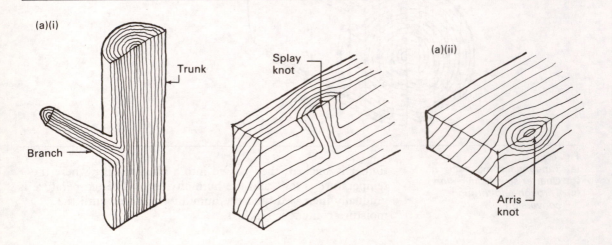

(a)(i)   Trunk   Branch

Splay knot   (a)(ii)   Arris knot

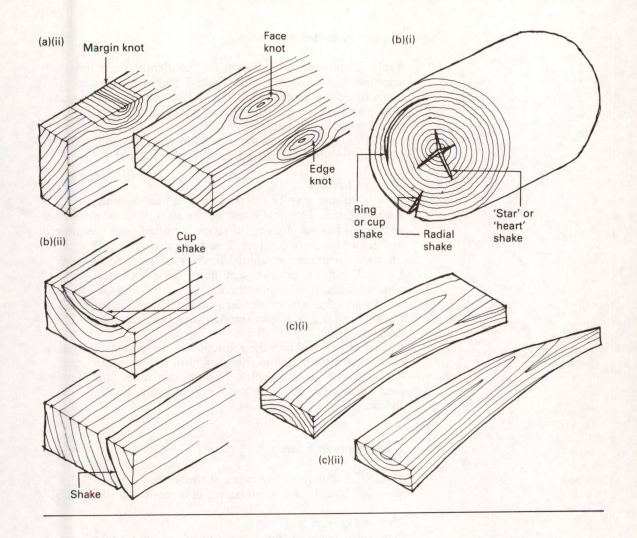

Sloping grain This occurs where the conversion of the timber is not carried out parallel to the axis of the tree, and consequently the grain will not run parallel to the edges of the board. Excessive slope of grain will reduce the strength of the timber.

Seasoning defects
Seasoning carried out with inadequate care can result in defects such as 'cupping' (see Fig. 7.2) resulting from tangential shrinkage, and 'bowing' and 'twisting' (Figs 7.4(c)(i) and (c)(ii) respectively) which result from shrinkage throughout the length of the timber.

Timber varieties and sizes

Table 7.2 provides information for the identification and selection of softwoods, whilst Table 7.3 provides similar information for hardwoods. It should be borne in mind that, as a generalisation, the strength of a particular timber is proportional to its density.

Table 7.4 gives a guide to the dimensional stability of some of the timbers referred to in Tables 7.2 and 7.3.

Sizes of timbers

Sizes of timber may be referred to as either 'nominal' or 'finished' sizes. Nominal sizes are the sizes of the sawn timber, which still has rough-sawn surfaces. Finished sizes are those of timbers which have been machined to provide smooth surfaces. Sometimes machined timbers are referred to as 'ex' 150 × 75 mm or 'ex' 100 × 50 mm. These 'ex' sizes are nominal sizes, prior to machining.

Table 7.5(a) shows standard nominal sizes for softwood, whilst Table 7.5(b) shows standard nominal sizes for hardwoods.

Table 7.5(c) shows the reduction in dimensions of a piece of nominal timber due to the planing of two opposite faces, for example a piece of 100 × 25 mm sawn timber for joinery would end up as 100 − 9 × 25 − 7 mm, that is, 91 × 18 mm.

Timber-based boards

Figure 7.5 illustrates five types of timber-based board: plywood, blockboard, laminboard, chipboard and fibreboard.

Plywood (Fig. 7.5(a))

Plywood is made of thin layers ('plies') of timber which are bonded together, face to face, resulting in a board material which is stiffer and stronger than solid timber of similar thickness.

Odd numbers of plies are always used, otherwise the plywood would be liable to warp. The direction of grain of adjacent plies run in opposite directions, which:

• provides the plywood with similar strength characteristics in its length and in its width (unlike solid timber where the strength parallel to the grain will be different to that perpendicular to the grain);
• results in moisture movement being very much less than that of solid timber.

Table 7.2   Characteristics of some softwoods

Name	Colour	Appearance	Approx. density (kg/m³)	Sources of timber	Typical uses
Douglas fir (British Columbia pine)	Pinkish/red-brown	Straight, even grain. Free from knots	530	USA/Canada	General structural work. Joinery, flooring
Hemlock	Light grey/brown	Slight sheen on surface	480	USA/Canada	General structural and joinery work
Larch	Heartwood: red/brown. Sapwood: yellow/brown	Straight grained, fairly free from knots	590	Europe	Fencing, gates and flooring
Parana pine	Heartwood: brown, sometimes with red streaks. Sapwood: almost white	Straight, uniform grain fairly free from knots	480–640	South America	Internal joinery and flooring
Pine (Scots pine)	Heartwood: dark brown. Sapwood: almost white	Straight grain, resinous. Difficult to obtain without knots	530	Europe	Structural work, cladding, joinery and flooring
Sitka spruce	Pale brown	Slight sheen on surface. Straight grain	450	USA/Canada	Internal joinery, kitchen furniture
Western red cedar	Heartwood: brown. Sapwood: light yellow	Straight grain, turns silver/grey after exposure to weather	380	Canada	External joinery and cladding. Internal joinery
Whitewood (European spruce)	Yellow/white	Has lustrous surface, tends to have knots	430	Europe	Internal joinery and furniture. Flooring
Yew	Straw pink	Wavy grain	700	Europe	Veneers, furniture and decorative joinery

Table 7.3 Characteristics of some hardwoods

Name	Colour	Appearance	Approx. density (kg/m³)	Sources of timber	Typical uses
Afrormosia	Golden brown	Close grained. Oily surface	700	West Africa	High-quality joinery. Furniture, veneers
Ash	White/cream	Rather coarse texture. Irregular grain	700	Europe	Good for steam bending. Decorative furniture, veneers
Beech	Light pinkish brown	Straight grain, fine texture	680	Europe	Good for steam bending. Furniture, veneers
Birch (European)	White/brown	Smooth surface	670	Europe	Large quantities used for plywood furniture
Chestnut	Straw brown	Similar appearance to oak	550	Europe	Good for steam bending. Internal woodwork. Panelling, veneers
Elm (wych elm)	Light brown (sometimes with greenish tinge)	Straight grain	670	Europe	Interior joinery. Veneers
Gaboon	Light pinkish brown	Straight grain, similar to mahogany	430	Equatorial Africa	Often used in plywood and other timber-based boards
Greenheart	Heartwood: dark brown with green tinge. Sapwood: pale yellow	Very hard. Free from knots	1000	South America	Flooring, stair treads
Iroko	Dark yellow, brown	Coarse texture	640	West Africa	High-class internal and external joinery. Flooring, veneers
Jarrah	Dark red	Very hard. Coarse texture, straight grain	850	Australia	External joinery. Cabinet-making

Name	Colour	Appearance	Approx. density (kg/m³)	Sources of timber	Typical uses
Mahogany (Honduras)	Yellow-brown/rich red-brown	Can have intricate grain pattern	550	South America	High-class joinery. Furniture, veneers
Mahogany (African)	Red-brown	Coarser grain than Honduras mahogany	560	West and East Africa	Internal and external joinery, veneers
Makore	Pink/red-brown	Close grained, lustrous surface	640	West Africa	Veneers for furniture. Flooring
Oak (European)	Yellow-brown	Coarse texture. Sometimes silver/grey grain. Pattern visible	720	Europe	Internal and external joinery. Furniture, veneers
Obeche	Light yellow	Dry open grain	380	West Africa	Cabinet-making
Padauk	Rich red	Coarse grain. Dark streaks	780	Burma	Panelling. Interior decorative joinery, veneers
Plane	Straw yellow to pale red	Straight grain, fine texture	640	Europe	Decorative veneers, furniture
Queensland walnut (Australian)	Variegated: pink-brown/dark brown	Often stripes of other colours	740	Australia	Fine interior joinery. Cabinet work, veneers
Ramin	Light yellow	Even texture, fairly straight grain	670	Borneo, Sarawak	Internal joinery. Furniture
Rosewood (American)	Variegated: rich purple/brown/yellow/black	Close texture	900	Brazil	Internal decorative joinery. Cabinet-making, veneers
Sapele	Reddish brown	Close texture	640	West Africa	Veneers for furniture and panelling
Teak	Golden brown	Fairly coarse texture	640	Thailand Burma	Internal joinery. External joinery. Furniture
Walnut (European)	Warm greyish brown	Darker markings	640	Europe	Veneers for furniture and panelling

Table 7.4

Timber		Tangential movement						Radial movement		
		As a percentage of the dimension of the timber								
		3.0	2.5	2.0	1.5	1.0	0.5	0.5	1.0	1.5
Timbers with small movement	Padauk									
	Western red cedar									
	Iroko									
	Rosewood (Indian)									
	Teak									
	Afrormosia									
	Obeche									
	Mahogany (American)									
	Mahogany (African)									
	Douglas fir									
	Hemlock									
	Makore									
Timbers with medium movement	Ash									
	Sapele									
	Scots pine									
	European walnut									
	Oak									
	Parana pine									
	Jarrah									
Large movement	Ramin									
	Beech									

Table 7.5(a)

Thicknesses (mm)	Widths (mm)								
	75	100	125	150	175	200	225	250	300
16	▓	▓	▓	▓					
19	▓	▓	▓	▓					
22	▓	▓	▓	▓					
25	▓	▓	▓	▓	▓	▓	▓	▓	▓
32	▓	▓	▓	▓	▓	▓	▓	▓	▓
38	▓	▓	▓	▓	▓	▓	▓		
44	▓	▓	▓	▓	▓	▓	▓	▓	▓
50	▓	▓	▓	▓	▓	▓	▓	▓	▓
63		▓	▓	▓	▓	▓	▓		
75		▓	▓	▓	▓	▓	▓	▓	▓
100		▓		▓		▓		▓	▓
150				▓		▓			
200						▓			
250								▓	
300									▓

Table 7.5(b) Nominal sizes for hardwoods

	Sizes (mm)
Thicknesses	16 19 25 32 38 50 63 75 100 125 150
Widths	75–300 in increments of 25; 450 also available
Lengths	1800–6300 in increments of 300

Table 7.5(c) Reduction of nominal timber sizes due to planing two opposite faces

	Sizes (mm)				
Purpose for timber		Nominal dimension of timber			
	Up to 22	Up to 35	Up to 100	Up to 150	Over 150
Constructional timbers			3	5	6
Floorings	3		4	6 over 100	
Tongued and grooved boards			4	6 over 100	
Joinery and cabinet work		7	9	11	13

Timber commonly used in plywood manufacture includes beech, birch, Douglas fir and gaboon. Adhesives used for bonding the plywood include animal and various synthetic glues.

Types of plywood Plywood is graded in Britain in accordance with the quality of the outer plies. Grade 1 is free from knots and other blemishes and is suitable in appearance for use with transparent finishes. Grade 2 may have occasional knots or blemishes, and is suitable for receiving paint or veneered finishes. Grade 3 has blemishes which would make it unsuitable for interior finishes, but may be used for construction which is unlikely to be visible.

Plywood is also graded in accordance with its intended use: 'Int.' – interior grade, which is suitable for only interior or dry locations; 'MR' – moisture-resistant grade, which provides only modest weather resistance; 'BR' – boil-resistant grade, which provides good weather resistance; and 'WPB' –

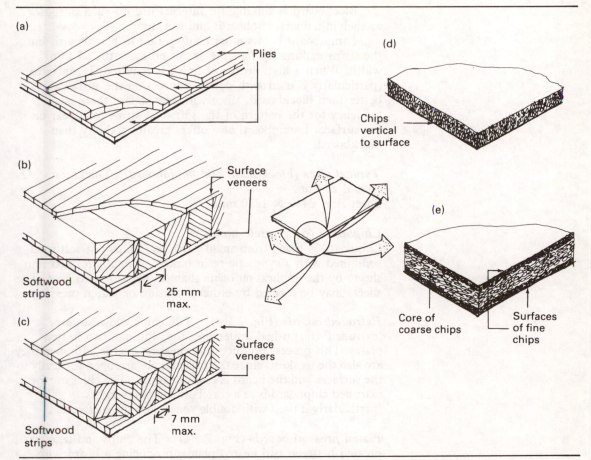

Fig. 7.5
(a) *Plywood*
(b) *Blockboard*
(c) *Laminboard*
(d) *Extruded chipboard*
(e) *Three-layer chipboard*

Labels in figure:
- (a) Plies
- (b) Surface veneers, Softwood strips, 25 mm max.
- (c) Surface veneers, Softwood strips, 7 mm max.
- (d) Chips vertical to surface
- (e) Core of coarse chips, Surfaces of fine chips

weather- and boil-proof grade, which has high resistance to weather, steam and heat.

Special types of plywood include decorative faced plywoods faced with wood veneers, metals or plastics.

Typical sizes Thicknesses: 4, 6, 9, 12, 15, 18 and 25 mm.
Sheet size: 2440 × 1220 mm.
Number of plies: 3, 5, 7, 9 and 11.

Blockboard and laminboard (Figs 7.5(b) and (c))
Blockboard is a sheet material which consists of a core of strips of softwood, each of up to 25 mm in width, with a sheet of veneer bonded to the top and bottom surfaces. The grain of the veneers runs in the opposite direction to the grain of the strips of softwood.

Blockboard is suitable for internal use for such purposes as shelving, doors, cupboards and worktops.

Laminboard is of similar composition to blockboard, but the strips making up the core are of no more than 7 mm width. When a high degree of finish is required, laminboard (particularly if used with double veneers on the surfaces) is better than blockboard, since with blockboard there is a tendency for the pattern of the softwood strips to appear on the surface. Laminboard also offers greater stability than blockboard.

Typical sizes (blockboard and laminboard)   Thicknesses: 12, 18 and 25 mm.
Sheet size: 2440 × 1220 mm.

Chipboard (or 'particle board')

Chipboard is made from small particles ('chips') of softwood combined with a synthetic resin binder, which is formed into sheets by the application of mechanical pressure and heat. The sheets may be formed by either extrusion or platen pressing.

Extruded boards (Fig. 7.5(d))   The chips and resin are 'extruded' (or pushed under pressure) through two heated plates. This process produces the cheapest boards, although they are also the weakest, since the particles are arranged vertically to the surfaces and the board is easily snapped. The chief use for extruded chipboard is as a core between two veneers (particularly if used with double veneers on the surfaces).

Platen pressed boards (Fig. 7.5(e))   The chips and resin are pressed between two heated plates, producing a board with particles arranged horizontally to the surfaces, which is consequently stronger than extruded boards.

Platen pressed boards are obtainable as either single-layer board consisting of a single layer of comparatively large particles, resulting in a coarse surface which is unsuitable for painting, or three-layer boards consisting of a core of large chips with a top and bottom surface of fine chips. The surfaces are sanded to a smooth finish.

Characteristics of chipboard   Chipboard is available in three densities – high density (640–800 kg/m^3), medium density (480–640 kg/m^3) and low density (less than 480 kg/m^3). Generally, the denser boards are stronger, high-density boards being suitable for a floor finish over timber joints whilst low-density boards would only be used where the chipboard was required to provide minimal support. The bending strength and rigidity of chipboard is less than timber plywood or blockboard, and consequently will require greater support. Chipboard will deform permanently if it is given inadequate support. It is little affected by movement due to changes in relative humidity, although unprotected edges of boards are vulnerable.

Special finishes   Boards are available with timber veneer surfaces, plastic laminate surfaces and boards may be obtained ready filled and sanded, suitable for applying a paint finish.

Typical sizes   Thicknesses: 12, 18, 22 and 25 mm. Sheet size: 2440 × 1220 mm.

Fibre building boards   These boards are made by the application of high pressure and heat to wood pulp. Medium-density fibreboard (MDF) is a recently developed material which is made by cured resin-impregnated wood fibres.

Hardboard

A dense material (800 kg/m^3) available in thicknesses of 1.3–6.5 mm suitable for purposes such as facing doors, lining walls and as a floor underlay. Special forms include:
- perforated hardboard (pegboard) for display and acoustic purposes;
- enamel finished, which has a hard gloss surface suitable for bath and other panels;
- decorative, with plastic laminate applied to the smooth face.

Medium hardboard   Available as low-density (350–560 kg/m^3) and high-density (560–800 kg/m^3) in thicknesses of 9–12 mm. It is suitable for ceiling linings and for pin display boards.

Insulation board (softboard)   A low-density material (less than 350 km/m^3) available in thicknesses of 12–19 mm, suitable for thermal and sound-insulation purposes.

Medium-density fibreboard (MDF)   Available in thicknesses of 2–35 mm, with a density of 600 kg/m^3. It is smooth on

Table 7.6 Characteristics of metals used in building interiors

Metal	Composition	Density (kg/m³)	Tensile strength (N/mm²)	Hardness (Brinell No.)	Melting point (°C)	Thermal conductivity (W/m K)	Colour	Typical uses
Ferrous metals								
Cast iron	Iron. Carbon content of 2–4%	7200	100–340	250	1150–350		Dark grey	Baths. Ornamental and garden furniture
Wrought iron	Iron. Carbon content of 0.02%	7850	355	100	1620		Dark grey	Gates. Decorative screens
Mild steel	Steel with carbon content of 0.15–0.25%. Also contains small amounts of other elements	7850	420–510	130	1370	52–62	Grey, but rusts rapidly if left untreated	Structural frameworks for all types of furniture
Stainless steel	Steel with at least 10% chromium. May have other elements such as Nickel	7700–7900	540	170		15	Natural finish light grey, may be polished to mirror finish	Sinks and industrial kitchen furniture. Structural frameworks for furniture
Non-ferrous metals								
Aluminium	'Pure' metal	2700	60–140	15–30	658	220	Whitish grey	Structural frameworks for furniture
Brass	Copper with zinc content of 3–50%	8380	370	75	904	129	Yellow, but darkens to dull brown if left untreated	Door furniture. Decorative trims to furniture
Copper	'Pure' metal	8940	215–250	50–60	1083	400	Pinkish brown, but turns pale green through oxidation if left untreated	Decorative table-tops

both surfaces and leaves a high-quality smooth finish when cut or shaped. It is suitable for internal joinery/furniture construction and available as standard and moisture-resistant grades.

METALS

Metals used inside buildings exist in nature in a chemically combined form known as 'ores'. The metal is extracted from the ore by either thermal reduction (a process involving heat and a 'reducing agent' such as carbon, which separates the metal from the ore) or 'electrolytic reduction' (a process often necessary for producing a purer form of metal). The principle of electrolysis is explained on p. 353 below.

Since the pure metal is very soft, it is often necessary for it to be further processed in order to increase its strength. This may be achieved by combining other chemical elements with the metal to produce an 'alloy'. The forming and joining of metals is considered in Chapter 6, p. 277. Characteristics common to all metals are:

- they are good conductors of heat and electricity;
- they are of high density;
- they have high 'melting points' (temperature at which the material melts). Mercury is the exception to this, being in liquid form at room temperature.

Table 7.6 compares some of the characteristics of metals used inside buildings and indicates some of their common uses. The figures given in the table should only be taken as a guide, since the properties of a particular metal will vary, depending on such factors as:

- whether the metal is in cast, rolled, extruded or other form;
- the precise chemical composition of the metal.

The notes below explain some of the terms used in Table 7.6:

(a) *Ferrous and non-ferrous metals*: ferrous metals are composed largely of iron, whilst non-ferrous metals contain little or no iron.

(b) *Hardness*: the degree of hardness is determined by a standard test, the higher the Brinell figure, the harder the metal. Hard metals are less easy to bend or cut.

(c) *Thermal conductivity*: these figures relate to the amount of heat that will pass through a 1 m^2 of material 1 m in thickness when the difference in temperature between the front and back surfaces of the material is 1 °C. The good conductors are indicated by high figures.

Ferrous metals

'Pig iron' is the basis of all ferrous metals, and is produced by the thermal reduction of iron ore. The thermal reduction takes place in a blast-furnace, where the iron ore is heated with other materials, including coke and limestone. The chemical reactions which take place in the blast-furnace result in molten iron being separated, whilst the coke and limestone combine with other materials in the ore producing gas and slag. The separated iron (pig iron) is impure, containing 3–4 per cent of carbon, and traces of other impurities.

Irons

Wrought iron Wrought iron is produced by melting pig iron, in a furnace, with iron oxide (basically rust). The impurities in the iron combine with the iron oxide to produce slag, leaving the iron in a relatively pure state, having a carbon content of about 0.02 per cent.

Wrought iron is strong, hard and 'ductile' (able to be stretched without fracturing) and is consequently easily shaped or bent either cold or with the application of heat.

Wrought iron is more resistant to corrosion than mild steel, although in exposed locations it will rust.

Cast iron Cast iron is produced by melting pig iron with steel scrap, resulting in an iron with a carbon content of 2–4 per cent. Various types and grades of cast iron are available, providing a wide range of tensile strengths and hardnesses. The cast irons used for such purposes as baths and garden furniture is known as 'grey cast iron'. Cast iron is suitable for casting and machining and is relatively resistant to corrosion.

Steels

Steel is, in effect, an alloy of iron and carbon, and is produced by heating pig iron and steel scrap in a furnace together with any necessary additional alloying materials.

The very wide range of steels available are classified as either:
(a) plain carbon steels; (b) alloy steels.

Plain carbon steels Plain carbon steels contain only small percentages of alloying materials, and are classified as either:
(a) Low-carbon steels (with carbon content of up to 0.15%) which are soft but tough and suitable for the production of rods, tubes and sheets.

(b) Mild steels (0.15–0.25% carbon) which are easily worked and welded. Mild steels are used for structural purposes in buildings in such forms as 'I' beams, channels, etc.

(c) Medium-carbon steels (0.20–0.50% carbon) which are harder and stronger, but less ductile than low-carbon or mild steels. These are used for forging and mechanical engineering items.

(d) High-carbon steels (0.50–1.50% carbon) are the hardest of plain carbon steels and are used for machinery items and tool manufacture.

Alloy steels Alloy steels may contain substantial proportions of alloying materials which provide different strength, wearing and working characteristics. Stainless steel is the best-known example, the chief alloying material being chromium. Stainless steel has excellent resistance to corrosion. The two stainless steels most commonly used in buildings are:

- '18/8' stainless steel (18% chromium and 8% nickel) which is used for internal purposes such as sinks;
- '18/8/3' stainless steel (18% chromium, 8% nickel and 3% molybdenum) which has better weather-resisting properties and is therefore more suitable for external uses.

Non-ferrous metals

Aluminium

Bauxite is the ore from which aluminium is extracted by electrolytic reduction. 'Pure aluminium' (containing at least 99% aluminium) is soft and weak, and therefore is restricted in use to non-structural purposes such as manufacture of foil or extruded trims. The strength of pure aluminium may be increased by 'cold-working' processes such as hammering and rolling, although alloying it with other materials such as magnesium and silicon will increase its strength further.

Aluminium alloys (which contain less than 99% of aluminium) are classified as either:

(a) 'Wrought alloys', which are suitable for forming by processes such as extrusion or rolling.

(b) 'Cast alloys', which are suitable for casting. Both wrought and cast alloys are obtainable as either:

- 'heat treatable' where the strength of the alloy may be increased by processes involving the controlled application of heat;
- 'non-heat treatable', where the strength is increased by cold-working processes such as hammering or rolling.

Copper

Copper is extracted from the ore as copper pyrites and is produced in several grades, including 'fibre-refined tough pitch copper' which is used for such purposes as sheet roofing, and deoxidised copper which is suitable for welding and is used for plumbing tubes.

'Annealing' (a process involving the heating and gradual cooling of a metal) results in increasing the strength of copper.

The best-known copper alloys are brass (copper and zinc alloy) and bronze (copper and tin alloy).

Brass

The zinc content of brass varies from 5 to 50 per cent. Other metals, such as tin, iron, zinc and aluminium may also be added in small quantities to improve particular qualities of the brass. Brasses with higher zinc contents are harder and have higher tensile strength.

The classification of brasses include:

(a) 'Alpha' brasses which have a zinc content of 10–38 per cent. They are strong, malleable and are golden yellow in colour. Alpha brasses include 'gilding brass' which is easily worked and is suitable for decorative purposes and 'cartridge brass' which may be shaped by pressing or spinning.

(b) 'Alpha-beta' brasses have a zinc content of 38–46 per cent; they are stronger but more brittle than alpha brasses and are light yellow in colour. 'Muntz metal' is an alpha-beta brass which is suitable for forming by casting or extrusion.

Bronze

'True bronze' is an alloy of copper and tin, the tin content ranging from about 5 to 40 per cent. Small quantities of other materials such as zinc and nickel might be added to improve certain qualities of the metal.

Bronzes are expensive metals, rich brown in colour, and are generally harder and stronger than brasses. Inside buildings, bronzes may be used for door furniture and ironmongery accessories.

Corrosion of metals

The corrosion of metals used in buildings results from an electrochemical process known as 'electrolysis'. Corrosion involves the destruction of metal, and in the case of ferrous metals (which tend to have little resistance to corrosion) may lead to severe weakening of the metal. In non-ferrous metals, a film of corrosion forms over the metal which tends to impair the development of further corrosion.

Electrolysis

Corrosion due to electrolysis requires:

(a) An 'electrolyte' which is a liquid, such as water, capable of conducting electricity. Impurities such as acids may speed up the rate of corrosion. Since air contains water (and in polluted localities, may also contain acids), the air surrounding a metal may act as an electrolyte.

(b) An 'anode' metal from which an electrical current will flow causing corrosion.

(c) A 'cathode' metal to which electrical current will flow and which is protected from corrosion.

Electrolysis in a single metal

Figure 7.6(a) illustrates the occurrence of corrosion due to electrolysis in a single piece of metal. The process is as follows:

- a part of the metal becomes anodic (areas of metal covered by dirt or scale may act as an anode);
- the air acts as an electrolyte, and an electric current will flow from the anodic metal through the air to a part of clean, exposed metal which becomes cathodic;
- the current passes from the cathodic metal, through the metal back to the anodic part of the metal;
- corrosion of the anode occurs as a result of electrically charged particles being removed from the anode, whilst the cathode is protected.

Fig. 7.6
(a) *Electrolytic corrosion in a single metal*
(b) *Corrosion of dissimilar metals*

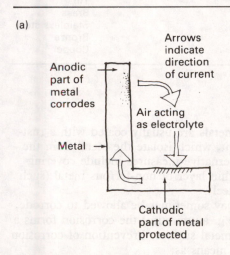

(a)

Anodic part of metal corrodes

Arrows indicate direction of current

Air acting as electrolyte

Metal

Cathodic part of metal protected

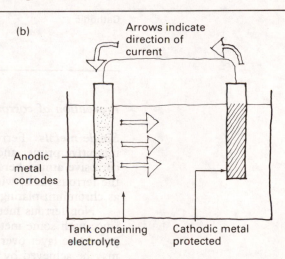

(b)

Arrows indicate direction of current

Anodic metal corrodes

Tank containing electrolyte

Cathodic metal protected

Corrosion in dissimilar metals

Electrolysis will also take place where dissimilar metals are in contact with an electrolyte. Figure 7.6(b) illustrates diagrammatically the process of electrolyte corrosion between two metals where:

- one metal acts as an anode;
- electrically charged particles are removed from the anode;
- an electrical current flows from the anode, through the electrolyte and back to the anode.

The list of metals in Table 7.7 shows metals that tend to be anodic at the top of the list, and metals that tend to be cathodic at the bottom of the list. The likelihood of corrosion is increased where metals from the top of the list are put in contact with metals from the bottom of the list such as copper with steel, or zinc with stainless steel.

Examples of corrosion of dissimilar metals occurring in a building might include:

- where different metals are used in a plumbing system and water flowing through the pipes acts as an electrolyte;
- where different metals are used in a roof construction, and rain falling on the roof acts as an electrolyte.

Table 7.7   Scale of anodic and cathodic metals

Anodic	Magnesium
	Zinc
	Aluminium
	Iron and steel
	Chromium
	Lead
	Tin
Cathodic	Brass
	Stainless steel
	Bronze
	Copper

Prevention of corrosion

Single metals   Ferrous metals are usually coated with a rust-inhibitive primer and paint, which isolate the metal from the corrosive atmosphere. Alternative measures include covering the ferrous metal with a thin layer of non-ferrous metal (such as chromium plating of steel).

Non-ferrous metals may sometimes be allowed to corrode, since with some metals (e.g. aluminium) the corrosion forms a protective layer over the metal surface. Prevention of corrosion may be achieved by such means as:

- forming a protective oxide layer over the metal (e.g. anodising aluminium, which is carried out by an electrolytic process);
- by painting the metal, after degreasing the surface of the metal and applying a suitable primer.

Dissimilar metals   Electrolytic corrosion of dissimilar metals may be avoided by the following measures:

(a) Careful design and selection of suitable materials to ensure that dissimilar metals are not in contact with water.

(b) Where dissimilar metals are in contact with water, providing protection by such means as installing a 'sacrificial metal'. Where, for example, a copper pipe enters a zinc-coated steel water tank, the zinc could become anodic, and corrode. By installing a piece of metal which is higher on the anodic scale (such as magnesium), the magnesium will become anodic instead of the zinc.

Electroplating

Electroplating is a process which utilises electrolysis in order to form a thin film of one metal over another metal. Chromium-plated steel is a well-known example of this process.

Electroplating is carried out by placing a suitable electrolyte in a tank and immersing the plating metal, and the metal to be plated, in the electrolyte. The plating metal (such as chromium zinc or tin) acts as an anode, whilst the metal to be plated acts as a cathode. When an electric current is applied, metal is released from the anode metal, and forms a uniform coating around the cathode metal.

CONCRETE

Concrete is a structural material which may be either:
- made and formed on the building site ('*in situ* concrete');
- made and formed in a factory ('precast concrete').

Concrete is made from a mixture of cement, water and aggregate. When cement and water are mixed to form a paste, the paste gradually changes from a liquid to a solid form. Aggregates (materials such as stones or sand) form the bulk of a concrete mix and when they are introduced to the water/cement mixture, the aggregates are gradually held together as the cement and water paste solidifies.

Cement and aggregates

Cement

When cement is mixed with water to form a paste, it initially 'sets' (becomes solid) then gradually 'hardens' (gains strength). During the process of setting and hardening (known as 'hydration') heat is generated, which may be of assistance in the manufacture of concrete in cold weather. The most widely used cement is 'ordinary Portland cement', the main constituents of which are limestone and clay. About 5 per cent of gypsum is added to prevent the cement from setting too rapidly.

'Rapid hardening Portland cement' is more finely ground than ordinary Portland cement, and although it sets in the same length of time, it hardens more rapidly. The rate of heat generation is high, which is useful in the prevention of damage that could occur in concrete due to frost.

Many different types of cement are available to provide properties such as increased water resistance, resistance to chemical attack or suitability for use in cold weather.

Aggregates

Aggregates form the bulk of the volume of concrete, and consequently will affect characteristics of the concrete such as density, strength and durability.

Aggregates used for making dense concrete include materials such as gravel, sand and crushed stone. Aggregates which will pass through a 5 mm sieve (e.g. sands) are known as 'fine aggregates', whilst those retained by a 5 mm sieve are known as 'coarse aggregates'.

Aggregates for making lightweight concrete include materials such as 'clinker' (the residue of burnt coal or coke from a furnace), 'foamed slag' (slag from a blast-furnace which is treated with jets of water, then ground to form a lightweight porous material), or 'exfoliated vermiculite' (vermiculite is a flaky mineral which, when heated, expands or 'exfoliates' to form a material consisting of expanded cylindrical particles).

Dense concrete

Dense concrete is the most widely used type of concrete for purposes such as cast *in situ* or precast roof and floor slabs, foundations, lintels and structural frameworks. The density of dense concrete is usually in the range of 2150–2500 kg/m^3, whilst the compressive strength (after 28 days) is in the region of 15–60 N/mm^2.

Different proportions of cement, aggregates and water in the concrete will affect both the strength of the concrete and its 'workability'. Workability refers to the ease with which the concrete can be compacted into the formwork.

The proportion of water in a mix is expressed as the water : cement ratio (the weight of water in the concrete divided by the weight of cement). When the quantity of water necessary to set and harden the mix is exceeded, the surplus water will evaporate leaving tiny pores in the concrete. These pores reduce the strength of the concrete.

Consequently, the lower the water : cement ratio, the greater the strength of the concrete. The water : cement ratio also affects the workability of the mix; the greater the ratio, the more workable the mix. In practice, ratios of between 0.45 and 0.60 are common.

The proportion of cement to aggregate also affects the workability of a mix; the 'richer' a mix (the greater the proportion of cement) the more workable it will be. The proportions of cement to aggregate are usually expressed as cement : fine aggregate : coarse aggregate. Mixes of 1 : 3 : 6 are common for purposes such as foundations and ground-floor slabs, whilst a mix of 1 : 2 : 4 is common for general r.f.c. work.

Lightweight concretes

Lightweight concretes vary in density from 300 kg/m^3 or less up to 1920 kg/m^3. The stronger, higher-density lightweight concretes might be used for structural purposes where the weight of the structure would be significantly less than a similar structure of dense concrete. The lower-density lightweight concretes are suitable for non-loadbearing structures, and for thermal insulation purposes.

Lightweight concretes may be produced by either: (a) using lightweight aggregates; (b) aerating the concrete; (c) excluding fine aggregate from the mix.

Lightweight aggregate concrete

These concretes are made in much the same way as dense concretes, although the aggregate used is of much lighter weight. Uses of this concrete include precast wall, floor and roof units, loadbearing and non-loadbearing blocks and for the provision of thermal insulation in roof and floor screeds.

Aerated (or 'cellular') concrete

Aerated concrete is made by either:
- introducing a 'foaming agent' into the mix, which causes bubbles to form in the concrete and remain when the concrete is set (this method is usually employed for manufacture of *in situ* aerated concrete);
- generating gas in the concrete by the inclusion of an aluminium powder in the mix (this method is more commonly employed for precast aerated concrete).

Aerated concretes are the lowest in density of lightweight concretes, and have better thermal insulation. The concrete, which is used for such purposes as concrete blocks, lintels and wall slabs, is easily cut with a saw. Screws or nails may be easily driven into aerated concrete.

'No fines' concrete

In no fines concrete, the fine aggregate is omitted from the mix. This results in voids being formed in the concrete, which reduces the weight. No fines concrete is usually cast *in situ* and is used for purposes such as wall construction.

Formwork

Since concrete is a fluid material when mixed, it requires some sort of mould to produce the final required shape. Where concrete is used below ground (e.g. for foundations) the soil is usually used to act as a mould. However, where concrete is used above ground, some form of temporary mould is usually necessary; this mould is referred to as 'formwork' or 'shuttering'.

Timber boarding and plywood are commonly used for formwork, although materials such as steel or plastic may be used in instances where a profiled surface is required, or where the formwork is to be reused many times.

Figure 7.7 shows an arrangement of formwork suitable for a simple beam or lintel. The surfaces of the formwork are of plywood which are supported and held together by timber bearers and props.

Casting large areas of concrete for such purposes as floor slabs or walls requires a complex system of formwork which is capable of supporting the wet concrete over a large area.

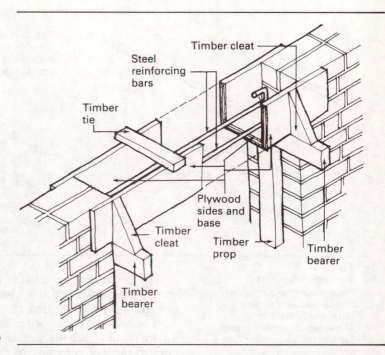

Fig. 7.7 *Formwork for simple lintel*

Reinforced concrete

As has been explained in Chapter 1 (p. 2), when loads are applied to a structure, the structure will be subjected to a combination of compressive, tensile and shear forces. Concrete is strong in compression, but its shear and tensile strengths are comparatively low. Consequently, if a structural member such as a beam or a column were made of plain concrete it could fail under comparatively light loading.

Steel is a material which is strong in compression, tension and shear and is suitable for reinforcing concrete. Glass and carbon fibres are also occasionally used.

Steel used for reinforcement is usually in the form of bars of between 10 and 32 mm diameter, which are placed in the concrete in the areas of shear and tensile stress. The steel bars may be of either mild steel, or of high-yield steel which has greater strength. Mild steel bars provide an adequate bond with the concrete and are of plain round section (Fig. 7.8(a)(i)), whereas high-yield steel reinforcement is usually of either twisted square section (Fig. 7.8(a)(ii)) or of indented round section, which increases the strength of the bond.

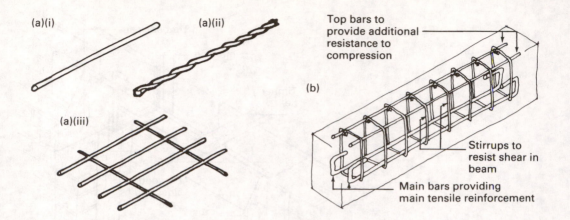

Fig. 7.8
(a) *Steel reinforcement*
 (i) *Plain round mild steel bar*
(ii) *Twisted square section high-yield steel bar*
(iii) *Mesh fabric reinforcement*
(b) *Typical reinforcement for a simple beam*

Reinforcement for slabs is usually provided by steel 'mesh fabric', which consists of a square or rectangular grid of small-diameter steel rods welded at the intersections (Fig. 7.8(a)(iii)).

The areas of shear or tensile stress in structural members usually occurs near the surface of the member. However, in order to prevent corrosion of the bars and failure of the bond between the steel and concrete, the bars are inset into the concrete.

The distance between the steel and the concrete surface is known as the 'cover'. Figure 7.8(b) illustrates typical reinforcement that might be used for a simple r.f.c. beam.

The main bars are of comparatively large diameter, and provide the main resistance to tensile forces. The stirrups are of smaller diameter and resist the shear that may occur diagonally between the top and bottom surfaces of the beam. The top bars are also of smaller diameter, and contribute to the compressive resistance in the beam.

PLASTICS

Introduction

Plastics are the most recent family of materials to be used in interior and building construction; the first commercial plastics were developed at the beginning of this century.

All plastics are 'polymers', that is, their molecular structure is in the form of long loose chains of molecules which result in lightweight, flexible, tough materials. Although some polymers are naturally occurring, most plastics today are derived from petroleum by a synthetic process known as 'polymerization'.

Many plastics are made by compounding other materials with the polymers to modify the characteristics of the plastic. Consequently, a very wide range of different plastics are used with widely differing properties. Plastics may be categorised as either 'thermosetting plastics' or 'thermoplastics'.

Thermosetting plastics, which tend to be rigid and hard, are chemically processed in such a way that when the plastic is heated, it will remain rigid. If heated excessively, the plastic will tend to char rather than soften. Examples of thermoplastics include phenol formaldehyde (the first commercially produced plastic, known as 'Bakelite').

Thermoplastics vary in their characteristics of hardness and softness, but all will soften if heated and harden again if cooled. This characteristic is useful for moulding and forming the plastic into complex shapes. Polythene is a familiar example of a thermoplastic.

Tables 7.8(a) and (b) compare some of the characteristics of common thermosetting and thermoplastics. In reading the table, it sould be borne in mind that many of the plastics are made in different forms, which will affect their characteristics.

In manufacturing a plastic item, three processes are usually involved:

Table 7.8(a)   Characteristics of some common thermosetting plastics

Plastic	Density (kg/m³)	Approximate tensile strength (N/mm²)	Appearance and characteristics	Typical uses
Phenolic (phenolic formaldehyde)	1400	80	Heavy. Hard. Usually opaque dark brown. Good resistance to heat, chemicals, etc.	Accurate mouldings for electrical fittings, door furniture, etc.
Melamine (melamine formaldehyde)	1500	50–90	Heavy. Hard. Better resistance to heat, chemicals, etc. than phenolic	Widely used in laminated form for surfaces of worktops, cupboards and decorative panels
Glass-reinforced polyester (g.r.p.)	1800	130	Strong. Rigid. Good resistance to heat, chemicals, etc. Fibres can often be seen on obverse side	External wall cladding panels. Corrugated sheets for roofing. Moulded into chair and other furniture shells

Table 7.8(b)   Characteristics of some common thermoplastics

Plastic	Density (kg/m³)	Approx. tensile strength (N/mm²)	Softening point (° C)	Appearance and characteristics	Typical uses
Polythene (high density)	950	20–30	90–100	Waxy feel to surface. Transparent to opaque, good resistance to chemicals and moisture	Containers and tanks
Polythene (low density)	920	7–15	80	As above, only translucent	In film form for membranes. Waste pipes and fittings. In foam form for upholstery
Polypropylene	900	35	100–120	Smooth. Very lightweight. Translucent	In fibre form for carpets. Chair shells and stools. Screws and bolts
Polystyrene	1040	35–60	80–100	Hard, transparent, brittle. Often toughened by mixing with other polymers	Internal moulded surface of refrigerator doors. In expanded form, used for thermal insulation

(a) 'Polymerisation', to produce the long-chain molecular material.

(b) 'Compounding', adding other materials to the polymers in order to modify the characteristics of the plastic. These compounding materials include:
   - fillers to either increase the bulk of the material or to improve the strength or wearing characteristics;
   - plasticisers to improve the flexibility of the product;
   - stabilisers to improve resistance to heat or sunlight;
   - colouring agents.

(c) 'Converting' or 'forming' to provide the plastic with its final shape. Forming processes include extrusion, calendering, thermoforming, blow moulding, injection moulding, transfer moulding, compression moulding, laminating and foaming.

Forming processes

Extrusion
Granular plastic is heated and forced through a die of the

Table 7.8(b)   continued

Plastic	Density (kg/m³)	Approx. tensile strength (N/mm²)	Softening point (° C)	Appearance and characteristics	Typical uses
Acrylic	1120	55–75	70–90	Transparent, like glass. Rigid. Good resistance to sunlight	Rooflighting. Signs/lettering. Sinks, baths, basins, etc. Glazing and furniture
Rigid p.v.c. (u.p.v.c.)	1350	55	55–85	Transparent to opaque. Rigid. Glossy, tough	Windows/door frames. External wall cladding. Skirtings and architraves. Corrugated roofing (sometimes wire laminated)
Flexible p.v.c. (plasticised polyvinyl chloride)	1250	10–25	70	Rubbery texture, transparent to opaque	As a coating for wallpapers, curtains, upholstery material
Nylon	1120	50–90	190–250 (melting point)	Very strong. Resistant to wear. Smooth waxy surface, translucent to opaque	Hinges, door furniture, curtain rails. In fabric form for upholstery
ABS	1000	17–60	85	Tough, rigid, strong, translucent	Baths, furniture, wall cladding

required final shape. Extrusion is used mainly for thermoplastics in the production of such items as tubes, rods, films and sheets.

Calendering
Calendering is a fast process for producing plastic sheets and films by heating thermoplastic and passing it between rollers.

Thermoforming
Sheet thermoplastic material is softened and formed over or into a mould. An example of thermoforming is vacuum forming, where:
- a mould is placed on a perforated bed;
- a sheet of plastic is held in place over the mould and bed;
- the sheet is heated and air is extracted through the perforated bed. This causes the sheet to form over the mould.

Items such as the profiled inner linings of refrigerator doors are produced by thermoforming.

Table 7.9

Plastic	Forming methods used						
	Extrusion	Thermo-forming	Blow moulding	Injection moulding	Compression moulding	Laminating	Foaming
Phenolic				■	■		■
Melamine					■	■	
Glass-reinforced plastic	■					■	
Polythene, HD	■		■	■			■
Polythene, LD	■		■	■			■
Polypropylene	■		■	■			■
Polystyrene	■	■	■	■			■
Acrylic	■	■		■			■
Unplasticised polyvinyl chloride	■		■	■			■
Polyvinyl chloride	■	■	■	■			■
Nylon	■			■			■
Acrylonitrile-butadiene	■	■		■			■

Blow moulding

Blow moulding is used mainly with thermoplastics for the production of such items as water cisterns and other containers. The procedure for blow moulding is as follows: a tube of soft plastic is placed in an enclosed mould. Air is then blown into the tube which expands to press against the sides of the mould.

Injection moulding

Granules of plastic are heated in a chamber and injected, under pressure, into a mould. This process is usually used for thermoplastic items such as crates and tanks; however, thermosetting plastics may also be injection moulded.

Compression moulding

Suitable for both thermoplastics and thermosetting plastics, compression moulding involves:
- placing granules of plastic on to a moulding plate;
- heating the granules;
- pressing a top moulding plate over the plastic.

Laminating

Plastic may be laminated as either: (a) fibre-reinforced plastic; (b) sheet laminates.

Fibre-reinforced plastics are formed by impregnating fibres with plastic resin. Glass-reinforced polyester is a well-known example, which is produced by pressing glass-fibre matting against a mould and impregnating it with resin. Several laminations of glass fibre and resin can be built up to increase the rigidity and strength of the material. Large-scale sections such as furniture shells may be fabricated by this method.

Foaming

Foaming of plastic is caused by such means as introducing air into the plastic or by the addition of chemicals. Foaming may result in either a 'closed'-cell material which is relatively rigid and suitable for use as thermal insulation or as an 'open'-cell material which is spongy and is suitable for furniture upholstery.

Table 7.9 indicates the forming methods which are usually associated with particular plastics.

CHAPTER 8

Internal finishes and materials

This chapter is concerned with some of the vast range of materials that are applied as interior surface finishes. These materials are considered under the headings: floor finishes; walls and joinery finishes; ceiling finishes; soft furnishings.

Table 8.1 summarises some of the commonly used applications of surface finishes. It is important to note that the table is by no means exhaustive, and many of the materials might be used as a finish for less typical surfaces. Glass, for example, could be used as a floor finish, carpet, as a wall finish, or marble as a ceiling finish.

The selection of a material for a particular surface involves comparing the performance required from the material, with performance data offered by manufacturers of surface finishes.

Table 8.1 Summary of some commonly used interior finishes

	Floors	Walls/joinery	Ceilings	Soft furnishings
Thick liquid	Jointless floors	Plaster	Plaster	
Thin liquid	Paints Clear finishes	Paints Clear finishes	Paints	
Thick solid	Timber based Clay based Stone based	Stone Ceramic tiles Mirror/glass Timber Metal	Panels and tiles Plasterboard	
Thin solid	Thin sheet and tile materials	Timber veneers Cork Plastic laminates		
Soft finishes	Carpet	Papers Fabrics	Papers	Fabrics Skins

In assessing the performance required from the material, the following factors should be carefully considered:

(a) Durability; taking into account:
- the facilities available for cleaning and maintenance of the finish;
- the likely wear and tear on the finish due to abrasion, water or chemical spillage, impact from foot traffic or movable furniture, etc.

(b) Economic considerations, including:
- the budget available for the cost of material and installation of the finish;
- the long-term budget available for cleaning and maintenance.

(c) Comfort and appearance:
- the softness of the finish to touch or underfoot;
- glare, caused by reflective surfaces or bright colours, or dullness caused by dark colours;
- the acoustic and thermal insulation properties required;
- the visual effects of colour, texture and pattern.

(d) Safety aspects; including:
- fire hazard;
- injury caused by hard or sharp edge of finishes.

This chapter aims to provide an introduction to the broad spectrum of commonly used surface finishes. In order to match the required performance with the actual performance of a surface finish, it is necessary to amass technical literature from manufacturers and to discuss the particular selection problem with manufacturers' technical experts.

FLOOR FINISHES

Durability is often the foremost consideration in the selection of a floor finish, since floors take heavy punishment from sources such as wear and dirt from foot traffic, spillage of liquids, moving or dropping of heavy objects, and sometimes wear caused by wheeled traffic in industrial buildings.

Other important considerations include:

(a) The means of maintenance and cleaning: some floors require sophisticated industrial machinery for cleaning, whilst others only require simple manual cleaning equipment.

(b) Cost: if a floor is required for long-term duration, it may be necessary to select a finish of high initial cost. Hardwood or marble floors, for example, have

Table 8.2 Jointless liquid floor finishes

Floor finish	Composition	Base for finish	Minimum thickness (mm)	Characteristics
Cement screeds (maximum bay sizes usually 15m²)(see also 'ground floors', Ch. 3, p.135)	1 : 3–4½ (cement : sand) or 1 : 1½ : 3 (cement : sand : aggregate	Usually concrete		Cement screeds are usually used as a base for other types of floor finish such as cork or carpet. Types of cement screed include:
			12	'Monolithic' screeds which are laid within three hours of the concrete base being laid. This minimises shrinkage
			40	'Bonded' screeds which are laid on a hardened concrete base, where the surface aggregate has been first exposed
			50	'Unbonded' screeds
	Expanded metal may be incorporated in the screed to reduce cracking	Concrete and insulation	65	'Floating' screeds which are used on top of thermal or sound insulation materials
Lightweight screed	Cement and lightweight or aerated aggregate	Concrete	40	Light in weight with improved thermal insulation
Granolithic (maximum bay sizes: monolithic 28 m², bonded 14 m²)	1 : 1 : 2 (cement : fine aggregate : coarse aggregate)	Concrete		Hard-wearing. Mainly used industrially. Slippery when wet unless special additive is included
			15	Monolithic
			40	Bonded
			75	Unbonded See cement screed characteristics for monolithic, bonded and unbonded characteristics

comparatively high installation costs but they are durable, and are capable of retaining their appearance.

(c) Appearance: the texture, pattern and colour of a floor finish should be selected in relation to:
- the style, function and size of the room;
- other finishes and furnishings in the room;
- the degree of lightness or darkness required in the room.

Table 8.2 continued

Floor finish	Composition	Base for finish	Minimum thickness (mm)	Characteristics
Cement Rubber-latex	Rubber latex, cement, aggregates (such as cork, sawdust, sand), fillers, pigment	Usually concrete with cement screed, or alternatively timber on which galvanised wire netting is first fixed	6	Hard-wearing. Surface less hard than granolithic. Non-slip. Smooth surface. Wide range of colours available
Terrazzo (specialist advice should be sought concerning bay sizes)	Cement and marble aggregate	Concrete	15	Hard-wearing. Slippery if polished or wet. Surface is usually ground, washed and polished. The marble aggregate provides a mottled mosaic appearance
Synthetic anhydrite screed	Screed using synthetic anhydrite in place of cement	Concrete	25	Rapid-hardening screed resulting in minimal residual moisture. This makes it suitable for use with underfloor heating. More expensive than sand/cement screeds
Magnesite	Burnt magnesite, magnesium chloride, fillers (such as wood flour, limestone, sawdust), pigments	Concrete or timber (when an underlay should be laid first)	15	Hard-wearing. Slippery if polished unless abrasive is incorporated. Range of colours available with speckled and mottled effects
Mastic asphalt	Bitumen, aggregate (such as limestone), fillers, and if colours are required, pigments	Concrete or timber (when an underlay should be laid first)	15	Hard-wearing. Slippery if wet or polished. Available in different grades with various wearing properties. Normally black in colour, but also available in red, brown, green and grey

(*d*) Comfort:
- the effect of the surface on the noise level in the room by either reflecting or absorbing sound;
- warmth;
- fatigue on feet: in rooms such as gymnasia or ballrooms, fatigue will be reduced if a resilient floor finish is used.

(*e*) Safety: accidents can be caused by slippery surfaces being

369

Table 8.2 continued

Floor finish	Composition	Base for finish	Minimum thickness (mm)	Characteristics
Pitch mastic	Coal-tar pitch, aggregate (such as limestone) and if colours are required, pigments	As above	15	As above, but has better resistance to water and fats, and is consequently more suitable for kitchens and washrooms than is mastic asphalt
Cement resin	Cement and polyester resin, aggregate (such as sand or crushed stone)	Concrete, timber	3	Hard-wearing. Non-slip. Slightly textured surface
Epoxy resin	Epoxide resin, hardener, aggregate, fillers	Cement screed, rigid plywood, other surfaces	2 if self-levelling, 3 if trowelled	Hard-wearing, can be made non-slip. Variety of colours and textures available. Self-levelling screed is poured on to the floor surface, spread out with a rake and allowed to level itself
Polyester resin	Polyester resin, catalyst, aggregates, fillers, glass fibres, pigments	Cement screed, rigid plywood	2	Hard-wearing. Wide colour range available
Polyurethane resin	Polyurethane resin, fillers	Cement screed, rigid plywood		Hard-wearing. Non-slip. Wide range of colours and textures available

installed in hazardous areas such as machine shops, or on stairs.

Jointless liquid floor finishes

The floor finishes considered in this section are laid in a liquid form which then hardens. The floors are laid without frequent joints, although granolithic, terrazzo and cement screed floors are laid in 'bays' which are separated by joints, in order to avoid shrinkage and cracking.

Jointless floors are often associated with industrial buildings, although materials such as mastic asphalt or terrazzo might be used for commercial, hospital, school and other buildings.

Cement screeds are widely used in many different types of building as a base for other floor finishes such as cork, carpet and p.v.c.

Table 8.2 provides information concerning these floor finishes.

Thick solid floor finishes

The floor finishes considered here are all solid, rigid and mostly over 15 mm in thickness. These finishes are either: (*a*) timber based; (*b*) clay based; (*c*) stone based.

Timber based
See 'Ground floors' and 'Upper floors' (Ch. 3, pp. 134 and 141); and 'Timber' (Ch. 7, p. 330).

Table 8.3(a)(i) provides details of the characteristics and fixing of timber-based floor finishes. Table 8.3(a)(ii) indicates some of the timbers suitable for flooring. Generally, the higher-density timbers are more suitable for heavier use.

Clay based
Clay-based floor finishes are summarised in Table 8.3(b), and include brick, brick paviors, ceramic tiles and quarry tiles. Clay finishes are hard, and consequently tend to be noisy and cold.

Ceramic and quarry tiles tend to expand due to the absorption of water. Where the width or length of tiling exceeds 6 m, an expansion joint should be incorporated around the perimeter of the floor area (at the junction with the skirting) in order to allow for this expansion. Expansion joints should also be incorporated at regular intervals in large areas of flooring. The expansion joints consist of compressible sealant material.

Stone based
Stones are derived from naturally occurring rocks which are classified in accordance with their method of geological formation.

(*a*) Igneous rocks were formed by the cooling of molten 'magma' (molten stratum lying under the earth's crust) of which granite is an example.

(*b*) Sedimentary rocks were formed from sediments laid down under either water or air. The sediments were composed of either:

Table 8.3(a)(i) Timber floor finishes

Type of floor	Description and characteristics	Base for finish	Typical sizes	Fixing
Boards	Usually softwood. Tongued and grooved boards provide greater rigidity and fire resistance. Resilient surface	Joists, or battens fixed into screed on concrete base	Widths: 100–175 mm Thicknesses: 19–32 mm	Boards nailed to either joists or to battens (as illustrated on page 373(a))
Strips	Hardwood and softwood. Characteristics as boards	As above	Widths: 50, 60, 75 mm Thicknesses: 19–22 mm	Strips secret nailed to joists or battens
Blocks	Usually hardwood laid in herringbone, brick-bond and basket patterns. Blocks jointed with tongued and grooved edges. Require dry, stable atmosphere	Screed on concrete	Lengths: 150–300 mm Widths up to 90 mm Thicknesses: 20–30 mm	Either (a) screed covered in hot bitumen then blocks laid in hot bitumen, or (b) blocks laid in cold latex bitumen adhesive (as illustrated on page 373(b))
Parquet	Made from specially selected hardwoods. Used mainly for high-class domestic work. Available as made-up square panels	Smooth level timber, possibly overlaid with plywood or hardboard	Thicknesses: 5–10 mm Widths: 50–90 mm Square panels 25–32 mm thick in squares of 300 or 600 mm	Parquet glued and pinned to base. Parquet panels either (a) glued and pinned to base, or (b) bedded in hot bitumen
Wood mosaic	Strips of hardwood. Usually arranged in square panels for laying. Used mainly for domestic work	Screed on concrete, plywood or hardboard	Usually 10 mm thick in 300 or 450 mm square panels	Panels are obtained with the hardwood ready bonded to bituminous felt or aluminium foil. The panels are bonded to the base with cold adhesive
Chipboard	Chipboard with tongued and grooved long edges. Moderate wearing ability	Timber joists	12 and 18 mm thick × 1220 × 2440 mm	Nailed to joists
Plywood	Plywood tongued and grooved. Moderate wearing ability	Timber joists	18 and 22 mm thick × 1220 × 2440 mm	Nailed or glued to joists

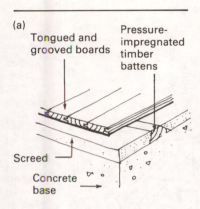

(a)
Tongued and grooved boards
Pressure-impregnated timber battens
Screed
Concrete base

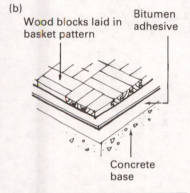

(b)
Wood blocks laid in basket pattern
Bitumen adhesive
Concrete base

Illustrations to table 8.3

Table 8.3(a)(ii)   Timbers suitable for flooring

Timber	Density (kg/m³)	Uses*
Afrormosia	700	L–M
Agba	500	L
Beech	680	M–H–G
Iroko	640	L–M
Mahogany (African)	560	L
Opepe	750	L–M
Sapele	640	L–G
Teak (Burma)	640	L–M
Teak (Rhodesian)	900	H

* Key for uses: G – gymnasia; L – light duty (e.g. domestic); M – medium duty (e.g. schools, offices); H – heavy duty (e.g. industrial buildings).

- broken-down particles of other rocks which were pressed and cemented together (such as sandstone);
- deposited material of organic origin (such as limestone).

(c) Metamorphic rocks were formed from igneous and sedimentary rocks subjected to influences such as heat and pressure. Marble, slate and quartzite are examples of metamorphic stones.

Stones are quarried and cut into slabs for flooring and other building purposes. Stone is a costly form of flooring which is hard and noisy, but a hard-wearing surface.

Stone floors are laid on concrete sub-floors on a bed of 1 : 3 cement : sand. Table 8.3(c) provides information concerning granite, limestone, sandstone, marble, quartzite and slate floors.

Thin solid materials

The thin solid materials considered in this section are mainly less than 15 mm in thickness. The materials are flexible, and all but vinyl asbestos and thermoplastic are available in both sheet and tile form. The finishes are all bonded with adhesive to the floor which is usually either of: (a) screed on concrete; (b) plywood or hardboard on timber boards. Thermoplastic, being less flexible than the other finishes, requires heating to soften the material prior to laying.

These finishes range from the low-cost linoleums and thermoplastics through vinyl-based finishes and cork to the less inexpensive rubber flooring.

Table 8.3(d) provides details of these finishes.

Table 8.3(b) Clay-based floor finishes

Finish	Composition and characteristics	Typical sizes (mm)	Typical laying methods
Brick and brick paviors	Characteristics will vary with the type of brick used (see 'Bricks', Ch. 2, p. 39). Engineering bricks are extremely hard-wearing and resistant to chemicals. Brick paviors are manufactured in a similar manner to bricks but are less deep. They may have special surfaces such as chequered, ribbed and roughened	Standard brick sizes (65 × 215 × 102.5) Paviors: 20–50 deep and of various widths and lengths	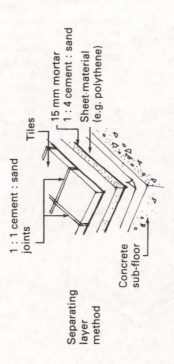
Ceramic tiles	Made from clay which is ground, pressed and tempered into shape, then fired at high temperature. Ceramic tiles are more hard-wearing and are more resistant to chemicals and fats than quarry tiles. The range of ceramic tiles include: colours: reds, buffs, brown, black and mottled and printed patterns; accessories (see illustration opposite); square and hexagonal shapes and chequered and raised dimple surfaces to provide anti-slip qualities	Thicknesses: 9, 12 Squares: 150 × 150, 200 × 200 Hexagons 175 × 175 Rectangles 200 × 100	

Quarry tiles | Made from clay which is pressed into shape and hard burnt. The range of quarries include: colours – reds, browns, buffs, blue; accessories (see illustration opposite and Fig. 5.8(d), Ch. 5); square and octagonal shapes and chequered and raised dimple surfaces to provide anti-slip qualities | Thicknesses: 12, 19, 25, 32 Squares: 150 × 150, 200 × 200 Octagons: 175 × 175 Rectangles: 150 × 75

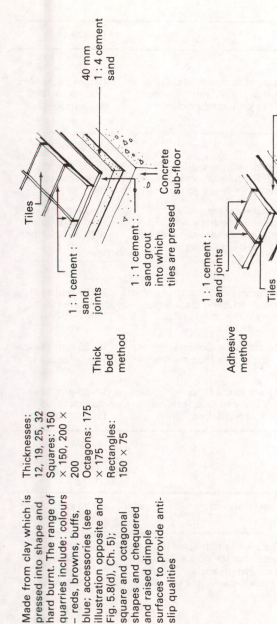

Thick bed method

Tiles

40 mm 1 : 4 cement sand

1 : 1 cement : sand joints

1 : 1 cement : sand grout into which tiles are pressed

Concrete sub-floor

Adhesive method

1 : 1 cement : sand joints

Tiles

Cement-based adhesive, max. thickness 5 mm

Screed and sub-floor

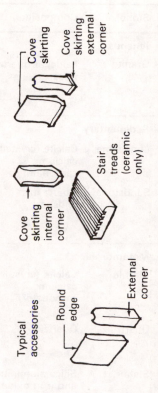

Typical accessories

Round edge

External corner

Cove skirting internal corner

Cove skirting

Cove skirting external corner

Stair treads (ceramic only)

Table 8.3(c)   Stone floor finishes

Stone	Composition	Characteristics	Typical sizes
Igneous			
Granite	Feldspar quartz mica	Hard-wearing, resistant to chemicals. Colours: black, blue, green, grey, pink, red. Slippery when polished. Used only rarely today, for prestigious interiors	40 mm thick. Max. area up to 1800 × 900 mm
Sedimentary			
Limestone	Calcite (crystalline calcium carbonate)	Some have very hard-wearing qualities. Colours: buff, grey. Slippery when worn. Rarely used internally for floors today	15–50 mm thick. Max. area up to 1500 × 600 mm
Sandstone	Quartz mica feldspar	Some have very hard-wearing qualities. Colours: buff, brown. Some have non-slip qualities. Tend to be used more for external paving rather than interior flooring	40–75 mm thick. Max. area up to 3000 × 1200 mm
Metamorphic			
Marble	Calcite (crystalline calcium carbonate)	Hard-wearing, but surface can scratch. Colours: wide variety including black/white, buff/brown, red/brown, red, black, white. Expensive, prestigious finish	15–30 mm thick. Max. area up to 1500 × 900 mm
Quartzite	Quartz	Good resistance to abrasion. Colours: green, grey, yellow. Less expensive than marble or slate	10–20 mm thick. Max. area up to 900 × 230 mm
Slate	Silica aluminium and iron oxide	Good resistance to abrasion. Colours: black, blue, green, grey. Variety of face textures such as sawn, riven, sanded. Expensive	12–38 mm thick. Max. area up to 1200 × 600 mm

Table 8.3(d)   Thin solid floor finishes

Material	Composition	Characteristics	Typical sizes (mm)
Cork	Derived from the bark of the evergreen oak (found in Mediterranean regions). Cork tiles are cut from blocks of granulated cork, the natural resins acting as a binder. Cork carpet is of similar material, but binders of linseed oil and resins make the cork more pliable. The material is then bonded to a jute canvas backing	The tiles in particular have good wear characteristics. Non-slip, quiet, warm surface. Colours: yellow-browns to deep dull browns	Tiles. Thicknesses: 5–12 × 300 × 300 square. Carpet. Thicknesses: 3–6. Sheet width 1800

Table 8.3(d) continued

Material	Composition	Characteristics	Typical sizes (mm)
Linoleum	A mixture of linseed oil, powdered cork, resin, fillers and pigments are rolled into sheets on either a hessian or a bitumen felt backing. Available in sheet or tile form	Ordinary grades suitable for domestic and light wear application. Hardened and toughened grades available for heavy wear use. Slippery if highly polished. Quiet, warm surface. Large range of colours and patterns available	Tiles. Thicknesses: 2.5, 3.2, 4.5, 6.0 × 300 × 300 square Sheet. Thicknesses as tiles. Sheet width: 1830
Rubber	Produced from natural or synthetic rubber, fillers and pigments. Available in sheet form or tiles either (a) cut from sheets, or (b) moulded with ribs on upper surface. Sheet available with foamed rubber backing	Good wear characteristics especially with thicker grades. Resilient, quiet surface (particularly in foam-backed sheet form). Tiles are non-slip; sheet may be slippery if wet. Wide range of colours, textures and patterns available	Tiles. Thicknesses: 3–6.5 225 × 225, 450 × 450 square Sheet. Thicknesses as tiles. Sheet width: 1830
Thermoplastic tiles ('Asphalt tiles')	Produced from non-vinyl resin binder, fillers and pigments. Originally, asphalt binders were used which restricted the colour range	Reasonably hard-wearing. Brittle compared with other thin solid finishes. Low-cost finish. Range of colours available, but tiles with asphalt binder base only available in dark colours	Thicknesses: 3, 4.5 × 225 × 225 square
Vinyl asbestos	Produced from p.v.c. resin binder, fillers, asbestos fibres and pigments	Characteristics similar to thermoplastic tiles, although use of p.v.c. resin binder means (a) tiles are more flexible, and (b) a brighter colour range is available than with thermoplastic tiles	Thicknesses: 1.6, 2.0, 2.5, 3.2 × 225 × 225 square
Flexible vinyl	Produced from p.v.c. resin binder, fillers and pigments. Available in tile and sheet form. Sheet available with p.v.c. foam backing	Fairly high wear characteristics. Fairly warm and quiet, especially with foam-backed sheets. Non-slip grade of sheet available. Wide range of colours and embossed designs available	Tiles. Thicknesses: 2.0, 2.5, 3.2 × 300 × 300 square Sheet. Thicknesses as above, or more if foam backed. Sheet width: 1830

Carpets

The bewildering variety of carpets makes their selection a difficult task. Carpets differ in their construction, the nature of their fibres, type of pile and backing, as well as in the usual differences of colour, pattern and texture. In an effort to simplify selection, the British Carpet Classification Scheme (BCCS) provides a labelling system for carpets in accordance with the suitability of their use. The BCCS classification and the use for each one is as follows:

Class 1   Light domestic use (rooms with light traffic, e.g. bedrooms).

Class 2   Medium domestic/light contract use (e.g. hotel bedrooms with light traffic).

Class 3   General domestic/medium contract use (e.g. hotel bedrooms, public areas of small buildings).

Class 4   Heavy domestic/general contract use (e.g. all heavy domestic uses, public areas subject to general use, hotels, shops, restaurants, offices, etc.).

Class 5   Heavy contract use (e.g. all heavy traffic areas of commercial buildings).

Class L   Luxury (carpets superior to grade 3; designed for comfort and visual appeal, but not necessarily suited to areas requiring high durability).

The following sections deal with: (*a*) carpet pile and backing; (*b*) carpet construction; (*c*) types of fibre; (*d*) factors affecting selection; (*e*) fixing methods; (*f*) matting.

Carpet backing and pile

Traditional carpets comprise a woven material consisting of 'pile' (the yarns visible on the upper surface) and a woven backing. The pile yarns are inserted simultaneously with the weaving of the backing. Although the traditional method is still widely used (particularly for better-quality carpets), in some modern manufacturing methods the pile tufts are inserted separately into a woven or non-woven backing.

Backing   The principal function of the backing is to provide a base for holding the pile tufts. In the case of woven carpets, the backing is made of either natural fibre yarns such as jute or linen, or polypropylene, which is used increasingly today.

Non-woven carpets use backings of synthetic materials (such as polypropylene) which increase wear and comfort

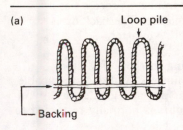

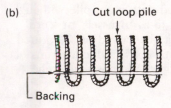

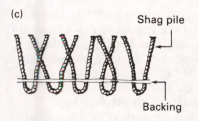

Fig. 8.1 *Woven carpets*
(a) *Loop pile*
(b) *Cut loop pile*
(c) *Shag pile*

characteristics. Non-woven carpets often have a secondary backing applied to them which may be either: (*a*) woven yarns of synthetic or natural fibres; (*b*) latex foam; (*c*) polymer-based foam.

A separate 'underlay' is often laid below the carpet to increase the life of the carpet. The underlay cushions the effect of traffic wear on the carpet, and also reduces wear caused by irregularities in the floor surface. Sound and thermal insulation properties will also be improved by the use of an underlay.

Materials used for underlays include felt (a fabric based on jute fibres), rubber and polymer-based foam.

Pile   Since the pile takes most of the impact of traffic on the carpet, it should be:
- of suitably strong resilient material;
- of sufficient 'pile density' (closeness of packing of the pile fibres).

If the pile density is low, each tuft is required to take more punishment from traffic, consequently the higher the pile density, the better the wearing characteristics of the carpet. Figures 8.1(a), (b) and (c) illustrate loop, cut loop and shag piles.

Carpet construction
Carpets may be made by the traditional method of weaving, using either the Axminster or the Wilton process. Axminster and Wilton carpets are made in 'body widths' (widths of up to approximately 1 m) or in 'broadloom widths' (widths of up to 4.5 m).

Modern carpet constructions not based on weaving processes include tufted carpets and needle-punch carpets which are available in widths of up to 4.0 m or in square tiles of 400 or 500 mm.

Axminster carpets   The pile tufts of Axminster carpets are inserted during the weaving of the backing. Large numbers of colours may be used to produce intricate patterns.

Spool Axminster   The spool Axminster process allows an almost unlimited number of colours to be used; however, it tends to be an expensive method, since the yarns have to be wound on to spools prior to weaving.

The backing of spool Axminster is rather limp, and a back coating of starch or latex is applied after the carpet has been woven.

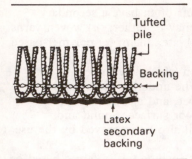

Tufted
pile

Backing

Latex
secondary
backing

Fig. 8.2 *Tufted carpet*

Gripper Axminster   The gripper Axminster process has a normal limitation of up to eight colours. The backing is much more firm than that of spool Axminster.

Wilton carpets   The manufacturing method of Wilton carpets is different to that of Axminsters. Wiltons are capable of being of heavier quality (heavier pile and greater density of tufts) although the number of colours is normally restricted to five.

Wilton carpet is usually finished with a cut pile, although a 'loop' weave (sometimes referred to as 'Brussels' or 'cord' carpet) is possible. The cut pile may be carved or sculptured to produce a textured effect. Where short production runs of carpet are required for special designs, Wiltons tend to be more economic than Axminsters.

Tufted carpets   Tufted carpets are made by inserting the pile tufts by needle into a woven or non-woven backing. The tufts are anchored into the backing by the application of a layer of latex, after which a further backing of either jute, synthetic yarn or latex is applied. This secondary backing improves the comfort of the carpet, and in the case of foam, obviates the need for a separate underlay (Fig. 8.2).

Tufting is a very rapid process which involves little labour, and consequently is a cheaper method of production than that of woven carpets. The pile may be cut, looped, or cut at different heights to produce a sculptured effect, and two- or three-colour designs are possible.

Designs may also be printed on to plain coloured carpet.

Needle-punch carpets (fibre-bonded carpets)   Needle-punch carpets are made by feeding a web of fibres into a machine, which interlocks and entangles the fibres by the penetration of barbed needles. The resulting carpet has a tough close finish without a pile, although loop pile and velour effects can be achieved.

Needle-punch carpets may be made with backings, patterned effects and in a range of colours.

Fibres

Fibres are the individual strands from which yarn used for carpets (or other fabrics) is made, and it is largely the fibre which determines the quality of the carpet. The most commonly used fibres for carpet yarns include acrylic, nylon, polyester, polypropylene, viscose rayon and wool. Often the fibres are blended in order to produce yarns that are: (*a*) sufficiently economical; and/or (*b*) able to provide specific

Table 8.4 Qualities of carpet fibres

Fibre	Qualities
Acrylic	The nearest synthetic fibre in quality to wool. Good resilience and wearing ability. Cleans well. Tends to soil easily, tends to melt with heat
Nylon	Tough, hard-wearing fibre with good abrasion resistance. Cleans well. Melts at high temperature. Carpets of 100% nylon are prone to problems of static electricity
Polyester	Good abrasion resistance. Easily cleaned. Poor pile recovery from the effects of heavy objects
Polypropylene	Good abrasion resistance and resistance to soiling. Durable fibre. Cleans well
Viscose rayon	Low cost, but low strength. Often blended with other fibres
Wool	Resilient, warm fabric with good abrasion resistance. Does not soil easily – easy to clean. Tends to be costly

performance characteristics. Table 8.4 summarises some of the qualities of these fibres.

Factors affecting selection
Besides cost and visual characteristics, some of the important factors in selecting a carpet are: soiling, dimensional stability, colour fastness, flammability, wear resistance, sound and thermal insulation, static electricity and resilience. Table 8.5 indicates some of the characteristics of carpets that will influence these factors.

Fixing methods
The floor on which carpet is to be laid should have a smooth, level surface. If a timber floor is badly warped, a layer of hardboard might be first laid, or if a concrete floor is cracked or badly pitted, a suitable screed might be first applied.

Where it is necessary to join carpets together, the carpets are either sewn together or joined with adhesive tape.

Carpets are best fixed by either 'loose laying' (in the case of carpet tiles), bonding to the floor with adhesives, or by using carpet grippers.

Table 8.5　Factors affecting carpet selection

Selection factor	Influences
Soiling (see also Table 8.4, p. 381 above)	Dense-pile carpets tend to retain their appearance better than low-density pile carpets. Rich patterns can conceal blemishes. Light, plain colours will soil more noticeably than dark, patterned carpets
Dimensional stability	Carpets can change dimensionally due to: (a) gradual relaxation of tensions induced during weaving, causing contraction of carpet; (b) contraction due to heat and moisture; (c) extension in the carpet caused by the effects of strains due to traffic
Colour fastness	Dyes vary in their colour fastness. Loss or change of colour can result from the effects of light, water, shampooing and abrasion
Flammability	Carpets are not usually a major cause of the initiation or spread of fires. Wool has the best flame-retardant properties, although some synthetic fibres melt, which extinguishes flames. Polypropylene will burn, but slowly. Fire-retardant properties can be improved by blends of fibres. Dense-pile carpets burn less readily; low-pile carpets allow flames to reach the underlay more easily
Wear resistance	Abrasion is the major cause of carpet wear (see also Table 8.4). Generally, short dense piles wear better than long dense piles; dense piles wear better than low-density piles. Latex coating on the carpet backing can increase retention of pile tufts, and underlays will spread the effects of impact on the pile resulting from traffic
Sound and thermal insulation	*Sound*: carpets with high, dense pile will improve sound absorption. High, dense pile and underlay will also improve resistance to impact sound *Thermal insulation*: increased pile weight and foam underlays contribute to the useful addition that carpets can make to the thermal insulation of floors
Static electricity	A combination of certain synthetic fibre carpets with low room humidity can result in people receiving electric shocks when metal objects resting on the carpet are touched. The problem can be reduced by: (a) anti-static treatments applied to the carpet; (b) incorporating metal particles or fibres in the yarn; (c) increasing the humidity in the room
Resilience	A carpet with high cushioning characteristics can be tiring to people using the carpet for long periods, although it provides a luxurious feel

Adhesives Adhesives are commonly used for fixing foam-backed and needle-punch carpets. The adhesives used include:

- rubber/resin emulsion adhesives, which are suitable for rubber-backed and certain fibrous-backed carpets;
- acrylic emulsion adhesives which are used with p.v.c.-backed carpets.

Adhesives may be used to provide either:

- a permanent bond, where the carpet may be lifted relatively easily after its useful life;
- a temporary bond, where the carpet may be easily lifted without damaging either the carpet or the floor.

Grippers Grippers, which are the usual way of fixing Axminster and Wilton carpets, consist of strips of plywood with two rows of projecting angled spikes. The strips are fitted around the perimeter of the carpet area (around the skirting in the case of wall-to-wall carpets) by either nailing the strips to timber floors, or bonding with strong contact adhesive to concrete or hard-surfaced floors.

The carpet edges are stretched over the gripper strips so that the spikes grip the underside of the carpet edge.

A range of accessories are available which include:

- threshold strips for installation of the carpet at doorways;
- edging strips for use at bare edges of carpet (where carpets are not fitted against a skirting board).

Matting

Mats are used for floors in areas such as entrance lobbies, where the floor finish is expected to receive particularly harsh treatment.

Entrance mats are designed to allow dirt from shoes to be scraped on to the mat surface, thus reducing the dirt carried through to the other floor finishes.

Thick matting is usually recessed into a 'mat well' such that the mat and floor finish surfaces are level. This avoids the danger of people tripping over the mat. Mat wells are often lined with frames, which protect the edges of the mat and floor finish (Fig. 8.3). Mat-well frames are usually of flat- or

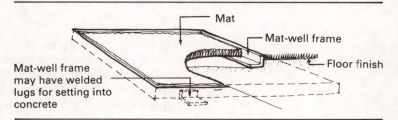

Fig. 8.3 *Mat-well frame*

angle-section galvanised steel, aluminium or brass. 'Coir' (or coco-fibre) matting is the traditional material for entrance mats, and is woven from fibres obtained from coconut husks. The matting, which is available in thicknesses of 25–50 mm, allows shoes to be both scraped and dried on the surface.

Rubber matting is available in several different forms including:

- rubber link matting, which consists of 19 mm thick blocks of rubber linked together by galvanised steel wire; dirt scraped from shoes falls into the spaces between the blocks;
- ribbed rubber matting, typically of 14 or 20 mm thickness;
- aluminium/rubber matting typically of 18 mm depth, which consists of rubber piled matting with aluminium intermediate strips which allow shoes to be scraped.

WALLS AND JOINERY

The factors affecting the selection of wall and joinery finishes are given under the headings below.

Durability, maintenance, cleaning and cost

Walls and joinery are rarely required to resist the degree of punishment that floor finishes take, and it is possible in some circumstances to use lightweight and delicate finishes. Such finishes may require replacement after a few years.

Wall and joinery finishes may become soiled and damaged from kicking and scuffing by people and objects, hence finishes below the level of about 2.5 m should be capable of receiving regular cleaning and maintenance. Finishes above this height are likely to require less attention since they are mostly subject to only airborne dust and dirt.

Generally, the heavier solid finishes will involve greater initial cost, but some may not require replacement during the life of the building.

The thinner finishes tend to be of smaller initial cost, but may require replacement many times during the life of the building.

Appearance

Since wall surfaces often constitute the largest visible surface area of an interior, the appearance of a room will be significantly affected by the choice of wall finish. In the selection of a finish (particularly if the finish is expected to remain unaltered for many years) it may be wise to compare the visual impact of various finishes by use of coloured perspective paintings, models or mock-ups of the interior.

Sound control (see also Ch. 5, p. 179)

Hard-surfaced finishes will reflect much of the noise hitting them. Where noise is likely to be a nuisance in a room, soft materials such as fabric or acoustic panels might be used which will absorb much of the sound hitting their surfaces.

Thermal insulation (see also Ch. 2, p. 36)

A wall finish could contribute to the thermal insulation of a wall by either:
- use of insulative materials such as fibre insulation board; and/or
- by the creation of a cavity between the structural wall and the wall surface.

Resistance to spread of flame

Approved document B of the Building Regulations 1991 includes requirements to restrict the spread of flame over wall and ceiling surfaces. This is discussed in Chapter 5 pp. 182–4 'Resistance to the spread of flame' and Tables 5.5 and 5.6.

Plaster

Plastering is a centuries-old craft which provides a comparatively uneven wall with a smooth, flat surface. In the past, plaster was based on lime, but today it is based on 'gypsum' (a naturally occurring material which is found in various parts of the world including the UK). Plaster is applied to wall or other surfaces as a wet mixture of gypsum plaster, aggregate (usually sand) and water.

Gypsum is a crystalline material, consisting of calcium sulphate and 'water of crystallisation' (water which chemically combines with the calcium sulphate to provide the crystalline structure of gypsum). The natural gypsum is extracted, then pulverised to a fine powder, and heated. Heating drives out the water of crystallisation; at 150–170 °C, approximately 75 per cent of the water of crystallisation is lost, and the gypsum is known as 'hemihydrate'; at 190–220 °C, the gypsum loses all of its water of crystallisation and is known as 'anhydrous'.

Types of plaster

British Standard 1191: 1973 classifies gypsum plasters. Part 1 lists:

Class A: hemihydrate gypsum plaster (plaster of Paris)
Class B: retarded hemihydrate gypsum plaster
Class C: anhydrous gypsum plaster
Class D: Keene's plaster
Part 2: 1973 is concerned with pre-mixed lightweight plasters:
Type a: undercoat plaster
Type b: finish plaster
(see 'Lightweight plasters', below)

Class A: plaster of Paris   Class A plaster has a very rapid
setting time (5–10 minutes). It is consequently unsuitable as a
wall plaster. It is used, in conjunction with jute or wire
backing, for fibrous plaster (see below) or as a 'filler' for
surface cracks in plaster surfaces.

Class B: retarded hemihydrate gypsum plasters   These
plasters have a retardant incorporated which delays the setting
time, thus making it suitable for wall plastering. Class B is
subdivided into type (a) which is mixed with sand and used
for undercoats, and type (b) a finer plaster used for applying a
smooth finish to undercoat plasters or plasterboard.

Class C and D plasters   These are no longer manufactured.

Class C: Anhydrous gypsum plaster had an accelerator which
resulted in a quick initial set. However, it could be
retempered and a high-quality finish produced due to its slow
final set time.

Class D: Keene's plaster was used as a finishing plaster
where a smooth hard surface was required.

Application of plaster

The mix proportion (plaster : sand), the number of coats and
the thickness of coats of plaster will depend largely on the
nature of the background to which the plaster is applied.
Backgrounds vary in their flatness, key and 'suction'.

Flatness   On firm flat backgrounds such as plasterboard, one
coat of plaster may be sufficient to provide a smooth, level
finish, whereas irregular surfaces such as brickwork will
require two coats. Where a surface is out of plumb, three
coats of plaster may be necessary.

Key   An irregular, rough background surface such as a brick wall, provides depressions into which the plaster will 'key' (or lodge). The key will provide support for the weight of the plaster. On smooth, flat, dense surfaces, a 'bonding agent' might be applied to the wall, prior to plastering. These are emulsions based on polymer resin which provide an adhesive surface suitable for the application of plaster.

Suction   Suction (the ability of a background to absorb water) will affect the drying rate of the plaster. Where the suction of a background is too high, the plaster will fail to set properly, resulting in a loss of adhesion of the plaster to the background. Where the suction of a background is too low, the plaster will retain excessive moisture, and drying shrinkage cracks could occur in the plaster.

Table 8.6 provides details of plaster mix proportions, and the number and thickness of plaster coats necessary for some common backgrounds.

Other plasters

Other plasters are available which are used for their specific functional properties. These include fibrous plaster and the premixed lightweight and acoustic plasters.

Fibrous plaster   Fibrous plaster consists of plaster of Paris reinforced by materials such as jute scrim or metal lathing. It may be moulded into complex ornamental forms for such purposes as cornices, covings and decorative work to walls and ceilings.

The moulding is usually carried out off site by specialist firms, then fixed on site by either:

- screwing or nailing the fibrous plaster sections to timber battens fixed to the wall;
- by attaching the sections to galvanised wire which is fixed to the wall.

Lightweight plasters   These are premixed Class B gypsum plasters with aggregates of either exfoliated vermiculite or expanded perlite. Expanded perlite is a form of volcanic glass which is expanded by heat.

When these materials are used as aggregate in plaster, the plaster is much lighter in weight than gypsum/sand plaster.

The advantages of lightweight plasters are:

- improved thermal insulation (Table 2.3, Ch. 2), hence condensation is less likely to occur;
- improved fire resistance (see Table 5.4, Ch. 5).

Table 8.6 Application of plaster to common backgrounds

	Background	Plaster mix proportions Undercoats Finish coat		Thickness of coats (mm)		
		Plaster : Sand		First undercoat	Second undercoat	Finish coat
Brickwork and blockwork	Porous bricks and blocks provide a good key. Some bricks and blocks have a special keyed surface for plastering. The key may be improved by raking out the mortar joints	1 : 2–3	Used neat or with up to 25% lime putty	11	—	2
	Dense bricks (such as engineering)	1 : 1½		8	—	2
Concrete	Dense concrete (a) where surface is flat and level; only one coat may be necessary;	— : —	Used neat	—	—	2–3
	(b) where a keyed surface is provided by for example use of rough formwork or use of a 'bonding agent'. Lightweight concretes: these shrink on drying and should be allowed to shrink prior to plastering	1 : 1½	Used neat, or with up to 25% lime putty	8	—	2
		1 : 1½–2		8	—	2
Metal lathing	Provides a good key for plastering. Problems may arise due to shrinkage of plaster applied to metal lathing; however, this is minimised by special metal lathing grades of plaster	1 : 1½–2	Used neat, or with up to 25% lime putty	8	3	2
Plaster-board	Some plasterboards (e.g. gypsum wallboards) are designed for direct decoration. Others (e.g. gypsum lath and gypsum baseboard, form a good base for plaster	— : —	Used neat	—	—	5
	Gypsum baseboard: joints should be 'scrimmed' (lined with jute-based fabric strip). Gypsum lath joints need not be scrimmed	1 : 1½–3	Used neat, or with up to 25% lime putty	8	—	2
Wood-Wool slabs	Have low suction and good key Joints and corners should be reinforced with expanded metal scrim	1 : 2	Used neat, or with up to 25% lime putty	11	—	2

Acoustic plasters These are premixed Class B gypsum plasters with a porous aggregate such as pumice, which absorbs sound.

Whilst most other acoustic finishes are available in flat panels, tiles, boards, etc. acoustic plaster may be applied to curved or irregular surfaces of suitable backgrounds (such as expanded metal lathing).

Paints and clear finishes

Paints and clear finishes are thin coatings applied to surfaces in liquid form, which gradually dry to become flexible solids.

Paints and clear finishes are used for such purposes as:

(*a*) Protection of a surface from the effects of abrasion, sunlight, dampness, dust, etc.

(*b*) Contributing to the cleanliness of a building, by providing surfaces with a finish which is easily cleaned.

(*c*) Contributing to the appearance of an interior by introduction of colours, lightness or darkness, and reflective or matt surfaces.

(*d*) Identification of pipelines: British Standard BS 1710: 1984 specifies colours to be used to either paint entire pipelines, or 150 mm bands on pipes, in order to identify the pipe contents. The colours are: green — water; silver grey — steam; brown – mineral, vegetable and animals oils and combustible liquids; yellow ochre – gases and liquid gases; violet – acids and alkalis; light blue – air; black – other fluids (including drainage); orange – electrical services (British Standards Institution).

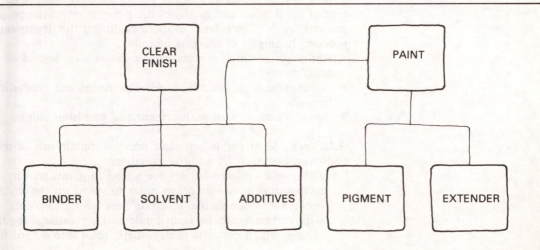

Fig. 8.4 *Composition of paint*

Composition of paints and clear finishes

Figure 8.4 shows diagrammatically the compositions of clear finishes and paints.

Clear finishes (sometimes known as 'lacquers' or 'varnishes') are transparent, and are used to protect and enhance the appearance of natural, or applied surface materials. A clear finish is composed of a binder, solvent and additives. Paints are composed of similar components, but also require pigments and extenders to provide colour and opacity.

Binders (also known as 'film formers' or 'mediums') The functions of binders are:
- to bind together the constituents of the paint or clear finish;
- to convert the liquid coating into a solid film;
- to provide the liquid with adhesive properties;
- to influence the degree of sheen;
- to provide paints with water-resistant properties.
  Binders today are based on either:
- drying oils such as linseed oil, soya-bean oil or tung oil;
- natural resins, which are either extracted from living trees or the fossilised remains of trees;
- oleo-resins, which are a combination of oil and resin;
- synthetic resins (which are very widely used today) and of which polyurethane resin, polyvinyl acetate (p.v.a.) resin, acrylic emulsions and phenolic resins are examples.

Solvents (or 'thinners') Solvents are colourless liquids which are incorporated in the paint or clear finish, and act on the binder to thin the quality of the paint. This both improves the ease of application and increases the penetration of the paint or clear finish. The solvent evaporates during the drying-out process. Examples of solvents include:
- white spirit, which is commonly used with oil and oleo-resin paints;
- ethers, which are used with cellulose resins and synthetic finishes;
- water, which is used with cement and emulsion paints.

Additives Most paints and clear finishes contain one or more additives. Some of the additives used are:
(a) *Driers and hardeners*: driers are added to paints which have a drying oil binder, in order to speed up the drying process. Hardeners (or 'catalysts') are added to paints with certain synthetic resin binders. They cause a chemical change which converts the synthetic resin into a hard film.

(*b*) *Flatting agents*, which reduce the gloss of the dried film of paint.

(*c*) *Anti-skinning agents*, which prevent formation of the skin which sometimes occurs in tins of oil-based paints.

(*d*) *Thixotropic agents*, which provide the paint with a gel quality. Such paints require no stirring, and drip less than other paints.

Pigments   Pigments are fine powdered materials which provide colour and opacity in paints, and may contribute to the durability of the paint. Pigments are derived from sources such as:

- chemically treated metals, for example titanium dioxide which is white, and is used for interior paints, or iron oxide which is yellow or red and used in all finishing paints;
- coloured earths, for example yellow ochre or burnt umber which are used for all paints;
- treated chemical compounds, for example prussian blue, which is used for oil/alkyd paints.

Extenders   Extenders are finely ground natural materials (mostly natural earths) which have no effect on the colour of paints, but may adjust certain of properties of the paint. Extenders may, for example:

- improve the ease of application of the paint;
- improve the adhesion of the paint to undercoats;
- give body to the paint;
- increase the hardness of the paint film.

Paint systems

Surfaces are usually painted with several coats of paints, each coat performing a specific function. A series of coats of paint is known as a 'paint system'. A typical paint system will comprise of: (*a*) primer; (*b*) undercoat(s); (*c*) finishing coat(s).

Sometimes different types of paint will be used for each coat whilst some paints, such as emulsions, will perform the function of more than one element in the paint system.

Primers   A primer is the first coat of paint applied to an unpainted surface, and its functions are:

- to adhere to the surface, and provide adhesion to subsequent coats of paint;
- to provide a smooth, stable surface for subsequent coats of paint;
- to prevent corrosion and chemical attack.

A wide variety of primers are available to suit different paint

Table 8.7 Primers and undercoats

Primer/under-coat	Composition	Properties	Uses
Acrylic primer/under-coat	White or pink pigment. Acrylic resin binder. Water thinner	Quick drying. Provides hard, flexible film. Good adhesion	May be used on many interior surfaces including brick, plaster and timber. Not suitable for bare metal
Alkali-resisting primer	Pigment. Oleo-resinous binder. White spirit thinner	Alkali-resistant. Resists water	For all surfaces of an alkaline nature
Aluminium primer	Grey-coloured aluminium-based pigment. Oleo-resinous binder. White spirit thinner	Seals surfaces. Resists water	Wood primer for resinous timbers. Used to seal timbers treated with coal-tar preservatives or bitumen
Calcium plumbate primer	White- or grey-coloured calcium-plumbate-based pigment. Linseed oil binder. White spirit thinner	Resists corrosion. Good adhesion. Toxic	For priming metals and composite metal timber components
Etch primer	Pigment. Oleo-resinous binder. Alcohol-based thinner. Phosphoric acid	Resists corrosion. Good adhesion. Lead free	For obtaining maximum adhesion to ferrous and non-ferrous metal
Oil/resin undercoat	White or coloured pigment. Drying oil, or oil-modified alkyd binder. White spirit thinner	Good adhesion. Good opacity	For use with gloss paints. May be used on any suitably primed surface
Lead-free primer	White- or pink-coloured pigment. Oleo-resinous binder. White spirit thinner	Good adhesion. Good opacity	General interior woodwork, particularly where non-toxic surfaces are important, e.g. nurseries, kitchens, etc.
Zinc chromate primer	Yellow-coloured zinc chromate pigment. Alkyd or oleo-resinous binder. white spirit thinner	Resists corrosion. Non-toxic	Suitable for ferrous or non-ferrous metals and composite wood and metal components

Table 8.8 Finishing paints

Paint	Composition	Properties	Uses
Oil-type paints (oil-modified alkyd paints)	Titanium white plus coloured pigment. Oil-modified alkyd resin binder (some gloss paints also contain other resins in the binder, such as silicone or polyurethane). White spirit thinner	Available in matt, eggshell and gloss finishes	The paints based on oil binders have been largely superseded by oil-modified alkyd binder paints (alkyd is a synthetic resin derived from alcohols and acids). These paints are used very commonly for woodwork and other surfaces
Polyurethane paints, which are available as either 'one pack' (ready mixed) or 'two pack' (which require mixing)	Titanium white plus coloured pigment. Binder, one pack: polyurethane-modified alkyd; two pack: polyurethane resin. Thinner, one pack: white spirit; two pack: special solvent produced by manufacturers	One pack: thixotropic (which gives the paint a 'non-drip' quality). Available as semi-gloss or gloss finish. Easily cleaned. More expensive than oil-modified alkyd paints. Two-pack: chemical, water and abrasion resistant. Good adhesion. Heat resistant	One pack: suitable for many interior surfaces, including woodwork, walls and metalwork. Usually, only one coat is necessary. Two pack: used where tough finishes are required, such as laboratory finishes or floors
Epoxy resin paints, which are available as either 'one pack' (ready mixed) or 'two pack' (which require mixing)	Titanium white plus coloured pigment. Epoxy resin binder. Thinner, one pack: white spirit based; two pack: hydrocarbon based	One pack: tougher and better resistance to water and alkalis than oil-modified alkyd paints. Two pack: highly resistant to abrasion, alkalis and solvent	One pack: used as two pack, but in less demanding locations. Two pack: used where a tough finish is required, such as furniture, machinery and floors
Water thinned gloss and eggshell paints	Titanium white and coloured pigment. Acrylic or other synthetic resin binder. Water thinner	Washable. Alkali resistant. The gloss of water-thinned gloss paints is not as high as oil-modified alkyd gloss paints	As emulsion paints (below), but where a greater degree of sheen is required
Emulsion paints	Titanium white and coloured pigment. Polyvinyl acetate (p.v.a.) or acrylic-based binder. Water thinner	Washable. Quick application, alkali resistant	Very widely used for internal wall finishes but may be applied over other surfaces
Cellulose paints	Titanium white, and a wide range of coloured and metallic pigments. Nitro-cellulose binder. Various thinners used	Quick drying. Highly inflammable. Hard, good abrasion resistance. Water resistant	Used as a metallic or coloured finish to metalwork such as balustrading and shop fronts
Metallic paints	Flakes of metal dispersed in a cellulose lacquer or varnish medium	Available in wide range of metallic colours. Available in powder form, or ready mixed	Decorative reflective finishes for metalwork

systems, surfaces and surface conditions. Table 8.7 provides details of some common primers and undercoats.

Undercoats   Undercoats are applied on top of primer coats, or alternatively on suitably prepared surfaces which have been previously painted. Undercoats are usually highly pigmented and of low gloss. Their functions are:
- to be sufficiently opaque to hide the colour of the primer or other surface;
- to be of a suitable colour for the more transparent finishing coat;
- to increase the thickness of the film.

Finishing coats   Finishing coats provide:
- the final colour and degree of gloss to the surface;
- protection against moisture, chemicals and abrasion.

The degree of gloss of a finishing coat may be either full gloss, semi-gloss, eggshell or matt.

Table 8.8 provides details of finishing paints.

Special paints

Anti-condensation paints   These are oil-based or emulsion paints containing small particles of cork, or other thermal insulative material, which gives the paint a textured finish. The paint acts as an insulative barrier between warm air in a room and a cold surface, and prevents the formation of condensation.

Anti-fungus paints   These paints contain toxic compounds which kill fungi attempting to feed on it. Some anti-fungus paints are formulated specially for their anti-fungus properties, whilst some ordinary paints may be obtained with a fungicide added to them. The fungicidal properties of the latter paints may not last for the duration of the paint film.

Fire-retardant paints   These paints may be either:
- 'intumescent', where the effect of heat causes the paint to swell up and form a barrier to insulate the painted surface from the effect of flames;
- paints which incorporate materials such as chlorinated paraffin, or chlorinated rubber which, when heated, fuse together to prevent combustion.

Relief texture paints   These paints have a heavy plaster-like consistency which may be applied and worked into a textured finish, or applied thickly then carved or modelled to form

Table 8.9 Clear finishes

Finish	Composition	Properties	Uses
Emulsion varnish	Binder: p.v.a. or acrylic. Thinner: water	Seals absorbent surfaces. Washable	When mixed with emulsion paint, may be used to seal surfaces such as brick or plaster. Used as a protective coating for surfaces such as wood and wallpaper
Oil–resin varnish	Binder: oleo-resin or oil-modified alkyd. Thinner: white spirit	Gloss, eggshell or flat. Hard, but brittle surface. Quick drying	Oil–resin varnishes used internally are known as 'short oil' varnishes since they have low oil and high resin content
Polyurethane varnish	Binder, one pack: polyurethane resin; two pack: urethane oil. Thinners. Hardeners	Water, abrasion, heat and chemical resistant. Two-pack varnishes more hard-wearing and flexible than one-pack	Available as one-pack or two-pack material. Used where hard-wearing timber surfaces are required, e.g. bar and counter tops, laboratories and floors
Scumble glaze	Extender: china clay or aluminium stearate. Binder: oleo–resin varnish and beeswax. Thinner: white spirit	Requires skilled handling for imitation marbling or graining. Coloured glazes may be applied	Transparent or semi-transparent material applied to painted surfaces for tone/colour effects, imitation marble and woodgrain, etc.
Spirit varnish	Binder: shellac (lac resin melted into flakes) or damar resin. Plasticiser. Thinner: methylated spirits	Quick drying. Brittle, and tends to crack if shellac content is high	(a) Spirit varnish used for bare interior woodwork. (b) Stained varnish incorporates dyes to stain and varnish woodwork in one operation. (c) Knotting: a form of spirit varnish used to seal wood knots and prevent discoloration of paintwork. (d) French polish: provides an excellent wood finish. A traditional finish which has been largely superseded by modern proprietary varnishes
Wood stain	Transparent coloured solutions of various compositions. Dyes or pigments are mixed with water or spirit	Water stains: cheap, easy to use, slow drying. Spirit stains: deep penetration of timber. Quick drying. Oil stains: low penetration, slow setting	Used to either colour whitewood or enhance the colour of other timbers

decorative borders or motifs. The paint is usually supplied as a powder containing materials such as plaster, china clay or exfoliated mica, and is mixed with water.

Heat-resistant paints   Ordinary oil-modified alkyd paints are suitable for applying to surfaces with temperatures approaching boiling point, although white and pale-coloured paints might discolour. Special paints are available which will resist temperatures in excess of 500 °C. The composition of the paint will be suited to a particular temperature range, although most of the high-temperature finishes include aluminium powder.

Multi-coloured paints   These paints contain globules of two, three or four different colours, which form a flecked, spotted or streaked textured film when sprayed on a surface. The paint has good adhesion, is hard-wearing and washable and is suitable for locations exposed to harsh treatment.

Clear finishes
A wide range of clear finishes is available for purposes such as:
- protection and preservation of surface materials;
- sealing porous materials;
- enhancing the appearance of surface materials.

Some of these clear finishes are described in Table 8.9.

Application of paints
Paint may be applied by either:
- brush, which provides good adhesion to the surface material;
- roller, which, being wider than a brush, is quicker in application;
- pads, which consist of a foam pad covered with a layer of hair pile, attached to a handle;
- mechanical spraying equipment, which is usually used for either complex sculptured or figured surfaces which would be difficult to paint by other means, or very large surface areas.

If paint is to adhere to a surface and be free of defects, care must be taken to ensure that the surface is suitably prepared.

Where paint is to be applied to an existing unsound surface such as cracking or flaking paint, it will be necessary to remove the existing paint. This may be done by:
- 'burning off' (using a blowlamp which heats and softens the paint sufficiently for it to be scraped off);

- applying a liquid paint remover which softens the paint sufficiently for it to be scraped off;
- by use of mechanical sanders.

Once the paint is removed, the surface may be treated in a similar manner to new surfaces.

Figure 8.5 illustrates the procedures necessary for preparing surfaces for the application of undercoats and finishing coats of paint.

Application of clear finishes

Clear finishes may be applied in order to enhance the natural appearance of attractive woods and metals (particularly hardwoods and non-ferrous metals). As with paints, it is important that the surface is properly prepared to ensure adhesion of the finish and lack of defects.

Preparation of a timber Figure 8.6 illustrates the necessary preparatory stages prior to applying a clear finish to previously coated, or new timber.

Preparation of metals Some metals, such as stainless steel and aluminium alloy, retain their bright appearance without the application of a protective clear finish. Copper-based metals will require either regular wax polishing or, alternatively, a clear finish such as cellulose or polyurethane lacquer may be applied.

The preparation of metals for receiving clear finishes involves thorough cleaning and degreasing of the surface prior to application.

Thick solid materials

The materials considered in this section are claddings which are fixed to internal wall surfaces, and are: (*a*) stone; (*b*) ceramic tiles; (*c*) mirror and decorative glass; (*d*) timber panelling; (*e*) metal cladding.

It should be noted that in certain instances, where a void is created between a wall structure and an internal cladding, the void should be filled with non-combustible materials such as mineral wool, in order to prevent the spread of fire.

Stone

Stones used in interiors are described on p. 376 above. Interior stone claddings are very durable, but expensive, heavy, hard and noisy. Igneous stones such as granite are

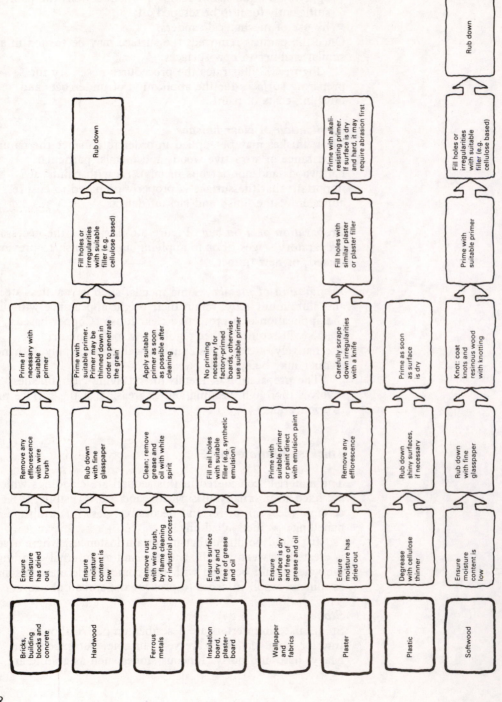

Fig. 8.5 *Preparation procedures for painting*

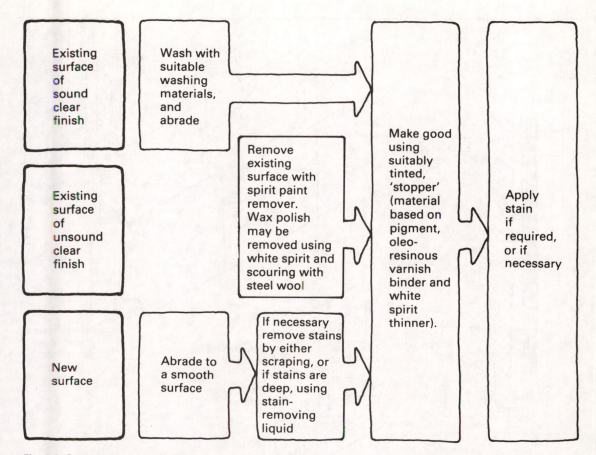

Fig. 8.6 *Preparation procedures for clear finishes to timber*

rarely used for internal cladding, sedimentary stones such as limestones and sandstones are sometimes used, but the metamorphic stones such as slates and marbles are the most frequently used.

The stones are available as either square or rectangular slabs, which may be fixed to the wall either: (*a*) in a simple square or rectangular grid pattern; (*b*) in a bond pattern. Typical thicknesses of marble and slate slabs might be 18 mm, whilst limestones and sandstones tend to be thicker.

Stone slabs are fixed back to the wall surface by means of non-ferrous metal cramps (such as phosphor bronze, brass or copper, as illustrated in Fig. 8.7(a)). One end of the cramp is built into the wall, whilst the other end hooks into a recess in the slab. The cramps may be built into the wall by such means as either:

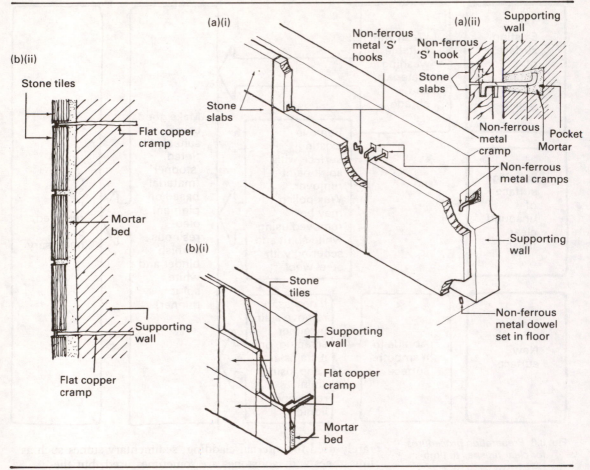

(a)(i)

Non-ferrous metal 'S' hooks

Stone slabs

(a)(ii)

Supporting wall

Non-ferrous 'S' hook

Stone slabs

Non-ferrous metal cramp

Pocket Mortar

Non-ferrous metal cramps

Supporting wall

Non-ferrous metal dowel set in floor

(b)(ii)

Stone tiles

Flat copper cramp

Mortar bed

Supporting wall

Flat copper cramp

(b)(i)

Stone tiles

Supporting wall

Flat copper cramp

Mortar bed

Fig. 8.7 *Interior stone cladding*
(a) *Stone slabs with air space between stone and wall surface*
 (i) *Pictorial view*
(ii) *Detail section*
(b) *Tiles bedded in mortar*
 (i) *Pictorial view*
(ii) *Detail section*

- setting the cramps into mortar joints of a brick wall;
- cutting a 'pocket' (or 'mortise') in a concrete wall, and bedding the cramp into the pocket with mortar.

Limestones and sandstones are usually backed with a bedding of mortar, whilst slabs of marble or slate are usually installed with a 12 mm air space between the slab and wall, and in the case of marble, the backs of the slabs are treated with an anti-staining solution to avoid stains appearing on the outer surface of the slabs. Marble and slate slabs are bedded on small 'dabs' of plaster located towards the corners of each slab. The plaster dabs position the slabs, but do not provide support.

Figure 8.7(a) illustrates a typical fixing system for marble or slate slabs. The top edges of the slabs are held by copper cramps which hook on to a recess in the top edge of the slab. The bottom of the slab is held by 'S'-shaped copper hooks which are housed at one end, in a slot at the bottom of the slab, whilst the other end hooks over the slab below.

Small slate tiles might be backed with a bedding of mortar and supported by flat metal cramps fitted to every third or fourth course of tiles (Fig. 8.7(b)).

Ceramic tiles

'Ceramic' (or clay) tiles have been used for centuries as a decorative finish for internal walls. Their hard, shiny surface makes them easy to clean, hence their association with bathrooms and kitchens.

Manufacture   Ceramic tiles are made in two stages:

(a) A 'biscuit' tile is made consisting of clays, flint, limestone and other materials, each of which is ground and mixed with water to form a slurry. The slurries are then mixed together, forming a liquid clay material known as 'slip'. Water is extracted from the slip, and the material is ground to a fine powder. This powder is moistened, pressed into the required shape and size, then fired in a kiln at high temperature.

(b) A coating of liquid glaze is applied to the top surface of each tile, and the tile is again fired at high temperature. This fuses the glazed coating to the ceramic tile.

Range   Most ceramic tiles are square, either 150×150 mm or 108×108 mm. Typical thicknesses are 5, 5.5 and 6 mm for the larger tiles, and 4 mm for the smaller tiles. Hexagonal and rectangular tiles, and tiles of greater thickness, are also made.

Many tiles incorporate 'spacer lugs' (2 mm projections) at their edges. These enable the tiles to be laid accurately with a neat 2 mm joint between each tile.

Some manufacturers also make tiles with one rounded edge ('RE' tiles) and tiles with two adjacent rounded edges ('REX' tiles). These tiles are used in order to form a neat rounded edge for small areas of tiles used for such purposes as splashbacks behind basins and baths.

Internal and external 'angle bead' tiles are also manufactured. These are designed to enable neat radiused corners to be made at the internal or external junctions of adjacent walls of tiling. Angle bead tiles are similar in length to square tiles and are typically 25 mm in depth.

Fixing   Ceramic tiles are suitable for fixing to most types of wall surface. However, since they are bonded to the wall surface with an adhesive, any movement occurring in the wall (such as moisture movement) could cause failure of the bond.

In domestic buildings, ceramic tiles are usually fixed with a proprietary mastic adhesive. If the wall is flat, the tiles may be fixed by the 'thin-bed' method, where the tiles are bonded by a thin layer of adhesive (typically 3 mm in thickness). Some mastic adhesives are suitable for applying over irregular wall surfaces using a technique known as the 'thick-bed' method. Here, a thick layer of adhesive is applied, which evens out any irregularities in the wall surface.

Mirrors and decorative finishes to glass

Mirrors Mirrors are made from either float glass (selected or special selected quality) or plate glass (silvering quality). A layer of silver, or other metal, is deposited on the back surface of the glass followed by a protective layer of copper, a layer of undercoat, paint and finally enamel, which is stoved. If the mirror is to be fixed in a damp environment, an additional protective layer of lead foil might be necessary.

Certain mirrored plastics are also made. These might typically consist of a layer of aluminium deposited on the back of acrylic sheet, protected by a coating of paint.

Fixing Large areas of mirror might be bonded to blockboard and screwed to battens fixed to the wall. Relatively small areas of mirror might be fixed directly to the wall by use of either clips or screws.

Clips Figure 8.8(a) illustrates a pair of clips suitable for fixing mirrors to walls. The clips have hooked ends which hook over the mirror edge. A nylon washer is placed between the screw and clip, and between the clip and wall. These create an air space between the wall and mirror, which prevents the formation of condensation on the back of mirrors in rooms such as kitchens and bathrooms.

The bottom clips are screwed rigidly to the wall. The side and top clips have a slotted screw hole which allow the clips to be pushed towards the mirror until the hooked ends locate over the edge of the mirror.

Screws Mirror screws are usually chromium plated, and have a screw thread tapped into the screw head. This allows a domed cap to be screwed over the screw head (Fig. 8.8(b)) which provides a neat appearance.

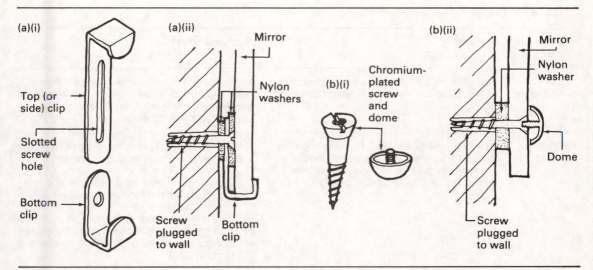

Fig. 8.8 *Mirror fixings*
(a) *Mirror clips*
(i) *Pictorial view of clips*
(ii) *Detail section*
(b) *Dome-headed screws*
(i) *Pictorial view of screw and dome*
(ii) *Detail section*

Nylon washers are placed between the mirror and wall to provide the necessary air space for prevention of condensation.

To prevent the mirror from cracking, screw holes should be drilled at not less than 25 mm from the mirror edge.

Curved mirrors   Curved glass mirror surfaces may be achieved by use of sheets of small pieces of mirror tile (typical sizes 25 × 25 mm or 25 × 50 mm) which are bonded to a cloth backing. Typical sheet sizes are 600 × 450 mm or 300 × 230 mm. The sheets may be bonded to flat, angled or curved surfaces by use of a contact adhesive.

Mirror tiles   Square mirror tiles are made in sizes ranging from 108 mm to 300 mm. These are fixed to the wall by means of self-adhesive pads.

Decorative finishes on glass   Decorative glasses may be used internally for such purposes as:
- Screens (where the glass will usually be held in rebated metal or timber frames);
- decorative wall panels (where the glass will usually be fixed to the walls either in a similar manner to that described for mirrors, or in rebated frames).

Table 8.10 describes different decorative glass treatments. Acid embossing, wheel cutting and sandblasting may also be used in conjunction with silvering to produce decorative mirrors.

Table 8.10　Decorative treatments to glass

Decorative treatment	Process	Decorative effect
Acid embossing	Hydrofluoric acid and an alkali are applied to the glass surface which decomposes the glass	A variety of textural effects may be achieved, including smooth shades which produce a grey appearance, sparkling stippled effects and almost obscured effects
Wheel-cutting	Revolving discs are used to cut and polish designs into the glass surface. Different kinds of cut may be achieved by cutting with discs of different edge profile	1. 'Brilliant cutting' produces precise cuts using a stone disc. Complex designs may be achieved. 2. 'Engraving' is used for more detailed, fine work using a hand tool with smaller-diameter discs. The visual effect of cut glass may be enhanced by arranging lighting at the side of the glass which emphasises the highlights and shadows of the cuts
Sand-blasting	A jet of sand is directed at the glass surface, which wears the surface away	Light sand-blasting produces an obscured effect, whilst 'peppering' (light sand-blasting which only partially removes the glass surface) provides a mottled effect. Sculptured effects may be produced by deep sand-blasting
Stained glass	Pieces of coloured or painted glass are set into strips of lead. Where stained glass is liable to bend under pressure, it may be supported by iron or copper rods. The rods might be incorporated into the design of the stained glass. 'Glass appliqué' consists of pieces of coloured glass stuck together with transparent adhesive	Rich coloured effects which may be enhanced by arranging light to shine through the glass
Gilding	Gold leaf is applied to the back surface of the glass by use of a transparent glue. Two layers of gold leaf are used for better-quality work	A range of colours varying from pale lemon to deep gold may be used to provide patterns or designs

The edges of glass may also be given a decorative treatment by grinding and smoothing the edge in order to produce either a bevelled, mitred or rounded edge profile.

Timber panelling

Timber, in the form of panelling, has been used as an internal wall cladding for centuries.

Traditional panelling (Fig. 8.9) consisted of a moulded timber framework infilled with solid panels of timber. Modern timber panelling (see Fig. 8.11) tends to be much more simple in detail, and might typically be little more than plain veneered plywood panels with no visible framework.

Both traditional and modern timber panelling are fixed to timber 'grounds', which are lengths of vertical and horizontal timber fixed to the wall. The grounds are fixed by means such as:

- nailing or screwing through the face of grounds into timber 'plugs' set into the wall;
- screwing through the face of the grounds into plastic plugs which are pushed into holes drilled in the wall (see 'Fixings to walls', Ch. 6, p. 284, and Fig. 6.6(b)).

The panelling is screwed to the grounds in such a way that the visual effect of the fixing is minimised. This might be achieved by either:

(a) Locating the screws in parts of the panelling framework which will be covered by other timber members such as skirtings or dado rails. The bottom, intermediate and top rails illustrated in Fig. 8.9 are fixed in this way.

(b) Sinking the screw heads below the face of the timber framing and inserting a pellet into the hole to conceal the screw head. The pellet will usually be made of similar timber to the framing. The dado rail illustrated in Fig. 8.9(c) is fixed in this manner.

(c) Using 'concealed fixing' techniques, which may be typically used for fixing modern panelling with no visible framework.

Traditional panelling (Fig. 8.9) At a first glance, there may appear to be little visible difference between the construction of traditional panelling and modern panellings which incorporate a visible framework. Two of the distinguishing features of traditional panelling concern: (a) the mouldings; (b) the panels.

Mouldings In traditional panelling, the mouldings tend to be more intricate in profile, and are cut into the framework members. In modern panelling, the framework members will usually be of plain profile, and any mouldings will usually be 'planted' (fixed on the face) of the framework.

Panels Traditional panels were of solid timber, with the exposed face usually being profiled, and the back face being left with a rough-sawn finish. Modern panelling is usually flat in profile, and made of veneered plywood.

Traditional panels were housed in grooves cut in the edges of the framework. Since solid timber is prone to expansion and contraction, resulting from changes in humidity, the grooves were made of sufficient depth to allow for expansion in the panels.

Modern plywood panels are much less subject to moisture movement and consequently the panels may fit tightly in the groove. Some modern panels fit in a rebate in the frame rather than a groove.

The grounds of traditional panelling were often of 19 mm thickness (corresponding to the thickness of plaster on the wall). Vertical and horizontal grounds were positioned to coincide with the positions of the vertical and horizontal members of the panelling framework.

The framework consisted of horizontal members (top, bottom and intermediate 'rails') and vertical members ('stiles' and 'muntins') which were typically of between 28 and 35 mm thickness. The stiles were full-length members forming the vertical outer edges of the framework, whilst muntins were the shorter lengths of vertical framework fixed between rails.

The skirting boards were usually screwed and pelleted to vertical sections of timber known as 'soldiers' (Fig. 8.9(b)) and the angled cornice would usually be screwed and pelleted to timber brackets projecting from the top horizontal ground (Fig. 8.9(d)).

Fig. 8.9 *Traditional timber panelling*
(a) *Elevation of traditional timber panelling*
(b) *Detail at skirting level*
(c) *Detail at dado level*
(d) *Detail at cornice level*

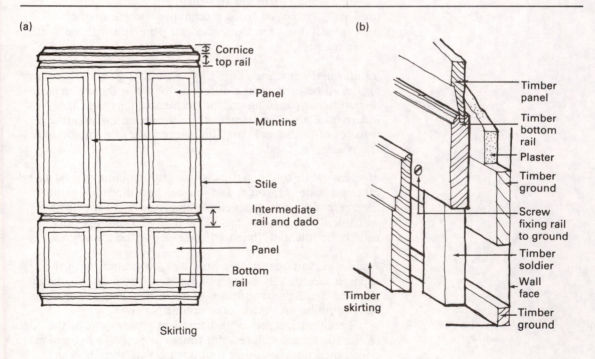

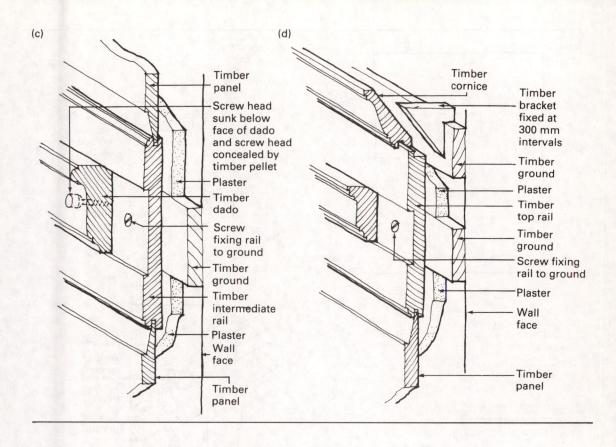

(c)

- Timber panel
- Screw head sunk below face of dado and screw head concealed by timber pellet
- Plaster
- Timber dado
- Screw fixing rail to ground
- Timber ground
- Timber intermediate rail
- Plaster
- Wall face
- Timber panel

(d)

- Timber cornice
- Timber bracket fixed at 300 mm intervals
- Timber ground
- Plaster
- Timber top rail
- Timber ground
- Screw fixing rail to ground
- Plaster
- Wall face
- Timber panel

Modern panelling with exposed framework (Fig. 8.10)

Figure 8.10 illustrates a more modern approach to timber panelling with an exposed framework, where the panels are much bigger than traditional panelling, and no mouldings are used either on the framework or on the panels.

The timber grounds are located in positions coinciding with the positions of the exposed framework.

The stiles, rails and muntins of the framework (which are rebated to receive the panels) are jointed together with mortise and tenon joints, and are screwed and pelleted to the grounds. A typical size for the framework might be 100 × 25 mm.

The panels are of veneered plywood (typically of 10 mm thickness) which are glued and pinned from behind into the rebates of the framework.

Where panels of large area are used, intermediate horizontal or vertical grounds might be first fixed to the wall.

407

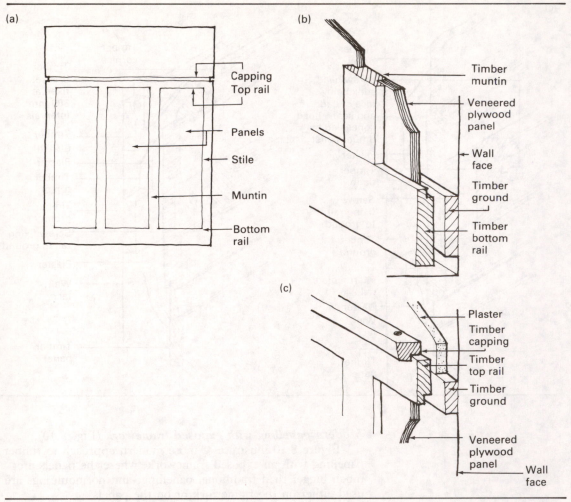

(a)

Capping
Top rail

Panels

Stile

Muntin

Bottom
rail

(b)

Timber
muntin

Veneered
plywood
panel

Wall
face

Timber
ground

Timber
bottom
rail

(c)

Plaster

Timber
capping

Timber
top rail

Timber
ground

Veneered
plywood
panel

Wall
face

Fig. 8.10 *Modern timber panelling
with exposed framework*
(a) *Elevation of panelling*
(b) *Detail at skirting level*
(c) *Detail at top rail level*

Modern timber panelling with no exposed framework
(Fig. 8.11)   Figure 8.11 illustrates a method of construction
that might be employed for blockboard panelling where no
framework is visible.

The grounds consist of a rebated skirting, and horizontal
rebated intermediate and top grounds. The panels (which
might typically be of 19 mm thickness) are of veneered
blockboard, and are fixed to rebated top, intermediate and
bottom rails. The panels are fixed by screwing through the
back of the rails into the blockboard panel.

With the blockboard panels attached, the rebated rails are hooked over the rebates of the grounds, and the top rail is screwed down into the top ground.

The vertical junction between adjacent blockboard panels might be made by forming a 'V'-jointed tongued and grooved joint (similar to the 'V'-jointed tongued and grooved boarding illustrated in Fig. 8.12(a)). The panels may then be secretly nailed to the grounds.

Fig. 8.11 *Modern timber panelling with no exposed framework*
(a) *Elevation of panelling*
(b) *Detail at skirting level*
(c) *Detail at top rail level.*

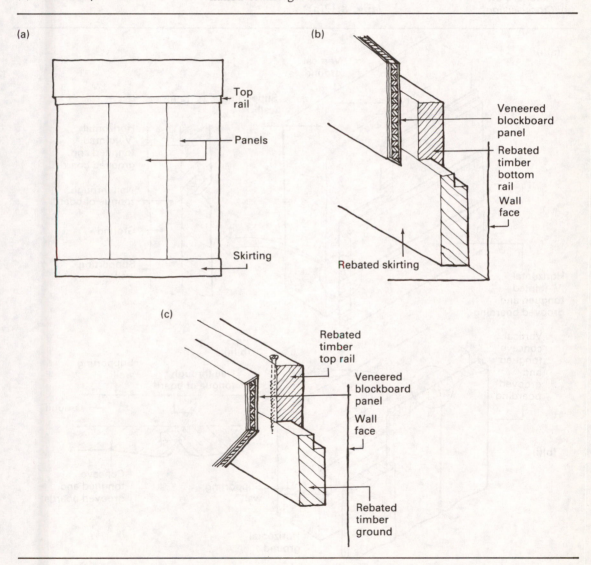

Fig. 8.12 *Timber boarding*
(a) *Horizontal 'V'-jointed tongued and grooved boarding*
(i) *Pictorial view*
(ii) *Cross-section*
(b) *Vertical concave tongued and grooved boarding*
(i) *Pictorial view*
(ii) *Cross-section*

Timber boarding (Fig. 8.12)   Timber boards may be nailed to grounds fixed to the wall. The boarding may be applied vertically (sometimes referred to as 'matchboarding'), horizontally or diagonally.

The most suitable boards for internal cladding are those of 'V'-jointed tongued and grooved profile. The boards may be secret nailed by driving nails through the tongue. The groove of the adjacent board will then conceal the nail heads (Fig. 8.12(a)).

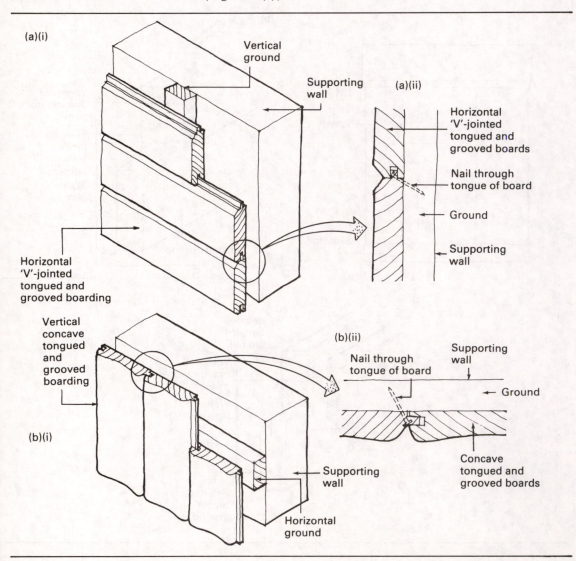

(a)(i)

Vertical ground

Supporting wall

(a)(ii)

Horizontal 'V'-jointed tongued and grooved boards

Nail through tongue of board

Ground

Supporting wall

Horizontal 'V'-jointed tongued and grooved boarding

Vertical concave tongued and grooved boarding

(b)(ii)

Supporting wall

Nail through tongue of board

Ground

(b)(i)

Supporting wall

Concave tongued and grooved boards

Horizontal ground

The bevelled edge of the board, which forms the 'V' profile with the adjacent board, ensures that any moisture movement that takes place in the boards will not be unsightly in appearance. Moisture movement occurring between straight-edged boards can result in noticeably irregular gaps. Other alternative profiles of tongued and grooved boarding include the concave sectioned tongued and grooved boarding illustrated in Fig. 8.12(b). Typical sizes for boarding are 16 and 19 mm thick and 75, 100 and 125 mm wide.

Metal claddings

Metals may be used for cladding internal walls or joinery, providing surfaces which are easy to clean and maintain. Metal cladding may be used because of functional qualities (e.g. for hygienic, hard-wearing surfaces in kitchens or bathrooms) and for visual qualities (e.g. decorative claddings of copper or chromium).

Two of the methods used for metal cladding are: (*a*) profiled metal cladding systems; (*b*) metal-faced boards.

Profiled metal cladding systems

Figure 8.13 illustrates a system of profiled aluminium sections which are screwed to timber battens fixed to the wall. The sections (which might typically be of 100 mm width) are designed with interlocking edges.

Standard accessories are usually available for forming details such as internal and external corners and edges.

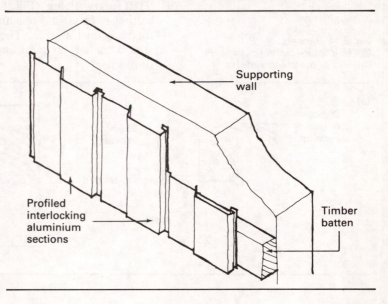

Fig. 8.13 *Aluminium wall cladding*

Metal-faced boards Metals may be used as surface finish to plywood and blockboard. Some manufacturers supply sheets of plywood with metals such as aluminium, copper or stainless steel bonded to one face.

If used as a wall cladding, metal-faced boards might be:
- fitted into a rebated framework similar to that illustrated in Fig. 8.10;
- fitted with secret fixings, similar to Fig. 8.11.

Thin solid materials

The materials considered in this section are: (*a*) timber veneers; (*b*) cork tiles; (*c*) plastic laminates.

Timber veneers

Since solid timber is expensive (particularly some of the hardwood varieties), veneered boards are frequently used as an alternative material for wall panelling and joinery. Veneers are thin slices of timber which are applied with glue to surfaces such as blockboard, plywood and chipboard. Some blockboard and plywood sheets are obtainable from manufacturers with veneers ready bonded to one or both surfaces.

Table 7.3 (Ch. 7) indicates some of the hardwoods from which veneers are cut. Veneers are available in various widths, and in lengths of up to 2 m.

Decorative veneers are usually cut by either:
(*a*) 'Half rotary slicing' (Fig. 8.14(a)) where a log of quadrant section is fixed in a machine which enables the log to rotate against a knife. The knife cuts off veneers which are almost 'radial' (at right angles to the growth rings of the timber).

Fig. 8.14 *Veneers*
(a) *Half rotary cutting of veneers*
(b) *Use of matching veneers*

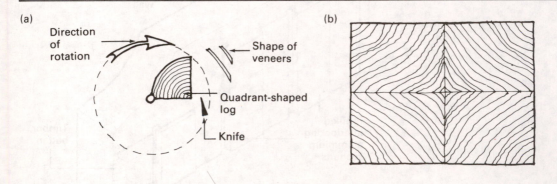

(b) 'Flat knife cutting' where a squared-off log (or 'fitch') is
fixed in position, and a knife slides up the log face to cut
the veneers.

Both of these cutting methods produce 'matching veneers'
where several adjacent veneers from a log will have almost
identical colouring and figuring. Matching veneers might be
used for panelling, doors, furniture, etc. by arranging similarly
figured veneers to form a pattern. Figure 8.14(b) shows the
use of four similar veneers to form a pattern. Since veneers are
thin, they may be glued over rounded edges of doors and
panels with curved surfaces. Some veneers are available with
backings such as cloth or aluminium, and may be used for
purposes such as cladding circular columns, by the application
of an adhesive.

Cork tiles

For further information concerning cork finishes see Table
8.3(d) and 'Cork-faced wallpaper', p. 418 below.

Cork is made in tile form, which is suitable for fixing to
smooth wall surfaces by use of an adhesive. The tiles vary in
texture from those with a coarse texture (which are composed
of a cork or hardboard base surfaced with rough-textured cork
stripped from the bark of the tree) to smooth tiles (which are
made from compressed granules). All but the roughest
textured tiles may be finished with a suitable matt or glossy
sealant (such as polyurethane).

The typical sizes of tiles are 300 × 300 × 2.5–13.5 mm in
thickness, and panels of 300 × 900 × 12 mm thickness are
available.

Decorative plastic laminates

Decorative plastic laminates (sometimes known as decorative
paper laminates) are thin sheet materials which are used as a
decorative finish on boards, for items such as worktops,
cupboards and wall panels. The material has become
immensely popular due to its advantages of cheapness and ease
of application (in comparison with timber veneers), its ease of
cleaning and the range of decorative effects available.

Decorative plastic laminates consist of:
- a core which is comprised of several layers of paper
  impregnated with phenolic resin;
- a surface which is composed of a decorative sheet of paper
  covered with a transparent layer of melamine resin.

Plastic laminates are bonded to board materials by use of
an adhesive. When plastic laminates are applied to thin
unsupported boards (such as cupboard doors) a 'balancing'

sheet of plastic laminate should be applied to the other surface of the board. This prevents the board from warping.

Typical sheet sizes are 3050 × 1220 mm or 2440 × 1220 mm with thickness of 1.2 and 1.5 mm. Sheets of these thicknesses may be bonded to curved surfaces with a radius of as little as 150 mm. 'Postforming' plastic laminates are also available, which are usually of 0.8 mm thickness. These sheets, if carefully applied, may be bonded to curved surfaces of as little as 12 mm radius.

The wide range of decorative effects available include plain colours, patterns, metals (such as aluminium, copper and stainless steel), and printed marble, wood grain, leather and other effects.

Also, manufacturers are able to incorporate artwork and prints into the laminated plastic, which are to the client's or designer's own specification.

Boards such as plywood, blockboard and chipboard are available with plastic laminates already bonded to their surfaces. These boards save time on site, particularly when large areas are involved. Typical board sizes are 3050 × 1220 mm and 2400 × 1220 mm with thickness varying from 14 to 18 mm.

Thin flexible materials

This section deals with thin flexible materials which are either:
- applied directly to a wall surface (surface coverings);
- applied to some form of panel fixed to a wall or to joinery (fabrics and skins).

Surface coverings
Surface coverings are thin materials which are 'hung' (applied to the surface of a wall) by use of an adhesive. They include: (*a*) lining papers; (*b*) decorative papers; (*c*) fabric and other surface coverings.

Lining papers Lining papers are surface coverings which are applied to a wall surface prior to hanging decorative papers or other decorative coverings.

Different lining papers are used for different purposes, such as concealing irregularities in a wall surface, resisting moisture, improving thermal insulation, providing even porosity, and facilitating even drying. Table 8.11 describes some of these lining papers.

Lining paper is hung with the joints running at right angles to the joints of the final surface covering. Where lining paper is used as a surface for receiving paint, the paper is usually hung with vertical joints.

The sizes of rolls of lining paper varies, but most are in the region of 530–550 mm in width and 10 m in length.

Decorative papers The decorative papers considered here are: (*a*) printed papers; (*b*) embossed papers; (*c*) other papers. The standard size of rolls of decorative papers is 10 m long by 520 or 530 mm in width.

Printed papers Most printed papers are printed by machines with rollers which are capable of printing up to 20 colours simultaneously. Some expensive papers are still printed by hand, where each colour is applied individually. The papers used are either 'pulps' or 'grounds'.

Pulps are papers where the pattern is printed directly on to the paper, and the natural colour of the paper forms the background colour of the design. Pulps tend to be cheapest decorative papers.

Table 8.11 Lining papers

Paper	Description and uses	Hanging
Cotton-backed lining paper	Heavy wood-pulp paper (see below) bonded to cotton. Used on walls where movement might be expected (e.g. timber boarding)	Cotton side applied to wall using starch paste or starch ether paste
Expanded polystyrene	Although not strictly a 'paper' expanded polystyrene (see Ch. 7) is available in rolls, and is used to increase the thermal insulation of a wall surface. This reduces the occurrence of condensation. Thickness of expanded polystyrene: 2 mm	Applied to wall with p.v.a. adhesive or fungicidal paste
Metal foil	Although not a 'paper', lead and aluminium foils are used to line walls either (*a*) to protect water-sensitive surface coverings from dampness, or (*b*) as a decorative surface	Epoxy-based adhesive
Pitch paper	Paper incorporating a facing of bitumen. Used for walls with small areas of temporary dampness	Bitumen side applied to wall using starch paste or starch ether paste. Joints overlapped to prevent moisture penetration through joints
Wood-pulp paper	Smooth white or coloured paper available in various weights. Used to (*a*) conceal irregularities in the wall surface, or (*b*) to provide a background of even porosity	Starch or cellulose paste

Grounds are papers which are coated with paint prior to printing. Some of the patterned and decorative effects obtainable are:

Chintz: patterns based on natural forms which resemble chintz cotton.

Flock: flock papers are produced by printing the design on the paper with adhesive, then blowing fibres of wool, rayon or silk across the paper surface. The fibres are retained on the adhesive to produce a pattern with a pile effect.

Jaspe: a printed effect of linear veins of intermingling colour.

Metallic: metallic powder is introduced either into the ground or the pattern which produces a shiny effect on the paper.

Moiré papers: these papers are of satin or satinette ground with a finely engraved embossing which produces a silky effect.

Satin: a sheen is produced on the paper by polishing or glazing the ground prior to printing the design.

Satinette: a sheen is produced by incorporating mica into the ground. Both satin and satinette papers are rarely used today.

Printed papers are hung with either starch or starch ether pastes, although cellulose pastes might be used for the lighter weight papers.

Embossed papers: Embossed papers have a raised pattern pressed into them by passing the paper through embossing rollers. Some of the embossed papers available are:

Ordinary embossed paper and duplex paper: Ordinary embossed papers are the cheapest and most commonly used embossed papers. The embossing is often used to provide textured papers with simulated fabric effects such as canvas. Duplex papers consist of two plies of papers bonded together. The resulting paper is stronger, and capable of deeper embossing. When hung, the embossed effect of both of these papers tends to be reduced.

Anaglypta: anaglypta papers contain cotton, and are produced by passing the paper through embossing rollers whilst the paper is still wet. The embossing (which may be deep) will not be reduced as a result of hanging.

Supaglypta: this heavy paper contains materials such as china, cotton fibres and alum. The material is passed through embossing rollers whilst still wet, producing a paper suitable for heavy embossing which is retained after hanging. Very heavily embossed papers are available as panels, where the depth of embossing may be up to 25 mm and the size of the panel up to 1 m².

Embossed papers are hung with starch or starch ether pastes. Embossed panels require a special adhesive and may require temporary fixing using steel pins.

The embossed, duplex and moiré papers are usually supplied in various colours, whilst the anaglypta and supaglypta papers are supplied in white only.

Other papers: Some of the other paper-based surface coverings include ingrains, and washable papers.

Ingrains: ingrain papers are made by sandwiching small particles of cork or wood between a stout backing layer and a lightweight top layer of paper. The coarser textured papers are known as 'woodchips' and the finer textured papers, 'oatmeals'.

Woodchips are supplied in natural colour suitable for receiving paint. The heavier texture of woodchip makes it useful for hiding surface irregularities in a wall. Oatmeals may be printed or embossed in the same manner as other ground papers.

Washable papers: ordinary embossed or ground papers may be factory finished with a clear p.v.a. glaze. Such papers may be washed by wiping with a damp sponge. Vinyl papers consist of a layer of p.v.c. applied to paper, printed with p.v.c. inks. These papers are more resistant to washing and abrasion than ordinary washable papers. Ingrains and ordinary washable papers are hung with starch, or starch ether pastes, whilst vinyl papers are best hung with special adhesives.

Fabrics and other surface coverings

Fabrics   Table 8.12 describes some of the very wide range of fabrics which might be hung directly on walls. Further information concerning fabrics is provided in 'Textiles' (p. 427 below).

Some of the fabrics are available with a paper backing bonded to them which makes them easier to handle and hang. Unbacked fabrics should be hung on walls lined with suitable

Table 8.12 Fabric surface coverings

Material	Description	Typical sizes
Felt	A fabric based on pressed wool fibres. The fabric is bonded to a paper backing and hung like wallpaper. A wide range of colours is available including bright colours	Width: 790 mm, length of roll: 40 m, or cut lengths are available
Glass-fibre fabric	Glass fibres are woven into a fabric which resembles coarse hessian. Adhesive is applied to the wall surface, and the fabric pressed to it. The fabric is supplied in white, suitable for finishing with emulsion paint	Width: 1 m, length of roll: 50 mm
Grass cloths	Strands of material such as grass, hemp, cane and bamboo are held together by fine threads and bonded to a paper backing. Hanging grass cloth is a skilled job requiring careful trimming at the edges	Width: 900 mm, length: 7 m, or longer rolls
Hessian	A woven fabric produced from jute fibres. Hessian is available backed for painting over, or backed in a wide range of colours. Hard-wearing fabric	Width: 1.3 m, length of roll: 32 m, or cut lengths are available
Linen	A fabric woven from flax fibres. Available as unbacked, or with a paper backing	Width: 700 mm, length of roll: 50 m, or cut lengths are available
Plaster and jute fabric	An open-weave jute fabric backed with plaster. The material is hung with an adhesive which crystallises the gypsum. Range of restrained colours available	Width: 1.2 mm, length: 3 m
Silk	A close-textured fabric woven from silk yarns bonded to a paper backing	Width: 900 mm, length of roll: 25 m, or cut lengths are available
Wool	A large number of fabrics are available based on wool. The wool might be woven, or arranged in linear fashion to produce a corded effect. The fabric is bonded to a paper backing	Widths: 1 m, length of roll: 25 m, or cut lengths are available

lining paper, whilst backed fabrics may be hung directly on the wall by use of a heavy-duty paste.

Manufacturers' instructions should be followed concerning suitable adhesives and hanging methods for particular fabrics.

Cork   (see also 'Cork tiles', p. 413 above). Cork is available in sheet form, consisting of a very thin layer of cork bonded to a coloured or metallic backing paper, such that the backing is partially visible. The material is available in rolls, typically of 760 mm width and 9 m length, and is hung in a similar manner to wallpaper.

Lincrusta-Walton Lincrusta is a material made from linseed oil and filling agents such as wood flour and is pressed at high temperature on to a stout paper backing. The material is available in a variety of figured effects including simulated timber and timber panelling, and decorative textural effects. It is hard-wearing, and available in its natural off-white colour suitable for painting, or in a range of colours.

Lincrusta is supplied in rolls of 530 mm width and is fixed with heavy-duty adhesive. Detailed fixing instructions are usually provided with the material.

Timber veneers (see also p. 412, above) Timber veneers are available which are bonded to a flexible backing, and may be hung in a similar manner to wallpaper.

A wide variety of veneers are available in sheet or roll form which, once fixed, may be finished with stains or clear finishes. Sizes vary from manufacturer to manufacturer.

Thin flexible materials applied to wall panels

Some thin flexible materials are suitable for covering panels which are fixed to walls or joinery. These materials fall into two categories: fabrics, and skin materials, both of which are considered in 'Soft furnishings', p. 424 below.

These materials could be bonded directly to a board material, then fixed to a wall or joinery surface in a similar manner to the veneered panels illustrated in Fig. 8.10 and 8.11.

However, since fabric and skin materials are flexible, they might also be used in conjunction with soft foam-covered panels. Figure 8.15 illustrates such a panel, which might be

Fig. 8.15 *Panel covered with soft foam and fabric or skin material*

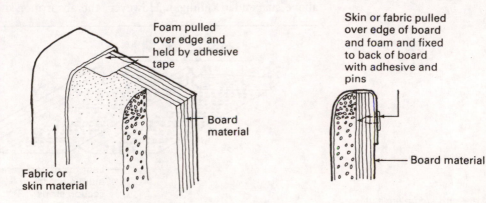

Foam pulled over edge and held by adhesive tape

Board material

Fabric or skin material

Skin or fabric pulled over edge of board and foam and fixed to back of board with adhesive and pins

Board material

fixed to wall or joinery surface, by means of secret fixings. The soft foam is bonded to the board with dabs of adhesive, and is pulled to a rounded edge by means of adhesive tape (details of soft foams and their use are considered in 'Upholstery, curtains and blinds', Ch. 6, p. 314). The fabric or skin material is pulled over the rounded edge of the foam and is fixed to the back of the panel with adhesive and pins or staples.

CEILING FINISHES

In Chapter 5, the constructional aspects of direct and suspended ceilings were considered. Here, ceiling finishes are considered. Of all the finishes in a building, the ceiling finishes usually receive the least wear. Some of the materials commonly used for ceiling finishes will be dealt with here.

The choice of a ceiling finish will be largely determined by such factors as: (*a*) acoustic properties; (*b*) appearance; (*c*) durability, maintenance and cost; (*d*) surface spread of flame characteristics; (*e*) thermal properties.

Acoustic properties

The acoustics of an interior will be considerably affected by the nature of a ceiling finish (see 'Providing acoustic properties', Ch. 5, p. 229). A hard, smooth material such as plaster, will tend to reflect sound hitting its surface, whilst a porous, textured material such as mineral fibre will tend to absorb sound hitting its surface. The absorption may be further improved by provision of an air space between the ceiling finish and the structural ceiling (such as the air space above suspended ceilings). However, the absorption of a

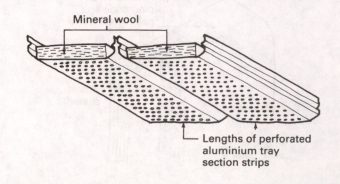

Mineral wool

Lengths of perforated aluminium tray section strips

Fig. 8.16 *Perforated aluminium strip ceiling*

420

porous material may be seriously impaired if it is decorated, since the decorative finish could cover the pores of the surface.

A wide range of proprietary ceiling finishes are available, which combine an absorbent porous material with a perforated surface. The perforated surface may then be redecorated without covering over the perforations. This allows redecoration to be carried out without serious loss of absorbency, since sound is still absorbed through the perforations in the surface into the porous material (see Fig. 8.16).

Appearance

The visual impact that a ceiling has on an interior might be affected by:

1. *Height*: By installing a suspended ceiling in a room, the atmosphere would be altered due to the modified proportions of the room. A low ceiling, for example, could contribute towards an atmosphere of warmth and intimacy, whilst a high ceiling could contribute towards an atmosphere of lofty dignity.

2. *Shape*: A suspended ceiling may also be used to alter the *apparent* proportions of an interior; for example, a suspended ceiling profiled in a series of deep corrugations could be installed in a long, rectangular room. If the corrugations were parallel to the long sides of the room, the apparent length is likely to be accentuated, whilst if the corrugations were parallel to the short walls, the apparent length of the room is likely to be foreshortened.

3. *Texture, colour and size of ceiling finish units.*

4. *Junction of wall and ceiling*: Figures 5.12(a)–(c) (Ch. 5) illustrate alternative treatments of cornice and coving details.

Durability, maintenance and cost

The durability of a ceiling finish will be influenced by:
- wear and tear caused by cleaning and maintenance of the ceiling and maintenance of electrical or other services housed in or above the ceiling;
- by stresses caused in the ceiling finish due to inadequate means of support, which could result in cracking or deformation of the finish;
- the harmful effects of dust or chemicals in certain industrial interiors, or the effects of condensation in rooms such as kitchens or bathrooms.

Some ceiling finishes, such as self-coloured metals and plastic laminates, require little more than periodic cleaning, whilst other finishes, such as plaster or timber, will require periodic redecoration.

As is the case with any finish, the initial cost should be weighed against the long-term maintenance costs. In the case of an interior with a high ceiling, for example, the long-term costs of an initially cheap finish could be considerable if the use of expensive scaffolding was necessary for frequent redecoration.

Surface spread of flame characteristics

Ceilings are an important element in the behaviour of fire in a building. 'Safety in fire' (Ch. 5, p. 231), together with Tables 5.5 and 5.6 (Ch. 5), outline the necessary surface spread of flame requirements for ceilings, and the surface spread of flame characteristics of certain materials.

Thermal properties

'Thermal insulation' (Ch. 5, p. 230) outlines the ways in which a ceiling might contribute towards the thermal insulation of a building.

The use of a suspended ceiling could reduce the effective volume of a room, and consequently reduce the quantity of energy necessary to heat the room.

Some of the materials suitable as ceiling finishes have already been referred to in this chapter. These include:

- Cork and fabrics: *see* 'Fabrics and other surface coverings', p. 417 above.
- Metals: *see* 'Metal claddings', p. 411 above.
- Papers: *see* 'Fabrics and other surface coverings', p. 417 above.
- Plaster: *see* p. 385 above.
- Plastic laminates: *see* 'Decorative plastic laminates', p. 413 above.
- Timber boarding: *see* p. 410 above.
- Timber veneers: *see* p. 412 above.

The following section will deal with (*a*) Panles and tiles; (*b*) Plasterboard.

Panels and tiles

Ceiling panels and ceiling tiles may be made of similar materials. Panels are large units, often of 600 × 600 mm or more, and are usually used in conjunction with some form of suspended ceiling. Tiles are smaller units, often 300 × 300 mm and are usually bonded with adhesive to a solid ceiling surface such as concrete or plasterboard.

Table 8.13 Suspended ceiling panels

Panel	Description	Typical sizes (mm)	Weight(kg/m²)
Mineral fibre	Panels composed of a mixture of mineral fibres, fillers and binders. Wide range of surface textures available. Available with square, rebated and grooved edges	Thicknesses: 12, 15 and 19 × 600 × 600, 600 × 1200	19 mm: 6.5 15 mm: 6 12 mm: 5
Wood fibre	Wood-fibre-based boards with series of slots or perforations in the surface. Available with square, rebated and grooved edges	Thicknesses: 13 and 19 × 600 × 600, 600 × 1200, 1200 × 2400	19 mm: 5.5 13 mm: 3.7
Glass fibre	Very lightweight panels composed of glass fibres with a p.v.c. surface. Wide range of surface textures available. Square edge detail	Thicknesses: 16 and 40 × 600 × 600, 600 × 1200	16 mm: 0.8

Panels

There are many proprietary forms of ceiling panel on the market, which include fibre-based panels and metal panels.

Table 8.13 compares the characteristics of some of the types of fibre-based panels. The figures are useful only as a guide, since they will vary from manufacturer to manufacturer.

The edges of fibre-based panels may be either:

- square; which are suitable for dropping into an exposed framework suspended ceiling (see Fig. 5.27(a), Ch. 5);
- rebated; these panels are also dropped into an exposed framework suspended ceiling, but the rebate causes the panel surface to be at a lower level than the level of the framework;
- grooved; for slotting into a concealed framework suspended ceiling (see Fig. 5.27(b), Ch. 5).

Metal-based panels are often tray-shaped in section and made of enamelled aluminium or steel. Typical sizes of such panels might be 600 × 600 mm or 1200 × 600 mm. Metal strips are also available, which might be of 100, 200 or 300 mm widths and several metres in length. The strips may be of similar material to trays, or of mirror effect, chrome or copper.

Figure 8.16 illustrates an aluminium strip ceiling. The tray sections might be held by some form of clip or panel carrier similar to that illustrated in Fig. 5.24(b) (Ch. 5). To provide increased sound absorption, the panels might be perforated and a 12 mm layer of mineral wool placed in the tray section.

Tiles

Some ceiling tiles are suitable for fixing direct to the ceiling by use of an adhesive. Materials commonly used for the manufacture of such tiles are wood fibre, mineral fibre and polystyrene.

Wood-fibre tiles are available with either a plain surface, or with a perforated patterned surface. The tiles are typically 300×300 mm with thicknesses of 12 or 18 mm.

Mineral-fibre tiles are available in a wide variety of textural finishes and are typically 300×300 mm with thickness of 15 or 19 mm.

Expanded polystyrene tiles are available with plain or textured surfaces and are typically 300×300 mm with a thickness of 9 mm.

Plasterboard

Plasterboard is a rigid flat sheet material which comprises of a core of gypsum plaster with the surfaces covered by stout paper. Plasterboard sheets are very commonly used for forming ceilings, by fixing the sheets to the underside of timber joists. Plasterboard is also used in conjunction with timber or metal studs in the construction of internal partitions (see Fig. 5.15, Ch. 5).

Table 8.14 outlines the four types of plasterboard available: baseboard, lath, plank and wallboard. The boards are fixed to the studs or joists with flat-headed plasterboard nails. Of the edge joints shown in the table, the bevelled joint (which may be used with bevelled-edged planks or wallboards) is usually only used in partition construction.

All four boards are available with a layer of aluminium foil applied to the back surface. This can improve the thermal insulation offered by a ceiling or roof construction, by reflecting heat back into a room.

SOFT FURNISHINGS

The materials discussed here are those used for 'soft furnishings'; that is, for items such as coverings for upholstered furniture, curtains, blinds, etc. These materials are basically categorised as either textiles or skin materials.

Selecting a soft furnishing material from the vast array of materials available will involve consideration of such factors as: (*a*) appearance and durability; (*b*) comfort; (*c*) cost; (*d*) flame retardance.

Table 8.14 Plasterboards

Plasterboard	Description and uses	Edge joint details	Sizes (mm)
Baseboard	Plasterboard designed as a surface suitable for application of plaster to ceilings or internal partitions. Joints should be 'scrimmed' (reinforced with 'scrim' – a strip of jute fabric). Boards have square edges	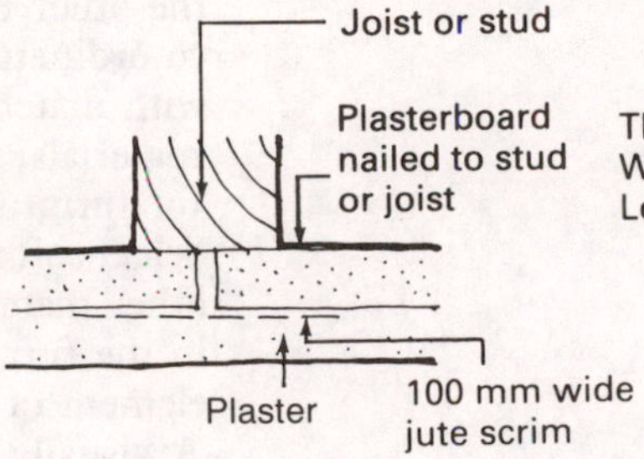	Thickness: 9.5 Width: 914 Length: 1200
Lath	Narrow-width plasterboards designed for application of plaster to ceilings or internal partitions. Because of the rounded edges, joints do not require to be scrimmed	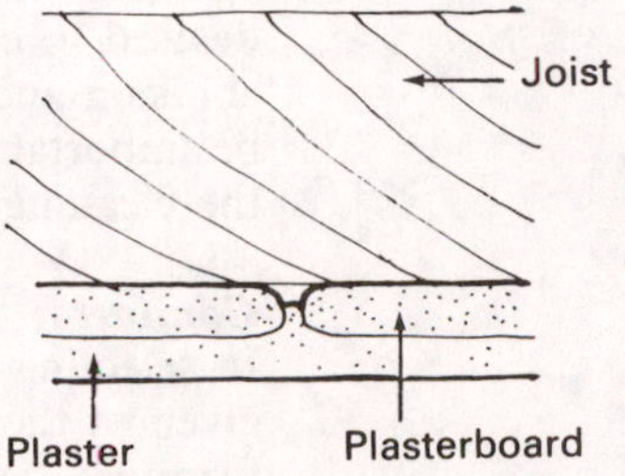	Thicknesses: 9.5 and 12.7. Width: 406 Length: 1200
Plank	Thick, narrow boards available either: (a) with square edges suitable for plastering (the edge detail being similar to baseboard detail); (b) with square, tapered or bevelled edges for direct decoration. Tapered boards (edge detail illustrated opposite) and bevelled edges (edge detail similar to wallboard) are used mainly for partitions. Planks may be used either for partition construction, or for ceilings where $\frac{1}{2}$ hr fire resistance is required	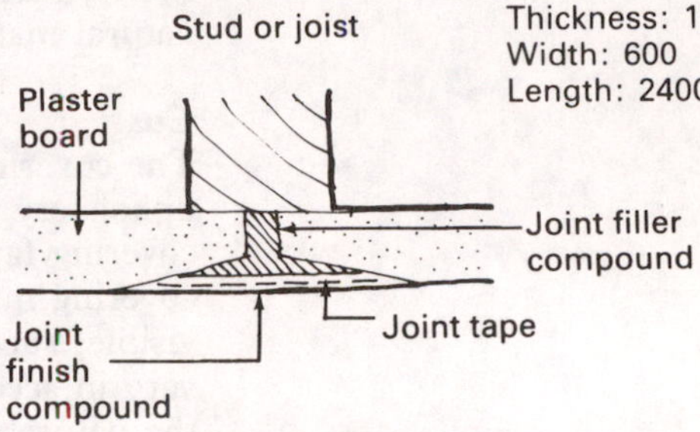	Thickness: 19 Width: 600 Length: 2400
Wallboard	Plasterboard designed for direct decoration, suitable for internal partitions and ceilings. Available with square edges and bevelled edges (illustration opposite) which are used primarily for partitions or with tapered edge where the edge detail is similar to planks	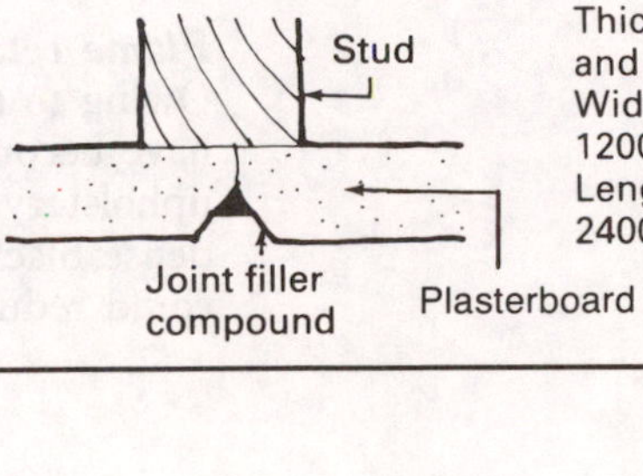	Thicknesses: 9.5 and 12.7 Widths: 600, 900, 1200 Lengths: 1800, 2400, 3000

Appearance and durability

The range of colours, textures and patterns available to the designer provides the opportunity to either:

(a) Select soft furnishing materials which sympathise with the other materials used in the interior. Ranges of co-ordinating textiles and wall coverings are available with matching colours and patterns. Alternatively, some materials, such as hessian, could be used on walls and also for furniture coverings.

(b) Select soft furnishing materials which contrast with the other materials used in the interior. This is likely to result in the furniture providing a more visually significant element of the interior scheme.

A visually well-selected material could soon become unsightly if inadequate thought was given to the anticipated wear and tear on the material. Standard tests have been devised to check such characteristics as colour fastness, abrasion and flame retardance. Other characteristics that may be important include crease, stain and water resistance, and the cleaning requirements of the material.

Comfort

In selecting upholstery coverings, consideration should be given to the length of time a person is liable to use the item of furniture. Some synthetic materials can become uncomfortably sticky, particularly in a hot environment, whilst certain woven natural materials can be an irritant to the skin.

Cost

The covering material is often the most expensive item in the upholstery of furniture. Consequently, when an expensive covering fabric is used, it might be worth using a cheaper covering material for parts of the furniture which are not visible, such as the underside of cushions. The cost of textiles vary in accordance with factors such as the type of yarn used, the nature of the weave construction, and special treatments applied to the fabric.

Flame retardance

Owing to their cheapness and ease of use, polyurethane foams have become very widespread as a cushioning material for upholstery. However, when the material is ignited it produces dense black smoke and toxic fumes. A fire-retardant covering could reduce the likelihood of this hazard by preventing a fire from starting and by resisting the spread of flames from a fire that has commenced. All foam fillings for domestic furniture must use CMHR foam, and upholstery coverings must pass the BS 5852 match test (see 'Fire safety' and 'CMHR foams', pages 314 and 319.

Textiles

Textiles are fabrics which are made from 'yarns' (threads) either by knitting, 'compacting' (a process used for producing materials such as felt) or by weaving. Most soft furnishing fabrics are woven on looms. In producing a woven fabric: (*a*) the yarn is manufactured; (*b*) the fabric is woven; (*c*) the fabric is finished.

Table 8.15 describes some fabrics commonly used in soft furnishing.

Yarns

Yarns are threads which are spun or twisted from 'fibres' (individual strands of vegetable-, animal- or synthetic-based material). Yarns may be produced by blending different fibres in order to provide yarns with particular qualities. Examples of blends are: 65 per cent wool, 35 per cent viscose; 93 per cent polyester, 7 per cent cotton; or 41 per cent cotton, 40 per cent acrylic, 19 per cent polyester.

Table 8.16 describes some of the fibres commonly used to produce yarns for soft furnishing fabrics.

Weaving

Woven fabrics are made on looms, where a series of threads known as 'warps' run lengthways, and a series of 'wefts' (which run crossways) interlace with the warps. There are three basic patterns of weaving: plain, twill and satin. From these simple patterns, many complex variations have been devised which allow intricate patterns to be woven. These complex designs are woven on 'Jacquard' looms.

Figure 8.17(a, b, c) illustrates the three basic weaves given under the headings below.

Plain weave   The weft passes alternatively over one warp then under the next, producing strong hard-wearing fabrics such as gingham or taffeta.

Table 8.15 Soft furnishing fabrics

Fabric	Description	Uses (B – bedcovers C – curtains U – upholstery)
Brocade	Finely woven patterned fabric usually made from silk, cotton or synthetic fibres	B, C, U
Chintz	Patterned cotton cloth usually glazed	B, C
Corduroy	Durable; thick ribbed fabric, usually made from cotton; rayon or wool	B, C, U
Gingham	Plain woven cotton fabric, often dyed in stripes or checks	C
Linen	Fabric woven from flax. May be plain or patterned	B, C, U
Moquette	Velvet-like looped, or piled fabric, made of cotton, synthetic fibres or wool	U
Taffetas	Plain woven fabric with silk-like surface. Usually made of silk or rayon	B, C
Velour	Cut-pile woven fabric usually made from cotton or rayon	C
Velvet	Closely woven, piled fabric made from rayon, silk or other fabrics	C, U

Fig. 8.17 *Patterns of basic woven fabrics*
(a) *Plain weave*
(b) *Twill weave*
(c) *Satin weave*

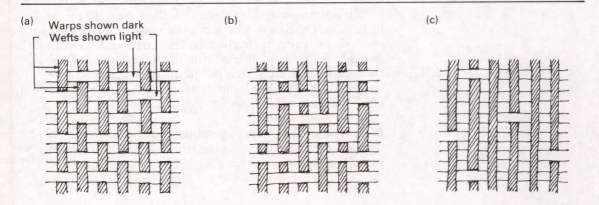

(a) Warps shown dark
Wefts shown light

(b)

(c)

Table 8.16 Fibres for soft furnishing

Fibre	Description	Properties
Vegetable		
Cotton	Taken from the fibrous substance surrounding the seeds of cotton plants. Often blended with other fibres	Economical; dyes well; durable; many different qualities of fabric may be produced from cotton
Flax	Taken from the stem of the flax plant to produce linen fabrics	Strong; resistant to moth and mildew; resistant to abrasion
Animal		
Silk	Very long fibres taken from the spun cocoon of the silkworm	Expensive; lustrous; resistant to abrasion; resistant to moth and mildew
Wool	Taken from fleece of sheep. Often blended with other fibres	Resilient; tends to soil easily; resistant to abrasion
Synthetic		
Acetate	A form of rayon. Derived from cellulose. Often blended with other substances	Less strong than viscose rayon; resistant to mildew; silk-like appearance; soils but washes easily
Fibreglass	Made from spun glass	Non-inflammable; poor resistance to abrasion; does not shrink or stretch
Nylon	Synthetic fibre often blended with other fibres	Water resistant; strong; easy to clean, does not burn
Polyester	Synthetic fibre often blended with other fibres	Does not shrink or stretch; resistant to moth and mildew; strong
Polypropylene	Synthetic fibre often blended with other fibres	Resistant to many chemicals; resistant to abrasion, strong
Viscose	A form of rayon. Derived from cellulose. Often blended with other fibres	Economical; reasonably durable; soils but washes easily

Twill weave The weft passes over one warp then under two warps. The same sequence repeats for each weft, although the sequence is staggered to produce a diagonal effect. Twill weave is hard-wearing, but stretches more than plain weave. Denim and serge are examples.

Satin weave The weft passes over one warp then under several warps, revealing much of the warp on the surface of the fabric. Sateen weave is similar, only much of the weft is visible on the surface. Satin and sateen weaves are smooth and lustrous, but tend to snag easily because of the large number of exposed threads.

Dyeing, finishing and printing

Woven fabrics may be dyed by either 'vat dyeing' where the yarn is dyed, prior to weaving, or by 'piece dyeing' where the woven fabric is dyed. Vat-dyed fabrics tend to have better colour fastness than piece-dyed fabrics.

Many woven fabrics require 'finishing' processes to modify the characteristics of the fabric. These processes include:

- 'bleaching', which whitens cotton and linen fabrics;
- 'calendering', where cotton or linen fabrics are passed between heated rollers which provides the fabric with a lustrous smooth surface;
- 'embossing', where the fabric is passed through embossing rollers, producing a raised design;
- application of fire-retardant solution;
- processes to improve the crease, stain and water resistance of the fabric.

After weaving, plain fabrics might be printed, either:

- by hand-printing methods such as hand block or screen printing, which might be used for small quantities of expensive fabric;
- by machine printing, where up to 14 colours may be applied to the fabric in one process. Machine printing is used for large quantities of fabric.

Skin materials

Skin materials are flexible, continuous materials which may be used for covering upholstered furniture. Traditionally, leather was widely used, but because of its high cost, it is only used today for high-quality upholstered furniture.

'Expanded vinyl', a synthetic skin material, has become very widely used, often as a low-cost substitute for leather.

Leather

Leather is produced from the skins of animals. The skins (known as 'pelts') are 'tanned', that is, soaked in a liquid containing tannic acid, which renders the skins firm and durable.

In the past, goat skins, known as 'moroccos', were used for high-quality work. Moroccos were very durable, lasting for 30–40 years, or more. Today, cow hide (or 'hide') is used, which is slightly less durable, but cheaper than morocco.

Hides are available in areas of 4–4.5 m². Two grades of hide are used – 'full-grain hides', which are of top quality, or 'buffed hides', which have some blemishes such as holes or abrasions, which are buffed over. Buffed hides are cheaper than full-grain hides.

Expanded vinyl-coated fabric

Expanded vinyl-coated fabric comprises of a tough knitted or woven cotton fabric coated with an expanded p.v.c. pigment and a protective coating of p.v.c. These fabrics are resistant to moderate abrasion and mild acids and alkalis, and are reasonably stain resistant. Some substances, such as ball-point ink, can become absorbed into the material and cause a permanent stain unless it is removed immediately.

Expanded vinyl-coated fabrics are available in a wide range of colours and textures including simulations of leather. The material is available in rolls of typically 1270 or 1370 mm width and lengths of 25–50 m.

CHAPTER 9

Environmental issues

The preservation of the environment has become a prominent issue in recent years due to the recognition of the harmful effects of pollution and unsustainable exploitation of natural resources. Human health and comfort are largely dependent on the maintenance of the delicate balance of natural forces and resources which provide stable climatic conditions and parity of atmosphere.

Since the building industry is a heavy user of synthetic and natural materials it could have an important impact on the preservation of the environment. While every building process and product could be evaluated to assess its environmental impact, this chapter concentrates on some of the more obvious issues:

- the use of chlorofluorocarbons (CFCs) in the manufacture of certain foamed insulants;
- the use of certain timbers and timber-related boards and the sustainability of tropical rain forests.

Chlorofluorocarbons and insulation materials

Introduction

The level of thermal insulation required for walls, floors and roofs has increased sharply in recent years due to the increased costs of fuel. Table 9.1 indicates the changing Building Regulation requirements for 'U' values of walls and roofs between 1976 and 1990.

While thermal efficiency could be increased by using a greater thickness of traditional insulant, new materials have been developed which are comparatively thin but of high insulation value. Two such materials are extruded polystyrene and polyurethane boards. Both materials rely on chlorofluorocarbon in their manufacture; as the blowing agent for foaming rigid polyurethane boards, and in the extrusion process of extruded polyurethane boards.

Table 9.1 Increases in Building Regulation requirements for thermal insulation

Date of publication	U value (W/m K)	
	External wall	Roof
1976	1.0	0.6
1978 Amendment	0.6	0.6
1985	0.6	0.35*
1991	0.45	0.25**

* (Dwellings) other buildings 0.6 or 0.7
** (Dwellings) other buildings 0.45

CFCs and the ozone layer

The 'ozone layer' refers to the comparatively high concentration of ozone occurring in the stratosphere (an atmospheric layer 20−50 km above the surface of the earth). The more damaging (short-wave) ultraviolet radiation from the sun is absorbed by the ozone layer. A depletion of the ozone layer (and consequent increase in short-wave UV light on the earth's surface) could result in:

- health problems;
- reduced crop yields;
- damage to marine life;
- degradation of certain building materials (particularly plastics).

Ozone is a highly unstable form of oxygen, the balance of which is maintained by its constant breakdown and regeneration. It is consequently vulnerable to destruction by certain chemicals, particularly chlorine.

Unlike ozone, CFCs are highly stable chemicals which release chlorine into the atmosphere and remain there a long time before being eventually broken up. The chlorine combines with the oxygen in the ozone layer to form chlorine monoxide and oxygen. The chlorine monoxide itself is then able to release its chlorine and the cycle is repeated.

CFCs are the principal contributors to the depletion of the ozone layer, and CFCs used in building account for approximately 10 per cent of the total released into the atmosphere.

The indications are that CFCs must be phased out very rapidly in order to avoid cumulative damage to the ozone layer. CFCs are also damaging since they are highly efficient 'greenhouse gases' contributing to the 'greenhouse effect' (see below).

CFCs and thermal insulation

The major uses of CFCs in buildings are in foamed plastics (described above) and as a refrigerant for refrigeration and air conditioning. While the use of CFC in refrigerants has increased slightly in the last decade, their use in foamed plastics has more than doubled.

The advantages offered by foamed plastic boards are
- very low thermal conductivity
- low water absorption.

Their uses as thermal insulants include
- cavity walls (pp. 45–46 and Fig. 2.13)
- roofs (pp. 81–2 and Fig. 3.5)
- floors (pp. 130–34 and Fig. 3.26).

In most cases alternative non-CFC insulants could be used such as
- mineral fibre
- glass fibre
- cork
- expanded vermiculite
- expanded glass.

At the time of writing many manufacturers of rigid polyurethane and extruded polystyrene boards are developing or have adopted production techniques that are either CFC free (e.g. using water or carbon dioxide as a foaming agent) or using HCFCs. HCFCs are hydrogen-containing CFCs which have a considerably shorter atmospheric lifetime. While these are less damaging to the environment, they still release chlorine. Due to the cumulative effect of chlorine on ozone and the high level of chlorine already in the atmosphere, HCFCs will still increase ozone depletion, although less rapidly than CFCs.

Timber and tropical rain forests

Trees are a vital component in the maintenance of a healthy global environment. They help purify the air, and conserve and preserve the fertility of the soil. In the developed world trees are suffering due to the effects of air pollution, particularly acid rain which results mainly from the effects of burning fossil fuels. The tropical rain forests, largely located in the third world, are being rapidly depleted by felling. This section considers

- tropical rain forests
- the greenhouse effect
- tropical hardwoods
- alternative timbers and products.

Tropical rain forests

Tropical rain forests form a belt of dense vegetation around the equator. The major geographical regions are Central and Southern America, Central Africa and South East Asia (Fig. 9.1). These forests are very rich in vegetation and make an important contribution to the reduction of carbon dioxide (see

Fig. 9.1 *Distribution of tropical hardwoods*

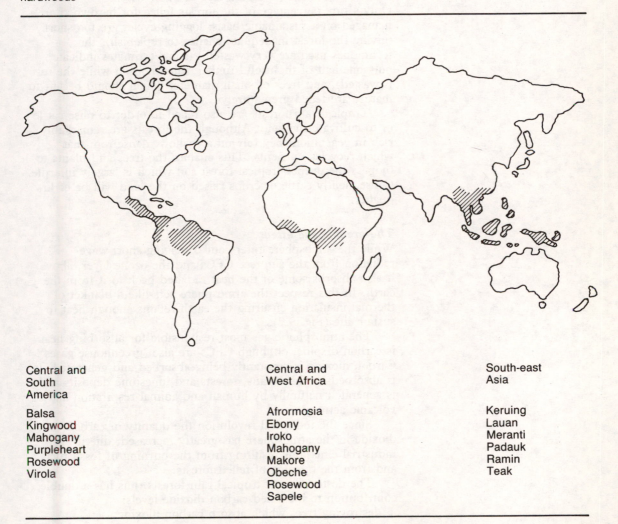

Central and South America	Central and West Africa	South-east Asia
Balsa	Afrormosia	Keruing
Kingwood	Ebony	Lauan
Mahogany	Iroko	Meranti
Purpleheart	Mahogany	Padauk
Rosewood	Makore	Ramin
Virola	Obeche	Teak
	Rosewood	
	Sapele	

435

greenhouse effect, below). They also have a significant effect on the world's climate. Water from the forests is absorbed into the air by evapotranspiration. The heat required to absorb this moisture has a cooling effect and forms clouds which move to the cooler, higher latitudes where the moisture is released as warm rain.

Tropical rain forests are being destroyed at an alarming rate: at the current rate of 2 per cent loss per year the forests could disappear entirely within 50 years. The primary reasons for the clearance of these forests are

- export of hardwood
- use of the land for agriculture or other uses.

Only a tiny percentage of the forests felled for hardwood are managed on a sustainable basis: logging cycles are too short (giving the forest inadequate chance to replenish), the techniques used are very wasteful (some estimates indicate only one-half of the useful timber felled is used while the rest is wasted), and trees of small diameter which should be left to mature are felled prematurely.

Tropical rain forests are also felled in order to raise cattle or to cultivate the land. Although the forests are remarkably rich in vegetation they rely on a shallow, dense root mat which recycles nutrients. This enables the trees and plants to by-pass the deeper tropical forest soil which is largely infertile. Consequently cattle or crops raised on the land will be of low quality.

The greenhouse effect

While the atmosphere filters out damaging short-wave radiation from the sun (see CFCs and the ozone layer, above) it also absorbs some of the heat radiated back to it from the earth. In this respect the atmosphere acts like a blanket of thermal insulation ensuring the earth retains enough heat to sustain life (Fig. 9.2).

The atmospheric gas most responsible for absorbing heat is carbon dixoide, although CFCs are also 'greenhouse gases'. Carbon dioxide is constantly being absorbed and generated. It is absorbed by the oceans, forests and limestone deposits and is generated naturally by human and animal respiration and volcanic activity.

Since the industrial revolution the quanity of carbon dioxide in the atmosphere has greatly increased, due to industrial emissions resulting from the burning of fossil fuels and from the burning of felled forests.

The destruction of tropical rain forests thus has a double contribution to increased carbon dioxide levels:

- destroying trees which absorb carbon dioxide;

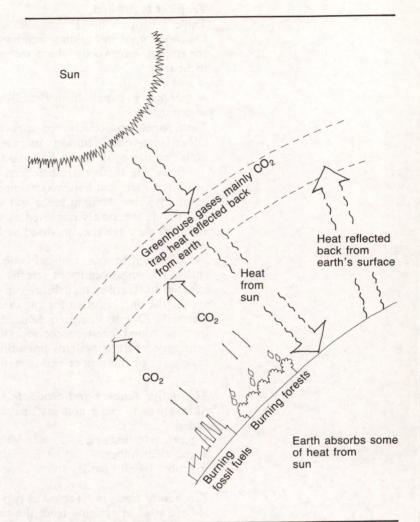

Sun

Greenhouse gases mainly CO_2 trap heat reflected back from earth

Heat reflected back from earth's surface

Heat from sun

CO_2

CO_2

Burning forests

Burning fossil fuels

Earth absorbs some of heat from sun

Fig. 9.2 *The greenhouse effect*

- burning rhe remains of the felled forest which creates carbon dioxide.

The immediate result of these rising levels of carbon dioxide is to increase the insulative effect of the atmosphere leading to a rise in temperature. Some scientists have predicted that by the mid-21st Century the global temperature will have increased by 1.5−4.5 °C resulting in a rise of sea level (due to thermal expansion and melting polar ice caps) of between 0.8 and 1.8 m. This would make immense areas of cultivated land arid while other areas and many major cities would become submerged by the sea.

Tropical hardwood

Table 7.3 (pp. 340–41) describes some of the species of hardwood used in building interiors. Of these, the following are tropical hardwoods which are imported in large quantities to Britain:

- iroko
- mahogany (largely from Brazil)
- ramin.

Teak, sapele and obeche are imported in smaller quantities. Other large-volume imports include keruing and red meranti both originating in South East Asia.

Since the 1950s there has been a massive increase in the quantity of tropical hardwoods imported due largely to their durability, fine, straight grain and strength. The imported hardwoods are mostly imported as sawn timber and plywood, although they are also imported as logs and, of course, as furniture.

Although in most tropical countries agreements exist to ensure the sustainability of forests, most are inadequately adhered to. Until sustainability can be guaranteed, it would appear the only solution for the designer is to seek alternative materials. Due to increased demand for tropical hardwoods from sustained forests, some suppliers are introducing voluntary labelling systems providing information concerning the source of the timber and evidence of sustainability.

Alternative timbers and products

Alternatives to using non-sustainable tropical hardwoods include:

- temperate timbers;
- recycled timbers;
- timber-based boards.

Temperate timbers   Table 7.3 (pp. 340–1) includes species of hardwood originating from the temperate forests in Europe and North America. While most of these species would be less durable than the more durable of tropical hardwoods, sweet chestnut, American white oak and European oak might be considered as substitutes.

For most interior purposes temperate hardwoods would be sufficiently durable. Many of the softwoods listed in Table 7.1 (p. 333) originate from temperate forests and would be suitable for most interior purposes. Yew is a durable European softwood and European Larch is reasonably durable.

To adjust the durability or colour characteristics of temperate timbers, preservatives and stains could be used (see

'Timber preservatives' Ch. 4, p. 159 and 'woodstain' Table 8.9 p. 395). It might be worth checking the environmental impact of such treatments since preservatives in particular can contain highly toxic chemicals.

While the use of temperate timbers will help towards the preservation of tropical forests, some temperate forests are themselves in danger. In particular some of the logging in the United States and Canada has been managed in a manner wasteful of timber.

Recycled timbers Some companies supply and manufacture components from timbers reclaimed from demolished buildings. The timber is cut and machined to provide either defect-free material or timbers with varying degrees of surface blemishes.

Timber-based boards (see also Timber-based boards, Ch. 7 pp. 338 and 345–7) Particularly in furniture construction, the use of timber-based boards as a lower-cost alternative to solid timber is increasingly widespread.

While some of these materials (particularly those using timber waste products) can assist in reducing the consumption of tropical timbers, many use formaldehyde in their manufacture.

Formaldehyde (which is also used in certain forms of cavity-wall insulation see p. 153) is a toxic chemical which is widely used in adhesives for plywood and veneers. The chemical is a potential health hazard and can lead to irritation of the eyes and nose. It is believed to have carcinogenic potential.

Plywood and blockboard

A large proportion of plywood used in Britain uses tropical hardwoods from South East Asia. Alternative plywoods and blockboards composed of temperate timbers such as ash, beech, birch, oak and Douglas fir might be used.

Chipboard (particle board)

Chipboard is made from chips, dust and waste wood bonded with natural or synthetic resin. However, since demand is currently out-stripping supply some chipboard is also manufactured from fast-growing softwood.

'Low-emission' grades of particle board are available which release lower levels of formaldehyde.

Fibreboards

Like chipboard, fibreboards use timber waste products, but may also contain waste paper.

Bibliography

Note: All titles published in the United Kingdom unless otherwise stated.

Structures generally
J E Gordon (1968) *The New Science of Strong Materials*, Penguin
J E Gordon (1978) *Structures*, Penguin

Building structure
Jack Stroud Foster (1973) *Structure and Fabric Part 1*, Batsford
W Morgan (1964) *The Elements of Structure*, Pitman

Building interior functions
Edward Allen (1980) *How Buildings Work*, Oxford University Press
E Schild *et al*, ed. M Finbow (1981) *Environmental Physics in Construction: its application in architectural design*, Granada

Building/interior materials
Building Research Establishment Digests (1977) *Building Materials*, Construction Press
A G H Dietz (1969) *Plastics for Architects*, MIT Press (USA)
Allan Everett (1986) *Materials* (4th edition), Batsford
A Fulcher, B Rhodes, N Stewart, D Tickle, and J Windsor (1975) *Painting and Decorating: an information manual*, Granada
The International Book of Wood (1976), Mitchell Beazley

Building/interior construction generally
Building Research Establishment Digests (1977) *Building Construction*, Construction Press
Francis D K Ching (1976) *Building Construction Illustrated*, Van Nostrand Reinhold
Common Building Defects (1979), Construction Press

A J Elder (1979) *Guide to the Building Regulations 1976*, *Architectural Press*

A J Elder and Maritz Vandenbergh (1974), *A J Handbook of Building Enclosure*, Architectural Press

W B and J K McKay *Building Construction*, Vols 1–4 (5th, 4th, 5th & 3rd editions respectively), Longman

Ivor H Seeley (1976) *Building Maintenance*, Macmillan

Specification 91 (1990), MBC, Architectural Press

Furniture, fittings and components

Dorothy Cox (1970) *Modern Upholstery*, G Bell

E V Foad (1982) *Purpose-Made Joinery*, Van Nostrand Reinhold (U.K.)

C Howes (1973) *Practical Upholstery*, Evans

Victor J Taylor (1977) *Modern Furniture Construction*, Evans, London

G Underwood and J Planck (1977) *Handbook for Architectural Ironmongery*, Architectural Press

Environmental issues

The Good Wood Guide (1990), Friends of the Earth

Philip Neal (1989) *The Greenhouse Effect*, Batsford

E Goldsmith (1990) *The Earth Report 2*, Mitchell Beazley

Index

Italics refer to illustrations